T0299306

Thin Film Shape Memory Alloys: Fundamentals and Device Applications

The first dedicated to this exciting and rapidly growing field, this book enables readers to understand and prepare high-quality, high-performance TiNi shape memory alloys (SMAs). It covers the properties, preparation, and characterization of TiNi SMA thin films with particular focus on the latest technologies and applications in MEMS and biological devices. Basic techniques and theory are covered to introduce new-comers to the subject, whilst various sub-topics, such as film deposition, characterization, post treatment, and applying thin films to practical situations, appeal to more informed readers. Each chapter is written by expert authors, providing an overview of each topic and summarizing all the latest developments, making this an ideal reference for practitioners and researchers alike.

Shuichi Miyazaki is a Professor at the Institute of Materials Science, University of Tsukuba, Japan, where he has worked since 1979 after obtaining his Ph.D. in Materials Science and Engineering from Osaka University. He has received numerous awards, including the Academic Deed Award from the Japan Institute of Metals in 1995 and the Yamazaki–Teiichi Prize from the Foundation for Promotion of Material Science and Technology of Japan in 2002, as well as co-authoring 25 books and over 400 technical and review papers in the field of materials science, especially on shape memory alloys.

Yong Qing Fu is a Lecturer in the Department of Mechanical Engineering at Heriot-Watt University, Edinburgh. He obtained his Ph.D. degree from Nanyang Technological University (NTU), Singapore in 1999 and his prior positions have included Research Fellow at Singapore–MIT Alliance, NTU, and Research Associate in EDM Group (Electronic Devices and Materials) at Cambridge University. He has extensive experience in thin films, surface coatings, microelectromechanical systems (MEMS), shape memory alloys, smart materials, and nanotechnology, and has authored over 100 journal papers.

Wei Min Huang is an Associate Professor at the School of Mechanical and Aerospace Engineering at Nanyang Technological University (NTU), Singapore. He was awarded his Ph.D. from Cambridge University in 1998 and has since published over 80 journal papers. His research interests include shape memory materials, actuators, advanced technologies and materials, surface patterning, and materials selection.

Thin Film Shape Memory Alloys

Fundamentals and Device Applications

Edited by

SHUICHI MIYAZAKI
University of Tsukuba, Japan

YONG QING FU
Heriot-Watt University, Edinburgh

WEI MIN HUANG
Nanyang Technological University, Singapore

CAMBRIDGE
UNIVERSITY PRESS

Shaftesbury Road, Cambridge CB2 8EA, United Kingdom

One Liberty Plaza, 20th Floor, New York, NY 10006, USA

477 Williamstown Road, Port Melbourne, VIC 3207, Australia

314–321, 3rd Floor, Plot 3, Splendor Forum, Jasola District Centre, New Delhi – 110025, India

103 Penang Road, #05–06/07, Visioncrest Commercial, Singapore 238467

Cambridge University Press is part of Cambridge University Press & Assessment, a department of the University of Cambridge.

We share the University's mission to contribute to society through the pursuit of education, learning and research at the highest international levels of excellence.

www.cambridge.org
Information on this title: www.cambridge.org/9780521885768

© Cambridge University Press & Assessment 2009

First published 2009

A catalogue record for this publication is available from the British Library

Library of Congress Cataloging-in-Publication data
Thin film shape memory alloys : fundamentals and device applications /
    [edited by] Shuichi Miyazaki, Yong Qing Fu, Wei Min Huang.
        p.   cm.
    Includes bibliographical references and index.
    ISBN 978-0-521-88576-8 (hardback)
    1. Shape memory alloys.   2. Thin film devices.   I. Miyazaki, Shuichi.
    II. Fu, Yong Qing.   III. Huang, Wei Min.
    TA487.T48 2009
    621.3815′2–dc22
    2009027554

ISBN   978-0-521-88576-8   Hardback

Contents

Contributors

P. J. Buenconsejo
University of Tsukuba, Japan

M. Cai
University of Houston, USA

G. P. Carman
University of Calfornia, Los
Angeles, USA

H. Cho
Oita University, Japan

C. Y. Chung
City University of Hong Kong,
China

Y. Q. Fu
Heriot-Watt University, UK

R. Gotthardt
Swiss Federal Institute of
Technology, Lausanne,
Switzerland

W. M. Huang
Nanyang Technological
University, Singapore

X. Huang
Yale University, New Haven,
USA

M. H. Hong
Data Storage Institute, Singapore

A. Ishida
National Institute for Materials
Science, Japan

A. D. Johnson
TiNi Alloy Company, USA

H. Y. Kim
University of Tsukuba, Japan

M. Kohl
Forschungszentrum Karlsruhe,
IMT, Germany

T. Lagrange
University of California,
Lawrence Livermore
National Laboratory, Livermore,
USA

H. -J. Lee
Sungkyunkwan University,
Korea

J. K. Luo
University of Bolton, UK

R. L. De Miranda
University of Kiel, Germany

S. Miyazaki
University of Tsukuba, Japan

K. P. Mohanchandra
University of California, Los
Angeles, USA

S. E. Ong
Nanyang Technological
University, Singapore

E. Quandt
University of Kiel, Germany

A. G. Ramirez
Yale University, New Haven, USA

Y. C. Shu
National Taiwan University,
Taiwan

M. Tomozawa
University of Tsukuba,
Japan

C. Zamponi
University of Kiel,
Germany

S. Zhang
Nanyang Technological
University,
Singapore

Preface

Shape memory alloys (SMAs) are materials that, after being severely deformed, can return to their original shape upon heating. These materials possess a number of desirable properties, namely, high power to weight (or force to volume) ratio, thus the ability to induce large transformation stress and strain upon heating/cooling, pseudoelasticity (or superelasticity), high damping capacity, good chemical resistance and biocompatibility, etc. These unique features have attracted much attention to the potential applications of SMAs as smart (or intelligent) and functional materials. More recently, thin film SMAs have been recognized as a new type of promising and high-performance material for microelectromechanical system (MEMS) and biological applications.

Among these SMA films, TiNi based films are the most promising ones. They are typically prepared by a sputtering method. Other technologies, e.g., laser ablation, ion beam deposition, arc plasma ion plating, plasma spray and flash evaporation, have also been reported in the literature, but with some intrinsic problems. It is well known that the transformation temperatures, shape memory behaviors and superelasticity of the sputtered TiNi films are sensitive to metallurgical factors (alloy composition, contamination, thermomechanical treatment, annealing and aging processes, etc.), sputtering conditions (co-sputtering with multi-targets, target power, gas pressure, target-to-substrate distance, deposition temperature, substrate bias, etc.), and the application conditions (loading conditions, ambient temperature and environment, heat dissipation, heating/cooling rate, strain rate, etc.).

The main advantages for MEMS applications of TiNi thin film include high power density, large displacement and actuation force, low operation voltage, etc. The work output per unit volume of thin film SMA exceeds that of all other microactuation materials and mechanisms. Application of SMA films in MEMS also facilitates the simplification of mechanisms with flexibility in design and creation of clean, friction free and non-vibrating movement. The phase transformation in SMA thin films is accompanied by significant changes in the mechanical, physical, chemical, electrical and optical properties, such as yield stress, elastic modulus, hardness, damping, shape recovery, electrical resistivity, thermal conductivity, thermal expansion coefficient, surface roughness, vapor permeability and dielectric constant, etc. TiNi thin films are sensitive to environmental changes such as thermal, stress, magnetic or electrical fields, and thus should be ideal for applications in microsensors as well.

Since TiNi films can provide a large force and/or large displacement in actuation, most applications of TiNi films in MEMS are focused on microactuators. Micro-pumps, microvalves, microgrippers, springs, microspacers, micropositioners and microrappers are typical among others that have been realized. TiNi based micro-pumps and microvalves are attractive for many applications, for instance, implantable drug delivery, chemical analysis and analytical instruments, etc. Grasping and manipulating small or micro-objects with high accuracy is required for a wide range of important applications, such as microassembly in microsystems, endoscopes for microsurgery, and drug injection micromanipulators for cells.

The good biocompatibility of TiNi films is promising for their biological applications, which is a huge market at present and is still growing. At present, increasing attention has been paid to the use of TiNi thin film for minimally invasive surgery, microstents and bioMEMS applications. Microactuators made of TiNi thin films may be used to infuse drugs, or be placed in strategic locations in the body to assist circulation. TiNi SMA thin films, in the superelastic state, are also promising as compliant elements in MEMS and biological devices.

The development of TiNi based SMA thin films and their microactuators has achieved considerable progress in recent years. This was largely driven by a fast expansion of, in particular, MEMS and biological communities, in which the demand for novel actuators and biological applications has been growing dramatically. As such, a timely review of the important issues pertaining to the preparation of high quality and high performance shape memory TiNi thin films and the technical applications of these films is necessary. This book aims to serve this purpose. We believe that this is the first book dedicated to thin film TiNi SMAs; it covers not only the state-of-the-art technologies for thin film SMAs (preparation and characterization), but also their applications, in particular, in MEMS and biomedical devices. This book should naturally serve not only as an introduction to those who want to know more about this exciting field, but also as a technical handbook for those who have some knowledge but want to know more. Hence, it is an essential reference book both for a better understanding of the fundamental issues in technical aspects and for catching up with the current developments in technologies and applications in new frontiers.

The book is naturally divided into two parts, namely, technologies (from Chapter 1 to Chapter 9) and applications (Chapter 10 to Chapter 19). The first part is focused on the fundamental issues of sputter-deposited TiNi based SMA thin films, covering a general overview (Chapter 1); the basics of martensitic trans-formation (Chapter 2); deposition technologies (Chapter 3), Ti/Ni multi-layer thin films (Chapter 4); microstructure, crystallization, mechanical properties and stress evolution in thin film SMA (Chapters 5 to 7); as well as advanced post-treatment of thin film SMA including ion implantation and laser annealing (Chapters 8 and 9). The second part is devoted to the device applications based on TiNi based SMA thin film, focusing mainly on MEMS and biological applications.

It covers: an overview of applications (Chapter 10); theory and simulation of shape memory microactuators (Chapter 11); MEMS devices of microvalves, micropumps, microcages, micromirrors, superelastic thin film for medical applications, and thin film composite microactuators (Chapters 12 to 19).

The editors express their heartfelt thanks to the distinguished international team of contributors whose scientific efforts unite to form this book. We also express special thanks to the staff, Dr. Michelle Carey and Miss Sarah Matthews, at Cambridge University Press for their assistance.

Abstracts of chapters

1 Overview of sputter-deposited TiNi based thin films

S. MIYAZAKI, Y. Q. FU AND W. M. HUANG

Abstract: The motivation for fabricating sputter-deposited TiNi base shape memory alloy (SMA) thin films originates from the great demand for the development of powerful microactuators, because actuation output (force and displacement) per unit volume of thin film SMA exceeds those of other microactuation mechanisms. Stable shape memory effect and superelasticity, which are equivalent to those of bulk alloys, have been achieved in sputter-deposited TiNi thin films. Narrow transformation temperature hysteresis and high transformation temperatures were also achieved in TiNiCu and TiNi(Pd or Hf) thin films, respectively. In the meantime, unique microstructures consisting of non-equilibrium compositions and nanoscale precipitates in the matrix have been found in Ti-rich TiNi thin films that were fabricated from an amorphous condition by annealing at a very low temperature. Several micromachining processes have been proposed to fabricate the prototypes of microactuators utilizing TiNi thin films. This chapter will review the recent development of the above-mentioned topics relating to sputter-deposited TiNi based thin films. Some critical issues and problems in the development of TiNi thin films are discussed, including preparation and characterization considerations, residual stress and adhesion, frequency improvement, fatigue and stability, and thermomechanical modeling. Recent developments in the microdevices based on SMA thin films are also summarized.

2 Martensitic transformation in Ti-Ni alloys

S. MIYAZAKI

Abstract: The basic characteristics of the martensitic transformation of Ti-Ni shape memory alloys are described. They include the crystal structures of the parent and martensite phases, the recoverable strain associated with the martensitic transformation, the transformation temperatures, the temperature and orientation dependence of deformation behavior, etc. Shape memory and superelasticity related to the martensitic transformation are also explained.

3 Deposition techniques for TiNi thin film

A. D. JOHNSON

Abstract: Direct current vacuum sputter deposition is the commonly used method of creating TiNi thin film. Polished silicon wafers are a preferred substrate. The limitations on composition and impurities are similar to those for bulk material. These limitations impose severe constraints on sputtering conditions for obtaining optimal performance of the resulting material. Obtaining material with desirable shape memory properties, uniform composition, and uniform thickness requires an understanding and control of the processes used. With sputter deposition it is possible to produce thin films with a range of transition temperatures from 173 K to 373 K. Superelastic thin film can be made without cold work. After deposition, photolithography and chemical etching are used to create shapes and combine thin film with other materials to produce microdevices. Producing thin film with shape memory properties is not difficult, but to obtain uniformity and high yield requires specialized equipment and great care in process control. This chapter introduces some specific recommendations for fabrication of TiNi thin film and its incorporation in useful devices. Applications of TiNi thin film are described elsewhere (see Chapter 10).

4 TiNi multilayer thin films

H. CHO AND S. MIYAZAKI

Abstract: The martensitic transformation temperatures of TiNi shape memory alloys are strongly affected by composition. However, the composition of the TiNi thin film fabricated by a conventional sputtering method using a TiNi alloy target is not easy to adjust precisely. In order to control the composition of thin films, a dual-source sputtering method using pure Ti and Ni targets can be an alternative candidate for fabricating TiNi multilayer thin films, which can be converted to TiNi alloy thin films by heat-treatment for alloying. This chapter explains the fabrication method and alloying process of the TiNi multilayer thin films in addition to the shape memory properties of the TiNi thin films made by alloying the TiNi multilayer thin films.

5 Crystallization and microstructural development

A. G. RAMIREZ, X. HUANG AND H.-J. LEE

Abstract: This chapter focuses on the crystallization of TiNi shape memory alloy thin films. These materials are commonly sputter-deposited in an amorphous form and require high-temperature thermal treatments to create their crystalline (actuating) form. The microstructures that emerge during crystallization depend on the nucleation and growth kinetics. This chapter briefly surveys crystallization theory and methods to determine these kinetic parameters such as calorimetry, X-ray diffraction and microscopy. Novel microscopy methods have also been developed to provide a robust description that can give rise to the prediction of microstructures. In addition to presenting these tools, this chapter will also survey various factors that influence crystallization and microstructural development, which include annealing temperature, composition, substrate materials and film thickness.

6   Mechanical properties of TiNi thin films
A. ISHIDA

Abstract: This chapter is devoted to the mechanical properties of TiNi thin films. In this chapter, the shape memory behavior, two-way shape memory effect, superelasticity, and stress–strain curves of sputter-deposited TiNi films are discussed with special attention to unique microstructures in Ti-rich TiNi films.

7   Stress and surface morphology evolution
Y. Q. FU, W. M. HUANG, M. CAI AND S. ZHANG

Abstract: The residual stress in a film is an important topic of extensive studies as it may cause film cracking and peeling off, or result in deformation of the MEMS structure, deterioration of the shape memory and superelasticity effect. In this chapter, the stress in thin film shape memory alloys is characterized. The recovery stress, stress rate and stress–strain relationship during phase transformation are introduced. The dominant factors affecting stress evolution during deposition, post-annealing and phase transformation are discussed. The mechanisms for the stress-induced surface relief, wrinkling and reversible trench have also been studied. New methods for characterization of stress-induced surface morphology changes have been introduced, including atomic force microscopy (AFM) and photoemission electron microscopy (PEEM).

8   Ion implantation processing and associated irradiation effects
T. LAGRANGE AND R. GOTTHARDT

Abstract: In this chapter, we describe the influence of ion implantation on the microstructural modifications in TiNi SMA thin films. We focus on investigations involving 5 MeV Ni ion irradiation since it can be used as a means to selectively alter the transformation characteristics and to develop NiTi based thin film actuator material for MEMS devices. The primary effects of ion implantation on microstructure are summarized.

9   Laser post-annealing and theory

W. M. HUANG AND M.H. HONG

Abstract: TiNi shape memory thin films have great potential as an effective actuation material for microsized actuators. However, a high temperature (above approximately 723 K) is required in order to obtain crystalline thin films either during deposition (e.g., sputtering) or in post-annealing. Such a high temperature is not fully compatible with the traditional integrated circuit techniques, and thus brings additional constraint to the fabrication process. Laser annealing provides an ideal solution to this problem, as the high temperature zone can be confined well within a desired small area at a micrometer scale. In this chapter, we demonstrate the feasibility of local laser annealing for crystallization in as-deposited amorphous TiNi thin films and present a systematic study of the theories behind this technique.

10   Overview of thin film shape memory alloy applications

A. D. JOHNSON

Abstract: This chapter discusses properties affecting thin film applications and resulting devices that have been developed since sputtered TiNi thin film with shape memory properties was first demonstrated in 1989. As the shape memory alloy technology has matured, the material has gradually gained acceptance. TiNi thin film shape memory alloy (SMA) exhibits intrinsic characteristics similar to bulk nitinol: large stress and strain, long fatigue life, biocompatibility, high resistance to chemical corrosion, and electrical properties that are well matched to joule heating applications. In addition, thin film dissipates heat rapidly so that it can be thermally cycled in milliseconds. These properties make TiNi thin film useful in making microactuators. Micro electro mechanical (MEMS) processes – specifically photolithography, chemical etching, and the use of sacrificial layers to fabricate complex microstructures – combine TiNi thin film with silicon to provide a versatile platform for fabrication of microdevices. A variety of microdevices has been developed in several laboratories, including valves, pumps, optical and electrical switches, and intravascular devices. Interest in thin film applications is increasing as evidenced by the number of recent publications and patents issued. Intravascular medical devices are currently in clinical trials. The future for thin film devices, especially in medical devices, seems assured despite the fact that to this day no "killer application" has emerged.

11   Theory of SMA thin films for microactuators and micropumps

Y. C. SHU

Abstract: This chapter summarizes several recent theoretical and computational approaches for understanding the behavior of shape memory films from the microstructure to the overall ability for shape recovery. A new framework for visualizing microstructure is presented. Recoverable strains in both single crystal and polycrystalline films are predicted and compared with experiments. Some opportunities for new devices and improvements in existing ones are also pointed out here.

12   Binary and ternary alloy film diaphragm microactuators

S. MIYAZAKI, M. TOMOZAWA AND H.Y. KIM

Abstract: TiNi based shape memory alloy (SMA) thin films including TiNi, TiNiPd and TiNiCu have been used to develop diaphragm microactuators. The TiNi film is a standard material and the ternary TiNi. The TiNiPd alloy films have their own attractive characteristics when compared with the TiNi films. The TiNiPd alloy is characterized by high transformation temperatures so that it is expected to show quick response due to a higher cooling rate: the cooling rate increases with increasing the temperature difference between the transformation temperature and room temperature, which is the minimum temperature in conventional circumstances. The martensitic transformation of the TiNiCu and the R-phase transformation of the TiNi are characterized by narrow transformation hystereses which are one-fourth and one-tenth of the hysteresis of the martensitic transformation in the TiNi film. Thus, these transformations with a narrow hysteresis are also attractive for high response microactuators. The working frequencies of two types of microactuators utilizing the TiNiPd thin film and the TiNiCu film reached 100 Hz while the working frequency of the microactuator using the R-phase transformation reached 125 Hz.

13   TiNi thin film devices

K. P. MOHANCHANDRA AND G. P CARMAN

Abstract: This chapter provides a brief review of TiNi thin film devices, both mechanical and biomedical, that have been studied during the last decade. Prior to reviewing the devices, we first provide a description of the physical features that are critical in these devices, including deposition, residual stresses and fabrication. In general, this chapter concludes that the main obstacle for implementing devices today remains the control of TiNi properties during manufacturing. If this can be overcome, the next issue is to develop acceptable micromachining techniques. While several techniques have been studied, considerable work remains on fully developing processes that would be required in mass manufacturing processes. Finally, if these issues can be resolved, the area with the most promise is biomedical devices, the argument here being that bio-medicine remains one of the major applications areas for macroscopic TiNi structures today.

14   Shape memory microvalves

M. KOHL

Abstract: Microvalves are a promising field of application for shape memory alloy (SMA) microactuators as they require a large force and stroke in a restricted space. The performance of SMA-actuated microvalves does not only depend on SMA material properties, but also requires a mechanically and thermally optimized design as well as a batch fabrication technology that is compatible with existing microsystems technologies. The following chapter gives an overview of the different engineering aspects of SMA micro-valves and describes the ongoing progress in related fields. Different valve types based on various design–material–technology combinations are high-lighted. The examples also demonstrate the opportunities for emerging new applications.

15   Superelastic thin films and applications for medical devices

C. ZAMPONI, R. L. DE MIRANDA AND E. QUANDT

Abstract: Superelastic shape memory materials are of special interest in medical applications due to the large obtainable strains, the constant stress level and their biocompatibility. TiNi sputtered tubes have a high potential for applica-tion as vascular implants, e.g. stents, whereas superelastic TiNi-polymer-composites could be used for novel applications in orthodontics and medical instrumentation as well as in certain areas of mechanical engineering. In orthodontic applications, lowering the forces that are applied to the teeth during archwire treatment is of special importance due to tooth root resorption, caused by the application of oversized forces. Furthermore, the use of super-elastic materials or composites enables the application of constant forces independent of diminutive tooth movements during the therapy due to the su-perelastic plateau. Superelastic TiNi thin films have been fabricated by mag-netron sputtering using extremely pure cast melted targets. Special heat treatments were performed for the adjustment of the superelastic properties and the transformation temperatures. A superelastic strain exceeding 6% at 36°C was obtained.

16   Fabrication and characterization of sputter-deposited TiNi superelastic microtubes

P. J. BUENCONSEJO, H.-Y. KIM AND S. MIYAZAKI

Abstract: A novel method of fabricating TiNi superelastic microtubes with a dimension of less than 100 µm is presented in this chapter. The method was carried out by sputter-deposition of TiNi on a Cu-wire substrate, and after deposition the Cu wire was removed by etching to produce a tube hole. The shape-memory/superelastic behavior and fracture strength of the microtubes were characterized. The factors affecting the properties and a method to produce high-strength superelastic TiNi microtubes are discussed.

17   Thin film shape memory microcage for biological applications

Y. Q. FU, J. K. LUO, S. E. ONG AND S. ZHANG

Abstract: This chapter focuses on the fabrication and characterization of microcage for biopsy applications. A microcage based on a free standing film could be opened/closed through substrate heating with a maximum temperature of 90 °C, or Joule heating with a power less than 5 mW and a maximum response frequency of 300 Hz. A TiNi/diamond-like carbon (DLC) microcage has been designed, analyzed, fabricated and characterized. The bimorph structure is composed of a top layer of TiNi film and a bottom layer of highly compressively stressed DLC for upward bending once released from the substrate. The fingers of the microcage quickly close because of the shape memory effect once the temperature reaches the austenite transformation point to execute the gripping action. Opening of the microcage is realized by either decreasing the temperature to make use of the martensitic transformation or further increasing the temperature to use the bimorph thermal effect. The biocompatibility of both the TiNi and DLC films has been investigated using a cell-culture method.

18   Shape memory thin film composite microactuators

E. QUANDT

Abstract: Shape memory thin film composites consisting of at least one shape memory thin film component are of special interest as microactuators since they provide two-way shape memory behavior without any training of the shape memory material. Furthermore, they allow the realization of novel concepts like bistable or phase-coupled shape memory actuators. The potential of shape memory thin film composite microactuators is discussed in view of possible applications.

19 **TiNi thin film shape memory alloys for optical sensing applications**

Y. Q. FU, W. M. HUANG AND C. Y. CHUNG

Abstract: This chapter focuses on the optical sensing applications based on TiNi films. When the TiNi film undergoes a phase transformation, both its surface roughness and reflection change; this can be used for a light valve or on–off optical switch. Different types of micromirror structures based on sputtered TiNi based films have been designed and fabricated for optical sensing applications. Based on the intrinsic two-way shape memory effect of free standing TiNi film, TiNi cantilever and membrane based mirror structures have been fabricated. Using bulk micromachining, TiNi/Si and TiNi/Si$_3$N$_4$ bimorph mirror structures were fabricated. As one application example, TiNi cantilevers have been used for infrared (IR) radiation detection. Upon absorption of IR radiation, TiNi cantilever arrays were heated up, leading to reverse R-phase transition and bending of the micromirrors.

1 Overview of sputter-deposited TiNi based thin films

S. Miyazaki, Y. Q. Fu and W. M. Huang

Abstract

The motivation for fabricating sputter-deposited TiNi base shape memory alloy (SMA) thin films originates from the great demand for the development of powerful microactuators, because actuation output (force and displacement) per unit volume of thin film SMA exceeds those of other microactuation mechanisms. Stable shape memory effect and superelasticity, which are equivalent to those of bulk alloys, have been achieved in sputter-deposited TiNi thin films. Narrow transformation temperature hysteresis and high transformation temperatures were also achieved in TiNiCu and TiNi (Pd or Hf) thin films, respectively. In the meantime, unique microstructures consisting of non-equilibrium compositions and nanoscale precipitates in the matrix have been found in Ti-rich TiNi thin films which were fabricated from an amorphous condition by annealing at a very low temperature. Several micromachining processes have been proposed to fabricate the prototypes of microactuators utilizing TiNi thin films. This chapter will review the recent development of the above-mentioned topics relating to sputter-deposited TiNi based thin films. Some critical issues and problems in the development of TiNi thin films are discussed, including preparation and characterization considerations, residual stress and adhesion, frequency improvement, fatigue and stability, and thermomechanical modeling. Recent development in the microdevices based on SMA thin films is also summarized.

1.1 Introduction

A shape memory alloy (SMA) is a metal that can "remember" its geometry, i.e., after a piece of SMA has been deformed from its original shape, it regains its original geometry by itself during heating (shape memory effect) or simply during unloading at a higher ambient temperature (superelasticity). These extraordinary

Thin Film Shape Memory Alloys: Fundamentals and Device Applications, eds. Shuichi Miyazaki, Yong Qing Fu and Wei Min Huang. Published by Cambridge University Press. © Cambridge University Press 2009.

properties are due to a temperature-dependent martensitic phase transformation from a low-symmetry (martensite) to a highly symmetric crystallographic structure (austenite) upon cooling and a reverse martensitic transformation in the opposite direction upon heating. Shape memory effects have been found in many materials, such as metals, ceramics and polymers. Among all these materials, TiNi based alloys have been extensively studied and found many commercial applications [1, 2, 3, 4]. For microelectromechanical system (MEMS) applications, thin film based SMAs possess many desirable properties, such as high power density (up to 10 J/cm³), the ability to recover large transformation stress and strain upon heating, the shape memory effect, peudoelasticity (or superelasticity) and bio-compatibility [5, 6, 7, 8]. The work output per unit volume of SMA exceeds those of other micro-actuation mechanisms. The phase transformation in SMA thin film is also accompanied by significant changes in the mechanical, physical, chemical, electrical and optical properties, such as yield stress, elastic modulus, hardness, damping, shape recovery, electrical resistivity, thermal conductivity, thermal expansion coefficient, surface roughness, vapor permeability and dielectric constant, etc. [9, 10]. These changes can be fully utilized in the design and fabrication of microsensors and microactuators [11, 12].

Since the early 1990s, several trials have been made in order to fabricate TiNi thin films using a sputter-deposition method [13]. Some of these results showed that conventional micromachining processes are applicable for making microstructures consisting of a silicon substrate and a TiNi thin film. If the films contain micro-defects, which are characteristic in sputter-deposited films, and other elements such as oxygen and hydrogen, they will become brittle. Application of stress to such films will cause them to fracture. Without information about the characteristics of materials, any trial for an improvement in the sputtering process will be ineffective. Therefore, it was very important both to establish fabrication methods for high quality thin films in enduring high stress applications and to develop mechanical testing methods to evaluate the shape memory characteristics of thin films.

The mechanical behavior of TiNi thin films can be characterized by damping measurement, tensile tests and thermomechanical tests [10]. The crystal structures of the austenite, martensite (M) and R phases were also determined to be B2, monoclinic and rhombohedral, respectively. The transformation and shape memory characteristics of TiNi thin films were shown to depend strongly on metallurgical factors and sputtering conditions. The former includes alloy composition, annealing temperature, aging temperature and time, while the latter includes Ar pressure, sputtering power, substrate temperature and so forth [14, 15]. Conventional mechanical properties such as yield stress, ductility and fracture stress have been investigated by measuring stress–strain curves at various temperatures. The maximum elongation amounted to more than 40% in an equiatomic TiNi thin film. The yield and fracture stresses of the martensite can be as high as 600 MPa and 800 MPa, respectively [6]. These mechanical properties provide good evidence to indicate that sputter-deposited TiNi thin films possess sufficient ductility and stable shape memory characteristics for practical applications.

The stability of shape memory behavior associated with both the R-phase and martensitic transformations against cyclic deformation was investigated in TiNi thin films [6]. The R-phase transformation characteristics showed perfect stability against cyclic deformation because of the small shape change which caused no slip deformation to occur, while the martensitic transformation temperatures increased and temperature hysteresis decreased during cycling because of the formation of internal stress which is due to the introduction of dislocations. However, no slip deformation occurred during cyclic deformation under 100 MPa so that perfect stability of the shape memory effect was also observed in the martensitic transformation. Besides, by increasing the number of cycles under higher stresses, the shape memory characteristics associated with the martensitic transformation were stabilized by a training effect.

TiNi thin films with Ni-rich composition, which were age-treated at intermediate temperatures, showed a strong aging effect on transformation temperatures and shape memory behavior [16]. The aging-treatment induced fine Ti_3Ni_4 precipitates, which are metastable. The size and density of the precipitates depend on aging temperature and time, hence the shape memory characteristics varied with changing aging conditions. The precipitates with a lenticular shape are formed on {111} planes of the B2 parent phase and have their intrinsic stress fields. The stress fields cancel each other by forming the precipitates on all four {111} planes when the film is aged without any applied stress. If the film is aged under elastic constraint, the precipitates are formed preferentially on one of the {111} planes, causing a specific stress field to be created. The stress field causes two-way shape memory behavior that is mainly associated with the appearance of the R-phase transformation. Since the critical stress for slip is increased by the Ti_3Ni_4 precipitates, stable superelasticity is achieved as well as a stable shape memory effect in Ni-rich TiNi thin films [17].

Bulk Ti-rich TiNi alloys are characterized by constant and high transformation temperatures. However, when sputter-deposited amorphous Ti-rich TiNi films were subjected to heat-treatment at various temperatures, they showed a strong dependence of the transformation temperatures on the heat-treatment temperature. The heat-treatment dependent characteristics originate from the formation of non-equilibrium Ti-rich plate precipitates that are several atomic layers in thickness on {100} planes [18]. Evolution of the internal structure as a function of heat-treatment temperature was systematically clarified [15], i.e., amorphous below a crystallization temperature, Ti-rich plate precipitates at temperatures a little higher than the crystallization temperature, a mixture of Ti-rich plate precipitates and equilibrium Ti_2Ni precipitates at intermediate temperatures, and Ti_2Ni precipitates at sufficiently high temperatures. The Ti-rich plate precipitates were found to be the origin of the stability of shape memory behavior.

If a large recovery strain is required for a microactuator, we need to use the martensitic transformation, which has a larger temperature hysteresis of about 30 K in TiNi binary alloys. In order to improve the response, it is necessary to decrease the temperature hysteresis. Addition of Cu is effective in decreasing the transformation temperature hysteresis without decreasing transformation temperatures themselves. TiNiCu thin films with Cu-contents varying from 0 to 18 at% were investigated [19, 20, 21, 22]. The martensitic transformation temperatures of all the thin films were around 323 K, decreasing only slightly with increasing Cu-content in the range below 9.5 at%, while slightly increasing in the range beyond 9.5 at%. The hysteresis associated with the transformation showed a strong dependence on Cu-content, i.e., it decreased from 27 K to 10 K with increasing Cu-content from 0 to 9.5 at%. In the 9.5 at% Cu thin film, a two-stage transformation appeared and it was determined as B2→orthorhombic(O) →M-phase by X-ray diffractometry. A perfect two-stage shape memory effect was observed corresponding to these transformations. The addition of Cu caused the maximum recovery strain to decrease from 3.9 % to 1.1 % and the critical stress for slip to increase greatly from 55 MPa to 350 MPa with increasing Cu-content up to 18 at%.

Additions of third elements, such as Pd, Au, Pt, Ag, Hf, etc., are effective for increasing transformation temperatures. In the case of sputter-deposited thin films, Pd and Hf have been added as a third element for this purpose [23]. However, Hf increases transformation temperature hysteresis, while Pd is effective in decreasing the transformation temperature hysteresis in addition to increasing transformation temperatures [24, 25]. For this reason, Pd addition is more promising for applications. The TiNiPd ternary alloy thin films basically show the M-phase and O-phase transformations but not the R-phase transformation in a narrow Pd-content region. The addition of Pd is effective in increasing the O-phase transformation temperatures, e.g., the O_s (O-phase transformation start temperature) of a 21.8 at% Pd thin film is higher than the M_s (M-phase transformation start temperature) of an equiatomic TiNi binary alloy thin film by about 50 K. It was found that Pd addition is also effective in decreasing transformation temperature hysteresis, e.g., 16 K in a thin film with 22 at% Pd content. The achievement of the small transformation temperature hysteresis in TiNiCu thin films or the high transformation temperatures in Ti-NiPd thin films is promising for achieving quick movement in microactuators made of TiNi base shape memory thin films.

Since TiNi films can provide large forces for actuation and large displacement, most applications of TiNi films in MEMS are focused on microactuators [26, 27], such as cantilevers, diaphragms, micropumps, microvalves, microgrippers, springs, microspacers, micropositioners, and microrappers, mirror actuators, etc., TiNi thin films are sensitive to environmental changes, such as thermal, stress, magnetic and electrical fields. Thus, they should be also ideal for applications in microsensors.

The main potential problems associated with TiNi thin film in MEMS applications include: (1) low energy efficiency, low dynamic response speed and large hysteresis; (2) non-linearity and complex thermomechanical behavior and ineffectiveness for precise and complex motion control and force tracking; (3) potential degradation and fatigue problems. Even with the above disadvantages, the TiNi based thin film is still considered as a core technology for actuation of some MEMS devices where a large force and stroke are essential in conditions of low duty cycles or intermittent operation, and in an extreme environment, such as radioactive, space, biological and corrosive conditions.

This chapter reviews the recent development of sputter-deposited TiNi base SMA thin films and fabrication of microsystems [6]. Successful implementation of the TiNi microactuators requires a good understanding of the relationship among processing, microstructure and properties of TiNi films. The required enabling technologies for TiNi films include [10]:

- low-cost, reliable and MEMS-compatible deposition methods with precise control of film composition and quality;
- reliable and precise characterization technologies for various properties (such as shape memory effect, superelasticity and mechanical properties, etc.);
- an appropriate post-deposition annealing (for film crystallization) or aging process compatible with the MEMS process;
- precise etching and patterning of TiNi film compatible with the MEMS process and the possibility of nano-size TiNi structures and actuators;
- prediction and modeling of the non-linear behavior of TiNi films as well as design and simulation of TiNi thin film microactuators.

Some basic requirements for TiNi films in MEMS applications are listed as follows [10]:

- large recovery stress and strain;
- low residual stress to prevent undesired deformation of MEMS structures;
- high actuation speed and fast response with precise control of deformation and strain;
- good adhesion on substrate (free of cracking, delamination and spallation);
- durable and reliable shape memory effects;
- wide range choice of working temperatures (from sub-zero to several hundred degrees C);
- good resistance to surface wear and corrosion;
- biocompatibility and good corrosion resistance (for instance in bio-MEMS applications).

1.2 Fabrication and characterization methods

1.2.1 Film deposition

TiNi based films are the most frequently used thin film SMA materials and they are typically prepared using a sputtering method. Laser ablation, ion beam deposition, arc plasma ion plating, plasma spray and flash evaporation were also reported but with some intrinsic problems, such as non-uniformity in film thickness and composition, low deposition rate, and/or non-batch processing, incompatibility with MEMS process, etc. Figure 1.1 shows a schematic drawing of a

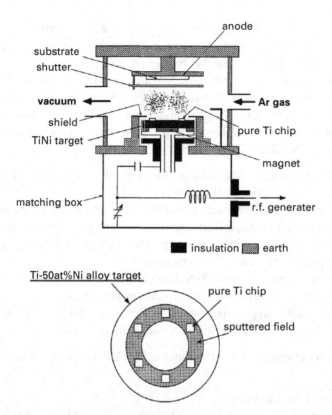

Figure 1.1   A schematic figure showing a radio frequency magnetron sputtering system [Miyazaki & Ishida, 1999 [6], with permission from Elsevier].

most common radio frequency (RF) magnetron sputtering apparatus [6]. Ar ions are accelerated into the target to sputter Ti and Ni atoms, which are deposited onto the substrate to form a TiNi film. Transformation temperatures, shape memory behaviors and superelasticity of the sputtered TiNi films are sensitive to metallurgical factors (alloy composition, contamination, thermomechanical treatment, annealing and aging process, etc.), sputtering conditions (co-sputtering with multi-targets, target power, gas pressure, target-to-substrate distance, deposition temperature, substrate bias, etc.), and the application conditions (loading conditions, ambient temperature and environment, heat dissipation, heating/cooling rate, strain rate, etc.) [28]. Systematic studies on the detailed effects of all the above parameters are necessary. The sensitivity of TiNi films to all these factors seems an intrinsic disadvantage but, at the same time, this sensitivity provides tremendous flexibility in engineering a combination of properties for intended applications.

Precise control of the Ti/Ni ratio in TiNi films is of essential importance, as has been documented since TiNi film studies started more than a decade ago. The intrinsic problems associated with sputtering of TiNi films include the difference in sputtering yields of titanium and nickel at a given sputtering power density, geometrical composition uniformity over the substrate and along the cross-sectional thickness of the coating, as well as wear, erosion and roughening of targets during sputtering [29]. To combat these problems, methods of co-sputtering of the TiNi target with another Ti target, or using two separate single element (Ti and Ni) targets, or adding titanium plates on a TiNi target are widely used [30]. Substrate rotation, optimal configuration of target position and precise control of sputtering conditions, etc. are also helpful. Varying the target temperature can produce a compositional modification: sputtering with a heated TiNi target can limit the loss of Ti, thus improving the uniformity of film properties [31, 32]. Good performance TiNi films can also be obtained by post-annealing of a multi-layer of Ti/Ni [33]. Since contamination is a big problem for good mechanical properties of sputtered TiNi films, it is important to limit the impurities, typically oxygen and carbon, to prevent the brittleness, deterioration or even loss of the shape memory effect. For this reason, the purity of the Ar gas and targets is essential, and the base vacuum of the main chamber should be as high as possible (usually lower than 10^{-7} torr). Pre-sputtering cleaning of targets before deposition effectively removes the surface oxides on targets, which thus constitutes one of the important steps in ensuring film purity. In order to deposit films without columnar structure (thus with good mechanical properties), a low processing pressure of Ar gas (0.5 to 5 mtorr) is essential. Application of a bias voltage during sputtering could modify the film microstructure, texture and stress, and is thus also important, but few studies have been reported on this topic so far. More information of this topic can be found in Chapter 3.

Important sputtering factors which will affect the quality of the films are r.f. power, Ar gas pressure, substrate–target distance, substrate temperature and alloy composition of the target used. Figure 1.2 shows the fracture surface of as-deposited thin films. The film prepared at a low Ar gas pressure exhibits a flat and featureless structure, while films prepared at high Ar gas pressure exhibit a columnar structure. This columnar structure suggests that the films are porous. This structure seems to be caused by the restricted mobility of deposited atoms on the surface of the growing film. A high Ar gas pressure is likely to decrease the energy of the sputtered atoms by collision with Ar ions, resulting in a decrease in their surface diffusion. Furthermore, under a high Ar gas pressure, Ar ions adsorbed on the film surface can interfere with the surface diffusion of sputtered Ti and Ni atoms. Of the films prepared at a high Ar gas pressure, the fracture surface of the film prepared at an r.f. power of 600 W seems to be less porous than the other films.

Depending on processing conditions, TiNi films can be deposited at room temperature or high temperatures. TiNi films sputtered at room temperature are usually amorphous, thus post-sputtering annealing (usually above 450 °C) is a must because the shape memory effect only occurs in materials of crystalline form. However, martensitic transformation and superelasticity of TiNi films are sensitive to post-annealing and/or aging temperature and duration [34, 35], thus post-sputtering annealing should be handled with care. It is suggested that the lowest possible

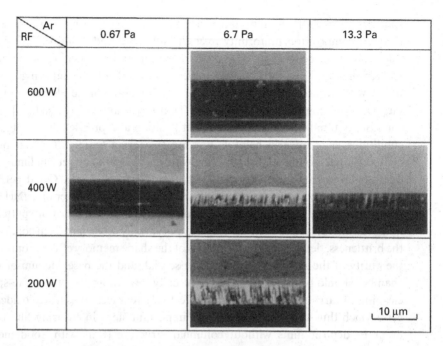

RF ＼ Ar	0.67 Pa	6.7 Pa	13.3 Pa
600 W			
400 W			
200 W			10 μm

Figure 1.2   Cross-section of TiNi thin films formed under various sputtering conditions [Miyazaki & Ishida, 1999 [6], with permission from Elsevier].

annealing or aging temperature be used in a bid to conserve thermal processing budgets and more importantly minimize the reactions between film and substrate [36]. A long-term post-annealing and aging process should be avoided since it could trigger dramatic changes in film microstructure (i.e., precipitation), mechanical properties and shape memory effects. Films deposited at a relatively high temperature (about 400 °C) are crystallized *in situ*, thus there is no need for post-annealing. Films can be deposited at relatively high temperatures (400 to 500 °C) during sputtering to form the crystallized phase, then at a relatively lower temperature (about 300 °C) to maintain a crystalline growth during the later sputtering process. Films can also be deposited at a low temperature (about 300 °C) to get partial crystallization, then annealed at a higher temperature (500 °C) for a short time to promote further crystallization.

Recently a localized laser annealing method was used for TiNi films [37], where only certain areas of a film are annealed by a laser beam to exhibit the shape memory effect, and the other non-annealed areas remain amorphous, thus acting as a pullback spring during the cooling process. This method, discussed in detail in Chapter 9, opens a new way for fabrication of microdevices [38]. The advantages of the localized laser annealing process include: (1) precision in selection of the areas to be annealed, down to a micron scale; (2) non-contact and high efficiency; (3) freedom from restrictions on design and processing; (4) ease in integration in MEMS processes; (4) ease in cutting of the final structure using the laser beam.

However, still some problems exist that include: (1) Energy loss. The TiNi film surface is usually smooth and reflection loss of laser beam energy is a big problem. Possible solutions include selection of an excimer laser beam, choice of suitable parameters (e.g. wavelength of the laser) and surface treatment or roughening of the film surface to improve laser adsorption. (2) Difficulty in duration control. Crystallization of the film structure is a thermodynamic process, and it is necessary to maintain sufficient treatment time for crystallization to complete. However, over-exposure easily causes surface damage of the thin films. (3) Need of a protection environment such as Ar gas or vacuum condition. This adds complexity and cost to the process.

1.2.2   TiNi film characterization

For freestanding TiNi films, conventional methods, such as differential scanning calorimetry (DSC) and tensile tests (stress–strain curves) are quite applicable to characterize the shape memory effects. The stress–strain and strain–temperature responses of freestanding films are commonly evaluated using tensile tests [39, 40]. Results show that the stress–strain–temperature relationship, elongation, fracture stress and yield stress are at least comparable to (if not better than) those of bulk materials, because of the grain size effect (micron or submicron size in thin films as compared with tens of microns for bulk materials) [41, 42]. The difficulties in tensile testing of TiNi thin films include: (1) to obtain free-standing films without pre-deformation; and (2) to clamp the films tightly on tester grips. For MEMS applications, the TiNi films are usually deposited on Si or other substrates. One of the important issues in characterization of TiNi films for MEMS applications is how to correctly evaluate the shape memory effects and mechanical properties of the constrained thin films on substrates. For this purpose, curvature and electrical resistivity measurements are widely used [43]. Some new methods based on MEMS techniques [44], such as the bulge test [45], TiNi/Si diaphragm [46, 47], cantilever bending or damping [48] are more appropriate for microactuator applications, which are compatible with small dimensions and high sensitivities. Nano-indentation testing with or without changes of temperature could reveal the different elastic and plastic deformation behaviors of austenite and martensite, which is also promising for characterization of superelasicity, phase transformation, the shape memory effect and mechanical properties of the constrained thin films [49, 50, 51]. Indentation of TiNi based films is strongly dependent on the materials' resistance to dislocation. Since the dislocation is closely related to the fatigue properties of films, indentation for material characterization is particularly useful for MEMS applications, where optimization of fatigue performance is critical.

Recently, an AFM based *in situ* testing method has been applied to characterize the phase transformation behavior of constrained films [10]. Figure 1.3 shows two micrographs of the AFM surface morphology of TiNi films on an Si substrate at a low temperature (martensite) and a high temperature (austenite), respectively. The surface roughness of the martensite phase is much higher than that of the

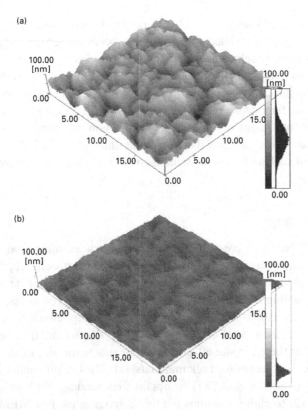

Figure 1.3   AFM surface morphologies of TiNiCu films: (a) low temperature in the martensite state and (b) high temperature in the austenite state [Fu [2004] [10], with permission from Elsevier].

austenite. With the change of temperature, the surface roughness changes drastically during transformation between the martensite and the austenite phases, thus clearly revealing the occurrence of phase transformation. The advantages of this method are its non-destructive nature and applicability to very small size films (down to nanometers). Moreover, the changes in optical reflection caused by the changes in the surface roughness and reflective index can also be used to characterize the transformation behaviors of TiNi films. The details relating to this issue can be found in Chapter 19.

There are usually some discrepancies in transformation temperatures obtained from different characterization methods [52]. The possible reasons include: (1) the phase transformation and mechanical behaviors of constrained TiNi films could be different from those of freestanding films, due to the substrate effect, residual stress, strain rate effect, stress gradient effect and temperature gradient effect; (2) the intrinsic nature of the testing method (thus the changes in physical properties will not start at exactly the same temperatures); (3) differences in testing conditions, for example, heating/cooling rate; (4) non-uniformity of the film composition over the whole substrate and along the cross-sectional

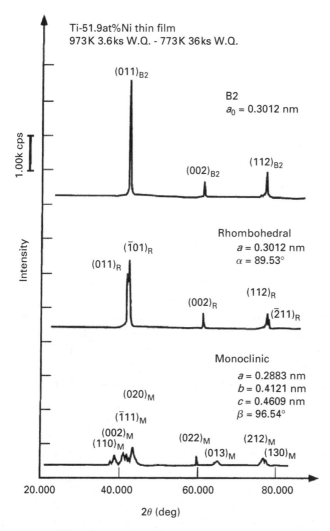

Ti-51.9at%Ni thin film
973K 3.6ks W.Q. - 773K 36ks W.Q.

$(011)_{B2}$

B2
$a_0 = 0.3012$ nm

$(002)_{B2}$

$(112)_{B2}$

Rhombohedral
$a = 0.3012$ nm
$\alpha = 89.53°$

$(\bar{1}01)_R$

$(011)_R$

$(002)_R$

$(112)_R$

$(\bar{2}11)_R$

Monoclinic
$a = 0.2883$ nm
$b = 0.4121$ nm
$c = 0.4609$ nm
$\beta = 96.54°$

$(020)_M$

$(\bar{1}11)_M$

$(002)_M$
$(110)_M$

$(022)_M$

$(013)_M$

$(212)_M$

$(130)_M$

1.00k cps

Intensity

20.000   40.000   60.000   80.000

2θ (deg)

Figure 1.4   X-ray diffraction profiles of the parent B2, R phase and monoclinic martensite phase in a Ti-51.9at%Ni thin film [Miyazaki & Ishida, 1999 [6], with permission from Elsevier].

thickness of the coating. Therefore, it is necessary to identify whether the application is based on the freestanding film or constrained film/substrate system, so that a suitable method can be chosen.

In the following sections, we will discuss the characterization results of TiNi films as well as Ti-rich and Ni-rich films.

1.2.2.1   TiNi binary alloy thin films
Basic characteristics

As mentioned above, the as-sputtered TiNi thin films are amorphous if the substrate is not heated during deposition. As a consequence, they should be crystallized by

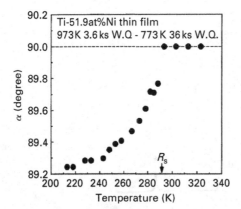

Figure 1.5   The rhombohedral angle of the R phase as a function of temperature. [Miyazaki & Ishida, 1999 [6], with permission from Elsevier].

heating at 973 K, which is higher than the crystallization temperature, followed by aging at 773 K for various times. The crystal structures of the Ti-51.9at%Ni alloy thin film age-treated for 36 ks were determined at three different temperatures, i.e., 300, 270 and 200 K, by X-ray diffraction. The parent (B2) phase, R phase and martensitic (M) phase exist independently at these temperatures, respectively, as shown in Fig. 1.4. The crystal structure of the parent phase was determined to be B2, while those of the R phase and M phase were rhombohedral and monoclinic, respectively. The lattice parameters of each phase are shown on the right side of the corresponding diffraction profile. The lattice parameters of the three phases are essentially the same as those measured in bulk specimens, although they differ only slightly depending on the alloy content. The rhombohedral angle a shows a unique dependence on temperature. Figure 1.5 shows the temperature dependence of a in the Ti-51.9at%Ni thin film. The angle starts to decrease at R_S, the temperature where the R-phase transformation starts, and gradually decreases with decreasing temperature. The lattice constant a of the R phase is almost constant irrespective of temperature.

Strain associated transformations

Figure 1.6 shows the transformation strains measured in the Ti-51.9at%Ni that was crystallized at 973 K for 3.6 ks followed by aging at 773 K for 3.6 ks. The recovery strain ε_A associated with the reverse martensitic transformation increases with the applied constant stress applied to the thin film during thermal cycling until a strain of about 2.6%. Then, the strain becomes a constant irrespective of the applied stress after reaching 2.6% at a critical stress of 300 MPa. The critical stress is the minimum stress required for the most preferentially oriented variant to grow almost fully in each grain. Therefore, stresses higher than 300 MPa are not effective in increasing ε_A. The strain ε_R associated with the R-phase transformation is also shown in Fig. 1.6. It also increases with applied constant stress up to about 0.26% which is one tenth of ε_A. After reaching a strain of 0.26% at a

Table 1.1   Calculated martensite transformation strains along specific orientations
[Miyazaki & Ishida, 1999 [6], with permission from Elsevier].

Crystal orientation	Calculated martensitic transformation strain (%)
[0 0 1]	2.63
[0 1 1]	8.20
[−1 1 1]	9.42
Average strain	8.42

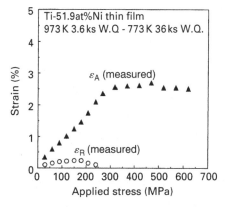

Figure 1.6   The effect of applied stress on the transformation strains in the Ti-51.9at%Ni thin film [Miyazaki & Ishida, 1999 [6], with permission from Elsevier].

stress of about 175 MPa, ε_R starts to decrease with a further increase in applied stress. The major reason for the increase in ε_R prior to the maximum point is that the most preferential variant grows with increase of the applied stress, while the major reason for the decrease in ε_R after reaching the maximum point is that the temperature range where the R phase can exist becomes narrower with increase of the applied stress, so that the lattice distortion of the R phase becomes less, as shown in Fig. 1.5.

Table 1.1 shows the transformation strains ε_A, which were calculated on the basis of the lattice distortion from the B2 phase to the M phase, along some specific crystal orientations and the average strain of ε_A. The average strain is 8.42%, which is about three times that of the observed strain. The reason why the observed strain is so small is not attributable to the texture, because the X-ray diffraction pattern of the B2 phase does not reveal a strong texture, as shown in Fig. 1.4, although the (011) peak intensity is a little larger than that of a TiNi film with randomly oriented grains. The possible reason is the strong internal structure consisting of fine Ti_3Ni_4 precipitates and fine grains of submicron size, since such an internal structure will suppress the growth of the most preferentially oriented M variant. On the other hand, Fig. 1.7(a) shows the recovery strain ε_A in a Ti-50.5 at%Ni thin film which

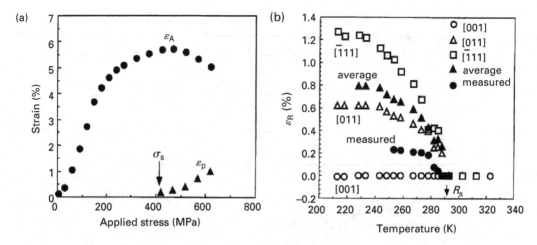

Figure 1.7   (a) The effect of applied stress on the transformation strain and permanent strain in a Ti-50.5at%Ni film. (b) The transformation strains associated with the R phase as a function of temperature [Miyazaki & Ishida, 1999 [6], with permission from Elsevier].

was solution-treated at 973 K for 3.6 ks. Since the solution-treated thin film does not contain fine Ti_3Ni_4 precipitates, the most preferentially oriented martensite variant can grow easily under low stresses. The maximum ε_A observed in Fig. 1.7(a) is 5.8% under a stress of 470 MPa, at around which the macroscopic plastic strain ε_p appears. σ_s is the critical stress for slip deformation.

In order to calculate the transformation strain ε_R, it is necessary to use the lattice constants of the R phase. The length of each axis of the R-phase unit cell is almost constant irrespective of temperature and is the same as that of the B2 phase. Therefore, the transformation strain ε_R depends on temperature. Figure 1.7(b) shows calculated strains along some specific orientations, and also the average strain as a function of temperature; the calculated average strain is also shown. The strain along each orientation increases with decreasing temperature in the temperature region below R_s: the strain along the orientation [$\bar{1}$11] shows the maximum, and that along the orientation [001] the minimum.

The observed strain is about one half of the calculated average strain, as shown in Fig. 1.6. Since the observed ε_A is one third of the calculated average strain, the suppressing effect against the growth of the preferentially oriented variant is less in the R phase than in the M phase. This can be understood in such a way that the transformation strains associated with the R-phase transformation is only one tenth of that associated with the M transformation. Details of the calculation methods for the strains associated with the M-phase and R-phase transformations are explained in Refs. [53, 54].

Superelastic behavior

Figure 1.8 shows stress–strain curves of a Ti-50.3at%Ni film at different temperatures. Curves (a) and (b) show the stress–strain curves obtained below A_s so

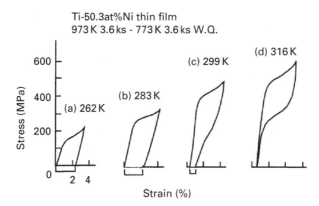

Ti-50.3at%Ni thin film
973 K 3.6 ks - 773 K 3.6 ks W.Q.

Figure 1.8 Stress–strain curves showing the shape memory effect and superelasticity in a Ti-50.3at%Ni thin film [Miyazaki & Ishida, 1999 [6], with permission from Elsevier].

that the shape change remains after unloading. The residual strain disappears upon heating to above A_f, revealing the perfect shape memory effect. Curve (c) is obtained by deforming the film at a temperature between A_s and A_f so that it shows a partial shape recovery upon unloading and further shape recovery upon heating. Finally, curve (d) shows a perfect superelasticity at a temperature above A_f. Since the superelasticity is accompanied by a stress hysteresis, it is necessary to apply a high enough stress to observe such superelasticity. In this case, the maximum stress applied to the film is higher than 600 MPa.

Texture and transformation strain anisotropy
The specific texture developed in materials causes anisotropic characteristics to appear in mechanical, electrical and magnetic behavior. The shape memory and mechanical properties of TiNi films depends significantly on the orientation of the crystal grains. Deposition conditions, film composition, and post-deposition thermo-mechanical treatment could have important consequences on the formation and evolution of the film texture. A strong film texture may lead to an anisotropic shape memory effect since the recoverable strain and deformation behavior is highly dependent on the film crystallographic orientation [55, 56, 57].

Figures 1.9(a) and (b) show pole figures obtained using diffraction from {110} and {200} planes in Ti-52.2 at%Ni and Ti-51.6 at%Ni thin films, respectively. The Ti-52.2 at%Ni thin film shows a considerably uniform orientation distribution of grains with a weak [302] fiber texture, the maximum axis density being only 3.9. On the other hand, the Ti-51.6at%Ni thin film shows a strong [110] fiber texture with a maximum axis density of 110. Since the poles of the fiber textures are both normal to film planes, the in-plane crystal orientation distribution is uniform so that transformation strain anisotropy is weak in both film planes. Calculated and experimentally measured transformation strains are shown in Figs. 1.9(c) and (d) for the Ti-52.2 at%Ni and Ti-51.6 at%Ni, respectively. The experimental results were obtained under various constant stresses by cooling and heating the films.

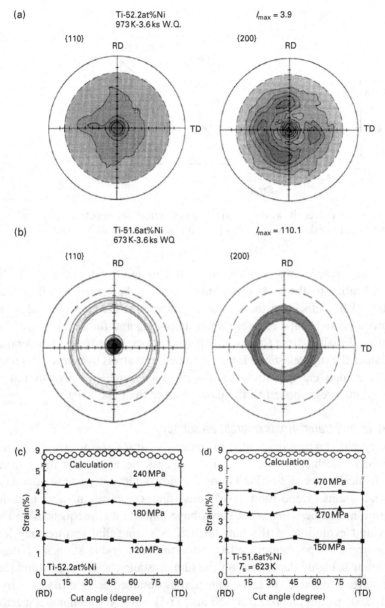

Figure 1.9   (a) {110} and {200} pole figures in a Ti-52.2 at %Ni thin film that was crystallized at 973 K for 3.6 ks. (b) {110} and {200} pole figures in Ti-51.6at%Ni thin films that were heat-treated at 673K for 3.6 ks The film was sputtered on the substrate at 623 K. (c) Comparison between measured and calculated transformation strains in Ti-52.2 at%Ni thin film that was crystallized at 973 K for 3.6 ks. The film was sputtered on the substrate at 623 K. (d) Comparison between measured and calculated transformation strains in Ti-51.6at%Ni thin film that was heat-treated at 673 K for 3.6 ks after being sputtered on the substrate at 623 K [Miyazaki & Ishida, 1999 [6], with permission from Elsevier].

The transformation strain increases with increasing stress. However, it is almost constant for each stress irrespective of direction. The calculated results were obtained by assuming that only the most favorable martensite variant grows in each entire grain, so that the calculated result gives the maximum recoverable strain, which is always larger than the experimental result. This isotropic nature of the transformation strain in the film plane is convenient for designing micro-actuators of TiNi thin films.

Some studies have been done to clarify the texture and the anisotropy in the deformation behavior in rolled TiNi thick plates [58]. However, few studies have investigated the texture and the anisotropy in shape memory behavior in sputter-deposited TiNi thin films with a thickness less than 10 μm. The post-annealed crystallized TiNi films normally have a strong fiber texture along austenite [110] [59], whereas at room temperature, martensite (200) and (022) peaks become dominant. Hassdorf et al. deposited a TiNiCu film on a SiO_2 substrate using molecular beam epitaxy (MBE) technology [60], and found that the film has a distinct austenite (200) diffraction peak. The crystallites are oriented within ±3° along the film plane normal. The authors pointed out that an intermediate Ti_2Ni layer is crucial for the formation of (200) texture. Under tensile load, [100] orientation is characterized as "hard" since it demonstrates small uniaxial transformation strain levels and begins transforming at a significant higher stress [61]. The film with (100) texture has a highest transformation stress compared with the [111] and [110] texture. The [111] orientation is characterized as "soft" since it demonstrates large uniaxial transformation strains and low critical transform-ation stress levels. The recoverable strain of TiNi film with (111) texture is much higher than that with the commonly observed (110) texture. However, so far, (111) dominant texture in TiNi based films is difficult to achieve. It clearly indicates an opportunity to improve shape memory effects by targeting special textures using novel processing techniques.

Strength and ductility

Strength and ductility were investigated by measuring stress–strain curves in three types of TiNi thin films, i.e., Ti-48.3at%Ni, Ti-50.0at%Ni and Ti-51.3at%Ni [6]. These films were heat-treated in order to form specific internal structure in each case. The Ti-48.3at%Ni thin films were annealed at 773 K for 0.3 ks (5 minutes) to form a non-equilibrium phase which has a shape of thin plates with a thickness of several atomic spacings on $\{100\}_{B2}$ planes. The Ti-50.0at%Ni is a typical alloy of TiNi and was annealed at 773 K for 3.6 ks. It was a single phase without any secondary phase or precipitates. The Ti-51.3 at%Ni was annealed at 973 K for 3.6 ks followed by aging at 673 K for 3.6 ks to form fine Ti_3Ni_4 precipitates, which have a lenticular shape locating on $\{111\}_{B2}$ planes. The average grain size of the Ti-51.5 at%Ni was 0.5 μm, while those of the Ti-50.0 at%Ni and Ti-48.3 at%Ni were 4.5 and 5.0 μm, respectively.

Figure 1.10(a) shows fracture strains of the three types of TiNi thin films at various temperatures. Each point was measured in an independent sample at each

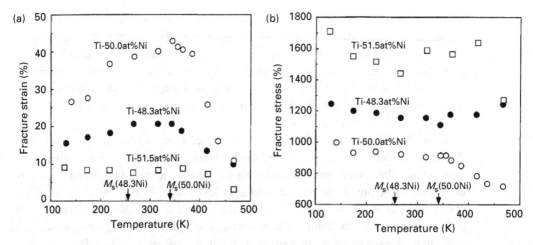

Figure 1.10   (a) Fracture strain and (b) fracture stress as a function of test temperature in Ti-48.3at%Ni, Ti-50.0at%Ni and Ti-51.5at%Ni thin films [Miyazaki & Ishida, 1999 [6], with permission from Elsevier].

temperature by deforming it until fracture. Fracture strains of the Ti-50.0at%Ni films are always highest, while those of the Ti-51.5at%Ni films are lowest. The fracture strains of both the Ti-48.3at%Ni and Ti-50.0at%Ni films show a specific relationship with the testing temperature. They increase with increasing temperature until the M_s point of each alloy films is reached: the M_s points of the Ti-48.3at%Ni and Ti-50.0at%Ni films are 253 K and 341 K, respectively. In this temperature region, the deformation mode is de-twinning in thermally induced martensites and reorientation of martensite variants. The reorientation is also essentially de-twinning, because the interface of martensite variants is a twinning plane. Therefore, the flow stress decreases with increasing temperature, resulting in the normal temperature dependence of fracture strain in this temperature region. On the other hand, above the M_s point, fracture strain decreases with increasing temperature. In this temperature region, the major deformation mode in the initial stage of deformation is stress-induced martensitic transformation at lower temperatures and slip deformation in the parent phase at higher temperatures. The stress for inducing the martensitic transformation increases with increasing temperature, causing the relaxation of stress concentration accumulated during deformation to be more difficult at higher temperatures. Larger fracture strain is usually expected in the martensite than in the parent phase, because many deformation modes such as slip deformation and twinning are available in the martensite phase, while slip deformation is the only one in the parent phase. Therefore, the fracture strain is lowest at a temperature where no martensitic transformation will be stress-induced. In the case of Ti-51.5at%Ni films, slip deformation is hard due to precipitation hardening so that the fracture strain does not exceed 10 %. Since the M_s point was not detected above 127 K, the fracture strains will be those of the parent phase.

Table 1.2 Mean diameter of Ti₃Ni₄ precipitates in Ti-51.3at%Ni thin films aged at various temperatures for various times after solution-treatment at 973 K for 3.6 ks [Miyazaki & Ishida, 1999 [6], with permission from Elsevier].

Aging temperature (K)	Aging time (ks)		
	3.6	36	360
773	92 nm	300 nm	460 nm
723	31 nm	87 nm	240 nm
673	10 nm	28 nm	74 nm
623	–	–	25 nm
573	–	–	–

Figure 1.10(b) shows the fracture stress of the three types of TiNi film as a function of testing temperature. The fracture stress is the highest in each stress–strain curve so that the fracture stress corresponds to the ultimate tensile stress. The temperature dependence of the fracture stress is different in each specimen. However, it is not easy at present to explain the difference in the temperature dependencies of these films. Since the fracture stress and strain are large enough and almost equivalent to those of TiNi bulk materials, sputter-deposited TiNi thin films are ductile and stable enough for practical applications. More discussions can be found in Chapter 6.

1.2.2.2 Ni-rich TiNi alloy thin films
Microstructure of age-treated thin films

Figure 1.11 shows the microstructures of Ti-51.3at%Ni thin films aged at various temperatures between 573 and 773 K for various times of 1, 10 and 100 h (3.6, 36 and 360 ks) after solution treatment at 973K for 1 h. As can be seen in this figure, the size of the precipitates increases with increasing aging temperature and aging time for all the age-treated thin films, while the grain size is almost constant, about 1 μm. All the precipitates in the figure are confirmed to be a Ti₃Ni₄ phase by electron diffraction. The bright field image and diffraction pattern of the precipitates in the thin film aged at 673 K for 10 h are shown in Fig. 1.12. The diffraction pattern is consistent with a mixture of [111]$_{B2}$ zone and [111]$_{Ti_3Ni_4}$ zones, and the same diffraction pattern is observed for all the age-treated thin films. The bright field image also shows the characteristic morphology of Ti₃Ni₄ precipitates. That is, the lenticular shape precipitates are observed along the three directions of $[\bar{1}10]_{B2}$, $[10\bar{1}]_{B2}$ and $[0\bar{1}1]_{B2}$, which are the traces of $(11\bar{1})_{B2}$, $(1\bar{1}1)_{B2}$ and $(\bar{1}11)_{B2}$ respectively. Another precipitated is located on the $(111)_{B2}$ plane, which is parallel to the photograph. The sizes of the precipitates were measured to describe the fineness of the microstructure. Since the precipitates are formed on {111} planes of the B2 phase in a lenticular shape, the longitudinal length was chosen to represent the size of the precipitates. The average sizes of the precipitates are given in Table 1.2. The sizes of the precipitates in the thin films aged at 573 K and 623 K are not listed here, since the precipitate contrast is too weak to measure the size.

Figure 1.11   Transmission electron micrographs of Ti_3Ni_4 precipitates in Ti-51.3at%Ni thin films which were solution-treated at 973 K for 3.6 ks followed by age-treatment at 773 K for (a) 3.6 ks; (b) 36 ks and (c) 360 ks, at 673 K for (d) 3.6 ks, (e) 36 ks and (f) 360 ks, and at 573 K for (g) 3.6 ks; (h) 36 ks and (i) 360 ks [Miyazaki & Ishida, 1999 [6], with permission from Elsevier].

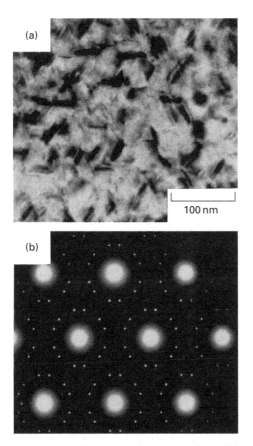

Figure 1.12   (a) Bright-field image and (b) electron diffraction pattern of Ti_3Ni_4 precipitates in a Ti-51.3at%Ni thin film which was age-treated at 673 K for 36 ks after solution-treatment at 973 K for 3.6 ks; the foil normal is nearly parallel to [111]$_{B2}$ in (a) and exactly parallel to [111]$_{B2}$ in (b) [Miyazaki & Ishida, 1999 [6], with permission from Elsevier].

Aging effect

Figure 1.13 shows the effect of aging time on the transformation temperatures of the Ti-51.9at%Ni alloy thin films that were aged at 773 K for different durations after solution treatment at 973 K for 3.6 ks. The solid lines show the DSC curves measured upon cooling, while the dashed curves show those upon heating. Solution treated film ($X = 0$) shows no transformation peak upon cooling in the DSC curve, indicating that the transformation temperatures are very low, hence again there is no reverse transformation peak upon heating. In age-treated thin films, there are two transformation peaks appearing on each solid curve: the two peaks denoted by R^* and M^* correspond to the R-phase and martensitic transformations, respectively. Upon heating the age-treated films show only one reverse transformation peak A^* except for the film aged for 3.6 ks which shows transformation peaks A^* and RA^*. A^* points of the films aged for 36 and 360 ks represent the reverse transformation from M directly to B2, while

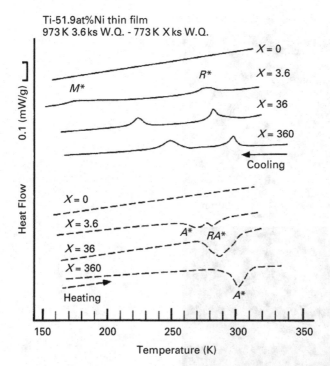

Figure 1.13   The effect of aging time on the transformation temperatures measured by DSC in the Ti-51.9at%Ni thin film [Miyazaki & Ishida, 1999 [6], with permission from Elsevier].

A^* and RA^* of the film aged for 3.6 ks represent the reverse transformation from M to R and from R to B2, respectively. Both R^* and M^* show an aging effect which causes them to increase with increasing aging time. However, M^* increases more sensitively to the aging time than R^*. The increase in M^* and R^* can be explained by the decrease in the Ni-content of the matrix of the film, because the growth of the Ti_3Ni_4 precipitates will consume excess Ni in the matrix and the transformation temperatures increases with decreasing Ni-content of the matrix.

The shape memory behavior of the thin films used in Fig. 1.13 is shown in Fig. 1.14. Strains were measured upon cooling (solid lines) and heating (dashed lines) under 240 MPa in the Ti-51.9at%Ni thin films which were aged at 773 K for 0, 3.6, 36 and 360 ks, respectively. Two-stage shape change appears both at R_s and M_s in the age-treated thin films, while a single stage shape change at M_s in the solution treated thin film. All the R_s and M_s in Fig. 1.14 are higher than those estimated by the DSC curves, because the DSC curves were measured under no load, while the strain was measured under stress: the stress increases the transformation temperatures following the Clausius–Clapeyron relationship. Since M_s (or M^*) increases more effectively by aging than R_s (or R^*), the temperature difference between R_s and M_s becomes smaller with increasing aging time. The strain induced by the transformations increases with increasing

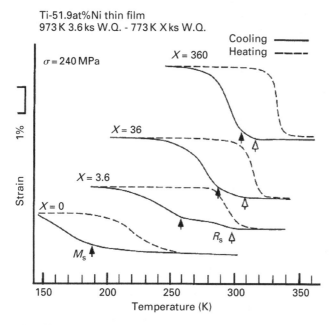

Ti-51.9at%Ni thin film
973 K 3.6 ks W.Q. - 773 K X ks W.Q.

Figure 1.14   The effect of aging time on the shape memory behavior in the Ti-51.9at%Ni thin film [Miyazaki & Ishida, 1999 [6], with permission from Elsevier].

aging time, indicating that as the Ti_3Ni_4 precipitates grow by aging they lose the suppressing force against the growth of the preferentially oriented martensite variants.

The aging effect was also investigated in the Ti-43.9at%Ni thin films. There is no aging effect observed in the transformation temperatures, as shown by the DSC results in Fig. 1.15. Two-stage transformation behavior is observed both upon cooling and heating. The appearance of the R phase is usually attributed to a fine internal structure consisting of dislocations and/or precipitates in bulk specimens. However, the R phase can not be attributed to such internal structure in the Ti-43.9at%Ni thin film, because there are neither dislocations nor Ni-rich Ti_3Ni_4 precipitates. The cause of the appearance of the R phase is considered to be another fine internal structure consisting of Ti_2Ni compounds and small grains. The grain size of sputter-deposited TiNi thin films generally ranges from 0.5 μm to several μm, while the grain size of solution-treated bulk TiNi alloys is up to several tens of μm. This fine internal structure in the thin films also suppresses the martensitic transformation more effectively than the R-phase transformation.

Figure 1.16 shows the shape memory behavior in the thin films corresponding to the samples used in Fig. 1.15. Again no aging effect is observed in the shape memory behavior. The transformation temperatures and transformation strains do not change with aging time. The reason why there is no aging effect in the Ti-43.9at%Ni thin film is that in Ni-poor TiNi alloy, the Ti_3Ni_4 precipitates can not be created, so that there is no variation in Ni-content of the matrix. The

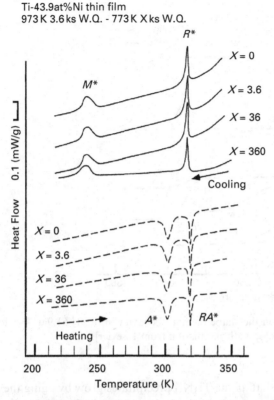

Ti-43.9at%Ni thin film
973 K 3.6 ks W.Q. - 773 K X ks W.Q.

Figure 1.15 The effect of aging time on the transformation temperatures measured by DSC in the Ti-43.9at%Ni thin film [Miyazaki & Ishida, 1999 [6], with permission from Elsevier].

strain vs. temperature curves clearly show two-stage deformation behavior both upon cooling and heating, corresponding to the two-stage transformation behavior in Fig. 1.15.

The M_s of both the Ti-51.9at%Ni and Ti-43.9at%Ni thin films are replotted against aging time in Fig. 1.17, where M_s measured by both DSC and mechanical tests are included. Controlling transformation temperatures is one of the important techniques needed in order to fabricate SMA thin film microactuators suitable for various purposes. The M_s increases with increasing aging time in the Ti-51.9at%Ni, while there is no aging effect in the Ti-43.9at%Ni film. These alloys are considered to possess an advantageous characteristic for fabricating microactuators: i.e., the Ni-rich alloy shows the aging effect so that the transformation temperatures are adjustable by heat-treatment even if the alloy content of the thin film can not be adjusted as one wishes, while the Ni-poor alloy does not show an aging effect so that the transformation temperatures are less sensitive to the variation of heat treatment condition. Besides, the Ni-content of the matrix is also constant irrespective of nominal composition, because the formation of Ti_2Ni

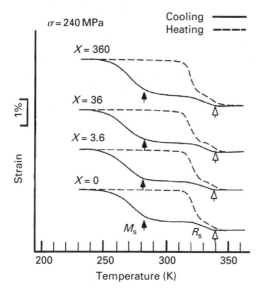

Figure 1.16 The effect of aging time on the shape memory behavior in the Ti-43.9at%Ni thin film [Miyazaki & Ishida, 1999 [6], with permission from Elsevier].

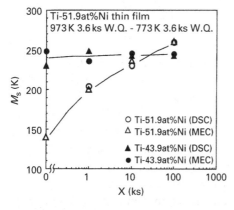

Figure 1.17 The effect of aging time on the martensitic transformation start temperature M_s in the Ti-51.9at%Ni and Ti-43.9at%Ni thin films [Miyazaki & Ishida, 1999 [6], with permission from Elsevier].

keeps the Ni-content of the matrix in equilibrium. Hence the transformation temperatures are insensitive to the nominal composition of the Ni-poor thin films.

However, since as-sputtered thin films are amorphous in Ni-poor (or Ti-rich) films, if the heat-treatment is not sufficient to achieve an equilibrium condition, a non-equilibrium condition can exist so that transformation temperatures and shape memory behavior will show sensitivity to heat treatment.

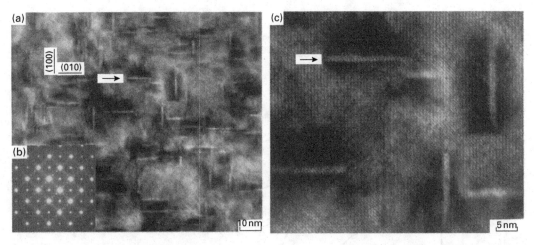

Figure 1.18   (a) High-resolution electron micrograph of a Ti-48.2at%Ni thin film heated at 745 K for 3.6 ks and (b) the corresponding diffraction pattern. (c) Enlarged micrograph of a part of (a), showing details of the plate precipitates [Miyazaki & Ishida, 1999 [6], with permission from Elsevier].

1.2.2.3   Ti-rich TiNi alloy
Non-equilibrium phase and composition

Since as-sputtered TiNi thin films are amorphous if the substrate temperature is not raised intentionally, the Ti-rich or Ni-rich TiNi thin films can contain excess Ti or Ni atoms, respectively, in the amorphous phase. Although equilibrium composition of TiNi varies depending on temperature, generally speaking it only ranges from 49.5 at%Ni to 50.5 at%Ni, where the B2 single phase is stable above M_s. Therefore, both Ti-rich and Ni-rich TiNi thin films may reveal non-equilibrium phases during a crystallization process.

It has actually been found that Ti-rich TiNi thin films show a non-equilibrium phase, which has never been observed in bulk TiNi alloys, when they are heat-treated at temperatures a little above the crystallization temperature [18]. Figure 1.18(a) shows a high-resolution electron micrograph of a Ti-48.2at%Ni thin film which was heated at 745 K for 3.6 ks. The electron diffraction pattern taken from this area is shown in Fig. 1.18(b). The electron micrograph in Fig. 1.18(a) was taken with all the diffraction spots in Fig. 1.18(b) passing through an objective aperture. The zone axis of the diffraction pattern corresponds to [001] of the B2 structure of the parent phase. Fairly strong streaks along the [100] and [010] directions are seen in Fig. 1.18 (b). It is considered that these have been caused by the very thin platelets observed in Fig. 1.18(a), which lie on {100} planes. Although the diffraction pattern in Fig. 1.18 (b) is of the R phase, it was confirmed by an *in situ* electron microscopy observation, using a heating stage, that these streaks also exist in the B2 parent phase. Each of these platelets is about 0.5 nm thick and has a disk shape with a radius of 5–10 nm. These platelets are distributed about 5–10 nm apart. Part of the micrograph of Fig. 1.18(a) is enlarged to show the detail of the platelets in Fig. 1.18(c). In this

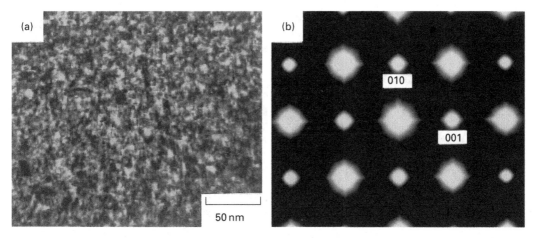

Figure 1.19 (a) Bright-field image and (b) the corresponding diffraction pattern of a Ti-48.2at%Ni thin film heated at 773 K for 0.3 ks. [Miyazaki & Ishida, 1999 [6], with permission from Elsevier].

enlarged high-resolution micrograph, we can see lattice images of {100} planes with an interplanar distance of 0.30 nm, and it is seen that each platelet consists of only two or three {100} lattice planes. These platelets have a Ti-rich composition and they are formed at the interfaces between crystallized domains during the crystallization process. Since the as-sputtered amorphous film includes 51.8at%Ti and the crystallized B2 phase can contain only 49.5at%Ti in an equilibrium condition, excess Ti atoms are carried away to the periphery of the crystallized domain, resulting in the formation of Ti-rich platelets at the interfaces between crystallized domains.

The same internal structure can be observed in the Ti-48.2at%Ni which was heated at 773 K for 0.3 ks (5 minutes) as shown in Fig. 1.19(a). The corresponding diffraction pattern in Fig. 1.19(b) shows a [100] zone pattern of the B2 parent phase. This observation indicates that the non-equilibrium Ti-rich platelets can be formed at the initial stage of heating even at considerably high temperatures. Figures 1.20(a) and (b) show a bright field image and the corresponding diffraction pattern of the same alloy film which was heated at 773 K for 3.6 ks. Figure 1.20(a) reveals spherical precipitates appearing as Moiré patterns in addition to the Ti-rich platelets. These spherical precipitates can be distinguished from the Ti-rich platelets by taking a picture in random orientation, as shown in Fig. 1.20(c). The formation of the spherical precipitates produces extra spots in the diffraction pattern in addition to the streaks due to the platelets (Fig. 1.20(b)). These extra spots could be indexed as a Ti_2Ni phase.

Figures 1.21(a) and (b) show bright field images of a Ti-48.2at%Ni thin film annealed at 773K for 36 ks. The corresponding electron beams are parallel to the [100] and [111] of B2, respectively. In both figures, the spherical precipitates are distinguished by Moiré patterns. Moiré fringes are parallel to the {110} planes of B2. The Ti_2Ni is an equilibrium phase. However, such a uniform distribution of

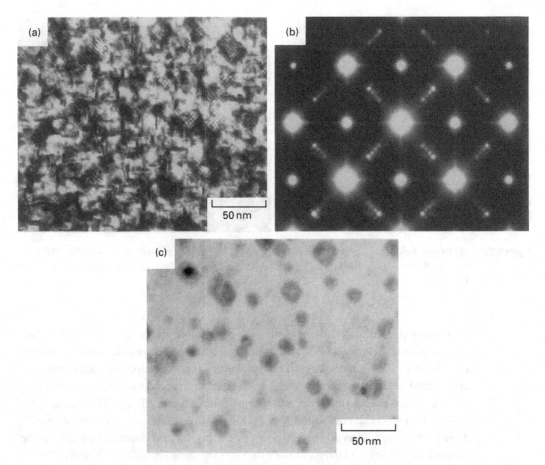

Figure 1.20   Microstructure of a Ti-48.2at%Ni thin film heated at 773 K for 3.6 ks: (a) bright-field
image; (b) the corresponding diffraction pattern and (c) bright-field image taken in
random orientation [Miyazaki & Ishida, 1999 [6], with permission from Elsevier].

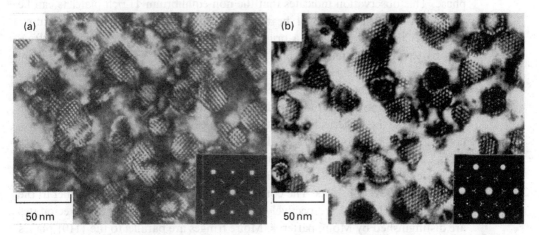

Figure 1.21   Bright-field images of a Ti-48.2at%Ni thin film heated at 773 K for 36 ks. The electron
beams are parallel to (a) [100] and (b) [111], respectively [Miyazaki & Ishida, 1999 [6],
with permission from Elsevier].

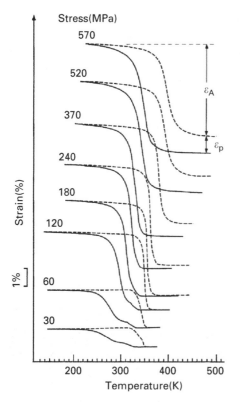

Figure 1.22 Strain–temperature curves under constant stresses for a Ti-48.2at%Ni thin film which was heated at 745 K for 3.6 ks [Miyazaki & Ishida, 1999 [6], with permission from Elsevier].

Ti$_2$Ni precipitates in grains is not of an equilibrium state. In the case of Ti-rich TiNi bulk materials, the Ti$_2$Ni precipitates preferentially distribute along grain boundaries. Such an equilibrium distribution of Ti$_2$Ni can be formed in the film when it is heated for a longer time or at higher temperatures. More detailed information can be found from Ref. [62].

Strengthening mechanism

As described above, the internal structure of the Ti-rich TiNi thin film evolves by means of changing the type of phases and their distribution. Therefore, it is expected that there will be an optimum heat-treatment condition for shape memory characteristics. Figure 1.22 shows a typical result of a mechanical test in which a Ti-48.2at%Ni thin film was thermally cycled between 140 K and 550 K under various constant stresses. The specimen was heat-treated at 745 K for 3.6 ks prior to the test. The test was performed in such a way that the magnitude of the applied stress was increased stepwise after each thermal cycle, starting from 30 MPa. One specimen was used throughout the whole test. It is revealed in Fig. 1.22 that the recoverable strain (ε_A) increases with increasing stress, up to 5.5% at 240 MPa without noticeable plastic strain involved. Above this stress level, the plastic strain ε_p in

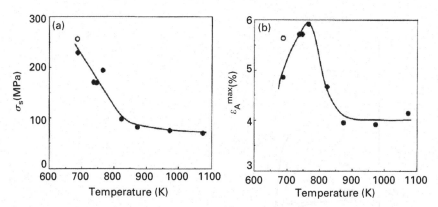

Figure 1.23   Variation in (a) critical stress for slip and (b) the maximum recoverable strain as a function of heat-treatment temperature [Miyazaki & Ishida, 1999 [6], with permission from Elsevier].

each thermal cycle gradually increases to nearly 1% at 570 MPa. However, it should be noted that a recoverable strain of about 5% can still be obtained under such a high level of stress.

The critical stress σ_s, below which any appreciable amount of ε_p could not be detected in the thermal cycle test, was measured in specimens heat-treated at various temperatures for 3.6 ks. The results are shown in Fig. 1.23(a), where σ_s is plotted as a function of heat-treatment temperature T_h. The critical stress σ_s rapidly increases with decreasing T_h below 820 K. Above 820 K, σ_s appears to be almost constant. This critical temperature of 820 K coincides with the heat-treatment temperature above which subnanometric nonequilibrium plate precipitates are no longer observable. This indicates that the formation of the nonequilibrium plate precipitates is an important factor for the remarkable increase in σ_s below 820 K. The maximum value of σ_s, 260 MPa, was obtained in the specimen heat-treated at 687 K for 6.4 ks (open circle in Fig. 1.23(a).

The maximum recoverable strain ε_A^{max} for each heat-treatment condition was obtained by the thermal cycling test and plotted as a function of T_h in Fig. 1.23(b). As shown in the figure, ε_A^{max} increases with decreasing T_h or increasing σ_s until T_h reaches down to 773 K. Below 773 K, however, ε_A^{max} decreases with further decreasing T_h. The ε_A^{max} decreases at 687 K, because the crystallization does not complete for this heat-treatment. Heating at this temperature for 6.4 ks caused ε_A^{max} to increase, as shown in Fig. 1.23(b), because further precipitation of the plate precipitates and further crystallization occur.

1.2.3   TiNiX ternary alloy thin films

1.2.3.1   TiNiCu films

Applications of microactuators require high frequency and fast response (narrow transformation temperature range and hysteresis). One of the challenges for the successful application of TiNi films is the effective reduction of hysteresis and

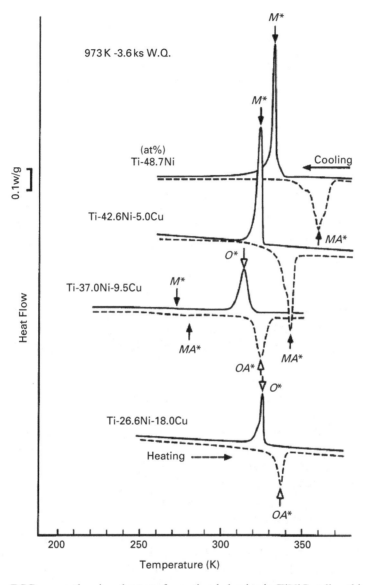

Figure 1.24 DSC curves showing the transformation behavior in TiNiCu alloy thin films [Miyazaki & Ishida, 1999 [6], with permission from Elsevier].

increase in operating frequency. The binary TiNi alloy films have a large temperature hysteresis of about 30 °C, and TiNi films with small hysteresis are preferred for faster actuation. Addition of Cu in TiNi films is effective in reducing the hysteresis [63]. Compared with TiNi binary alloy, TiNiCu alloys show less composition sensitivity in transformation temperatures, lower martensitic yield stress, and superior fatigue property, etc., which make them more suitable for microactuator application [64].

Table 1.3 Lattice parameters of the Ti-37.0Ni-9.5Cu(at%) [Miyazaki & Ishida, 1999 [6], with permission from Elsevier].

	a (nm)	b (nm)	c (nm)	α (°)	β (°)	γ (°)
B2	0.3032	0.3032	0.3032	90.00	90.00	90.00
Orthorhombic	0.2875	0.4198	0.4508	90.00	90.00	90.00
Monoclinic	0.3031	0.4254	0.4828	90.00	96.96	90.00

Figure 1.24 shows DSC curves measured in the TiNiCu thin films [6]. The solid lines indicate the transformation upon cooling, while the dashed lines the reverse transformation upon heating. The Ti-48.7at%Ni binary and the Ti-42.6Ni-5.0Cu (at%) ternary alloys show a single peak associated with the transformation from the B2 (parent phase) to M (monoclinic martensite) upon cooling and associated with the reverse transformation from M to B2 upon heating, respectively. The sharpness of these peaks indicates that the distribution of alloy composition is uniform. The DSC curve of Ti-37.0Ni-9.5Cu(at%) film shows two peaks. The first peak is very sharp and high, but the second one is very diffuse and almost invisible. However, when the ordinate of the curve is magnified, the second peak becomes visible. Based on the X-ray diffraction results, it is confirmed that the former transformation is associated with the transformation from B2 to O (orthorhombic martensite) upon cooling and the reverse transformation from O to B2 upon heating. The lattice parameters of these three phases (B2, O, M) are shown in Table 1.3. The Ti-26.6Ni-18.0Cu(at%) shows only a single-stage transformation which is associated with the transformation from B2 to O upon cooling and the corresponding reverse transformation from O to B2 upon heating, respectively. The transformation temperature from O to M is supposed to decrease significantly with an addition of Cu, so that it becomes un-measurable.

Figure 1.25 shows the transformation temperatures of all the specimens as a function of Cu-content, M^* and O^* being the transformation peak temperatures of the B2 to M (or O to M) and B2 to O, respectively, while A^* and OA^* are the transformation peak temperatures of the corresponding reverse transformations, respectively. These temperatures are as high as those of bulk specimens, implying that the thin films contain few impurities. The M^* decreases slightly with increasing Cu-content until 9.5 at%. The 9.5at%Cu alloy shows a two-stage transformation, the first transformation temperature O^* being 314 K and the second one M^* being 270 K. By adding more Cu, M^* decreases drastically, while O^* increases slightly. The transformation hysteresis (A^*-M^*) or (OA^*-O^*), shows a strong dependence on Cu-content, as shown in Fig. 1.26. A stronger Cu-dependence of the hysteresis is observed in the single-stage transformation region than in the two-stage transformation region. The hysteresis decreases from 27 K to 11 K with increasing Cu-content from 0 at% to 9.5 at%, and this property is comparable to that of bulk specimens.

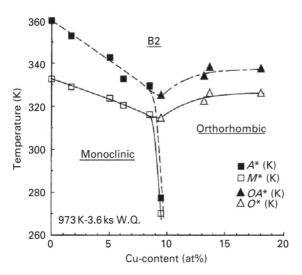

Figure 1.25 Cu-content dependence of transformation temperatures (M^*, O^*, A^*, OA^*) in TiNiCu thin films [Miyazaki & Ishida, 1999 [6], with permission from Elsevier].

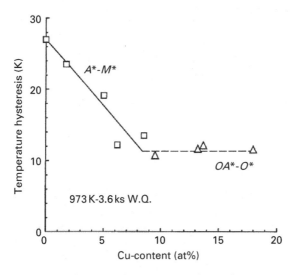

Figure 1.26 The effect of Cu-content on the transformation temperature hysteresis in TiNiCu alloy thin films [Miyazaki & Ishida, 1999 [6], with permission from Elsevier].

Figure 1.27 shows the strain vs. temperature (ε–T) curves for TiNiCu alloy thin films measured during cooling and heating under a variety of constant stresses. The ε–T curves measured under the same stresses (60, 120, 180 MPa) in each specimen are shown, in order to clarify the effect of the Cu-content on the deformation behavior. The 0at%Cu (binary) alloy thin film shows a single-stage shape change associated with the transformation from B2 to M upon cooling. The deformation starts at M_s

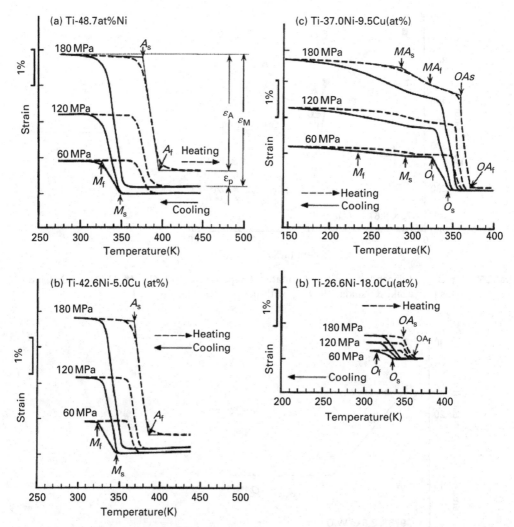

Figure 1.27   Strain versus temperature curves measured during cooling and heating under a variety of constant stresses in TiNiCu thin films. [Miyazaki & Ishida, 1999 [6], with permission from Elsevier].

and finishes at M_f; the strain induced here is estimated as ε_M, which generally consists of recoverable transformation strain and unrecoverable plastic strain. The elongated specimen contracts toward its original shape due to the reverse martensitic transformation upon heating; the recovery strain is estimated as ε_A. The unrecoverable strain ε_P is the permanent strain due to slip deformation which occurs during the transformation. Both the transformation strains ε_M and ε_A increase with increasing stress until the stress reaches a critical value under which the most preferential martensite variant will occupy the most part of the specimen.

The 5.0at%Cu specimen also shows a single-stage deformation associated with the transformation from B2 to M, similarly to the 0at%Cu specimen. The strain

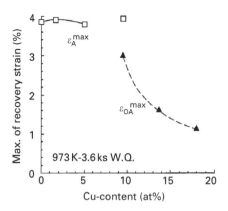

Figure 1.28   Cu-content dependence of maximum recovery strain in TiNiCu thin films [Miyazaki & Ishida, 1999 [6], with permission from Elsevier].

and the shape of the ε curve are similar to those of the 0at%Cu specimen except for the temperature hysteresis. This indicates that the addition of Cu is effective in reducing the hysteresis even through the structural change of the transformation is the same as that of the binary alloy. By adding further Cu, a two-stage deformation is observed for 9.5at%Cu. The first shape change is associated with the B2–O transformation, while the second shape change is associated with the O–M transformation. The first shape change occurs in a narrow temperature region, and the second in a broad temperature region; these shape changes correspond well to the transformation behavior measured by DSC, as shown in Fig. 1.24. The temperature hysteresis of the first stage transformation is smaller than that of the 5at%Cu specimen. The ε–T curve of the 18.0at%Cu shows only a single-stage deformation again, although it is associated with the transformation B2–O. The strain induced by the transformation is very small.

The maximum of the recovery strain ε_A^{max} is shown as a function of Cu-content in Fig. 1.28. Open squares show the ε_A^{max} associated with the B2–M transformation in the specimens with 0, 1.7 and 5.0at%Cu or the B2–O–M transformation in the 9.5at%Cu specimen. Closed triangles show the ε_A^{max} associated with the B2-O transformation in the Cu-rich specimens. The ε_A^{max} is almost constant irrespective of Cu-content if the Cu-content is less than 9.5 at%, while it decreases from 3.9 % to 1.1 % with increasing Cu-content from 9.5 at% to 18.0 at% in the Cu-rich region where the transformation only occurs from B2 to O. This Cu-dependence of the ε_A^{max} in the thin films is similar to that in bulk specimens. However, the ε_A^{max} is a little smaller than that of the bulk specimens in the Cu-poor region. This is supposed to come from the grain size effect; i.e., the grain size of thin films is less than that of bulk specimens.

The permanent strain ε_P due to slip deformation appears when a specimen is subjected to thermal cycling under a constant stress which is higher than the critical stress for slip. The critical stress for slip, σ_s, can be estimated by extrapolating the data of ε_P to zero strain in a diagram showing the ε_p vs. constant applied

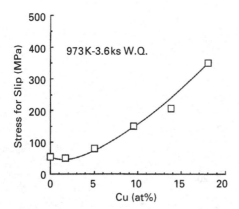

Figure 1.29   Cu-content dependence of the critical stress for slip in TiNiCu thin films. [Miyazaki & Ishida, 1999 [6], with permission from Elsevier]

stress relationship. Values of σ_s estimated in this way are shown in Fig. 1.29. It is found that σ_s increases with increasing Cu-content. For example, the σ_s of the 0at%Cu specimen is only 55 MPa and that of the 18at%Cu specimen increases to 350 MPa, showing that the addition of Cu is also effective in increasing the stress for slip.

Two factors are considered to be the causes for the effect of the addition of Cu on the critical stress for slip; i.e., (1) solid–solution hardening due to the third element and (2) small transformation strain. Therefore, the addition of Cu is quite effective not only to decrease the temperature hysteresis but also to stabilize the shape memory effect.

1.2.3.2   TiNiPd films

The working principle of TiNi microactuators renders them very sensitive to environment. The maximum transformation temperature of binary TiNi thin films is usually less than 100 °C. However, a lot of MEMS applications require higher temperatures. For example, in automobile applications, the transformation temperature required is up to 150 °C, and in high-temperature gas chromatography the operation temperature is up to 180 °C, etc. A ternary system is a solution. By adding a varying amount of a third element, such as Pd, Hf, Zr, Pt, Au, etc., into the binary alloys, one can easily adjust the transformation temperatures from 100 °C to 600 °C. TiNiPd and TiNiHf films are also effective in decreasing the temperature hysteresis, thus promising quick movement at higher temperatures [65]. The potential problem is that all these high-temperature ternary thin films are high cost with poor shape memory effect and thermal stability, as well as brittleness [66]. A small amount of Pd or Pt addition (less than 10 at%) could reduce martensitic transformation temperatures rather than increase them [67]. A slight increase in Ni content in the film can dramatically decrease the phase transformation temperatures. However, Hf increases transformation hysteresis, while Pd is effective in decreasing the transformation hysteresis apart from

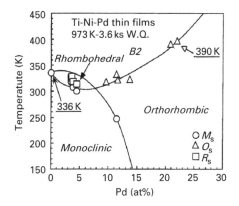

Figure 1.30   Pd-content dependence of transformation temperatures (M_s, R_s and O_s) in TiNiPd thin films [Miyazaki & Ishida, 1999 [6], with permission from Elsevier].

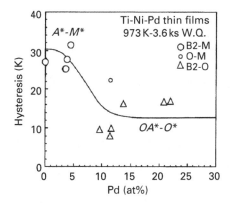

Figure 1.31   The effect of Pd-content on the transformation hystereses (A^*-M^*, OA^*–O^*) in TiNiPd thin films [Miyazaki & Ishida, 1999 [6], with permission from Elsevier].

increasing transformation temperatures. For this reason, Pd addition is more promising for applications. Figure 1.30 shows the effect of Pd-content on the transformation temperatures (M_s, O_s, R_s) in TiNiPd thin films which were annealed at 973 K for 3.6 ks, Ti-content being kept between 50.0 and 51.0 at%. M_s decreases slightly with increasing Pd-content up to around 6 at%, then increases up to 390 K, which is 54 K higher than that of the binary TiNi film, with further increasing Pd-content up to 22 at%. On the other hand, it is also found that the transformation temperature hysteresis is strongly affected by Pd-content, i.e., it decreases with increasing Pd-content down to around 10 K as shown in Fig. 1.31. This is an unexpected phenomenon, but both effects of Pd-addition are useful for quick response in actuation.

Figure 1.32 shows the strain vs. temperature relationships of a Ti-26.4Ni-21.8Pd (at%) thin film. Upon thermal cycling, the shape changes occur in a higher temperature region than the binary TiNi and ternary TiNiCu alloys, although the transformation temperatures are a little lower than those measured by DSC because

Table 1.4 Transformation and shape memory characteristics of TiNi, TiNiCu and TiNiPd alloy thin films. [Miyazaki & Ishida, 1999 [6], with permission from Elsevier]

	TiNi (M phase)	TiNiCu (O phase)	TiNiPd (O phase)
$M^*(O^*)$ (K)	332	313	385
$A^*(OA^*)$ (K)	359	324	401
Hys (K)	27	11	16
ε_A^{max} (%)	3.8	3.0	2.5
σ_S (MPa)	90	173	200

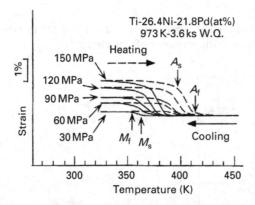

Figure 1.32   Strain versus temperature curves measured during cooling and heating under a variety of constant stresses in a TiNiPd thin film [Miyazaki & Ishida, 1999 [6], with permission from Elsevier].

of the possible variation in alloy composition in one film. TiNiPd thin films with different Pd-contents were investigated. The transformation temperatures during shape change decrease with increasing Pd-content until reaching 7at%Pd, then start to increase and become higher than those of the TiNi binary alloy when the Pd-content is more than 17 at%. The Ti-26.4Ni-21.8Pd(at%) alloy thin film has not only higher transformation temperatures but also a higher resistance against slip, because an almost perfect shape recovery exhibits even under a stress of 200 MPa. The maximum recoverable strain is about 2.5% and it is the smallest among TiNiCu, TiNiPd ternary and TiNi binary alloy films. The small transformation strain is also another reason for high critical stress for slip in addition to a solid–solution hardening effect by the third element of Pd. A summary of the transformation and shape memory characteristics of these three alloy thin films is shown in Table 1.4. $M^*(O^*)$ and $A^*(OA^*)$ are abbreviations of the transformation peak temperatures for the martensitic (orthorhombic) and reverse-martensitic (reverse-orthorhombic) transformations, respectively, measured by DSC. Hys is an abbreviation of the temperature hysteresis, i.e., the temperature difference between $M^*(O^*)$ and $A^*(OA^*)$. ε_A^{max} and σ_S stand for the maximum recoverable transformation strain and the critical stress for slip, respectively.

1.2.4   Residual stress and stress evolution

Residual stress and stress evolution in the films could pose potential problems in applications, as it may influence not only adhesion between film and substrate, but also deformation of the MEMS structure, mechanics and thermodynamics of transformation and superelasticity effects, etc. [68]. Large residual stress could lead to either film cracking or decohesion under tension, or film delamination and buckling under compression. Deposition conditions, post-deposition thermo-mechanical treatment and composition of the TiNi films could have important consequences with respect to the development of residual stress. These have been studied in detail and reported in Ref. [69]. In crystalline TiNi films, large tensile stress is generated during heating due to the phase transformation from martensite to austenite, while during cooling, the martensitic transformation occurs and the tensile stress drops significantly due to the formation and alignment of twins. The stress generation and relaxation behaviors upon phase transformation are significantly affected by film composition, deposition and/or annealing temperatures, which strongly control the formation and evolution of intrinsic stress, thermal stress and phase transformation behaviors [70].

Stress evolution could have a significant effect on the film surface morphology evolution. Significant surface relief (or surface upheaval), caused by the martensitic transformation, is commonly observed in TiNi bulk materials and has recently also been reported in sputtered TiNi thin films [71]. During the martensitic transformation, atomic displacement introduces stacking faults that lead to surface relief morphology on the film surface. A flat surface in austenite transforms to twinned martensite upon cooling and becomes rough, without macroscopic shape change, and vice versa. Reference [72] reported a phenomenon of film surface morphology evolution between wrinkling and surface relief during heating/cooling in a sputtered TiNiCu thin film. *In situ* optical microscopy observation upon heating revealed that the interweaving martensite plate structure disappeared. However, many radial surface wrinkles formed within the original martensitic structure. Further heating up to 300 °C did not lead to much change in these wrinkling patterns. On subsequent cooling to room temperature, the twinned martensite plates or bands reformed in exactly the same wrinkling patterns as those before thermal cycling.

After post-annealing a partially crystallized TiNiCu film at 650 °C, optical microscopy (at room temperature) revealed an interconnected network structure of trenches on the film surface [73]. *In situ* observation using optical microscopy, interferometry and AFM during heating/cooling showed that these trenches gradually disappear upon heating, and the film surface becomes smooth and featureless. On subsequent cooling the trenches re-appeared, with almost the identical surface morphology as was present before heating, and the surface became slightly opaque and cloudy. The details are discussed in Chapters 7 and 19.

The substrate effect is also significant in stress generation and evolution, because the difference in thermal expansion coefficients between the substrate and

TiNi films significantly affects the thermal stress. The film intrinsic stress is also critically dependent on the mismatch between film and substrate. So far, most studies have been focused on Si-based substrates for MEMS applications. TiNi deposited on other substrates (with different coefficients of thermal expansion) could result in different stress states (compressed or tensile) and stress–temperature evolution behaviors, thus detailed studies of substrate effect on stress state, shape memory effect, phase transformation and mechanical properties of TiNi films deserve more systematic effort [74].

In order to minimize the residual stress in TiNi films, it is necessary to: (1) precisely control the Ti/Ni ratio; (2) deposit films at a possible lower pressure; (3) select a suitable deposition temperature or annealing temperature, with a compromise between thermal stress and intrinsic stress; (4) use interlayers (with possible compressive stress) to reduce large tensile stress in TiNi films; (5) perform post-annealing, ion beam post-modification, or *in situ* ion beam modification during sputtering in order to reduce intrinsic stress; (6) select a suitable substrate to reduce thermal stress. More information can be found in Chapter 7.

1.2.5 Frequency response

Applications of microactuators require not only large recovery stress and large transformation deformation, but also high frequency and fast response (narrow transformation hysteresis). One of the challenges for the successful application of TiNi films is effective reduction of hysteresis to increase operating frequency. External heat is necessary for generating phase transformation and actuation, and the response speed of TiNi microactuators is mainly limited by their cooling capacities. The binary TiNi alloy films have a large temperature hysteresis of about 30 °C. TiNi films with smaller hysteresis are preferred for faster actuation. The hysteresis could be slightly reduced by decreasing the cyclic temperature amplitude and/or increasing working stress. R-phase transformation usually has a much smaller temperature hysteresis, which is useful for MEMS applications [75]. However, the problem in R-phase transformation is that the strain and stress (or force) generated are too small to be of much practical use. Addition of Cu in TiNi films is effective in reducing the hysteresis. Compared with a TiNi binary alloy, TiNiCu alloys also show less composition sensitivity in transformation temperatures, lower martensitic yield stress, and superior fatigue property, etc., which make them more suitable for microactuator application. However, the transformation temperatures of TiNiCu films decrease slightly, and the transformation becomes weaker with the increase of Cu content, in terms of recovery stress, maximum recovery strain and heat generation, etc. Also the film becomes brittle when Cu content is more than 10 at%.

Generally speaking, the constrained films have smaller hysteresis than freestanding films, and the films with larger compressive stress could have much smaller (even almost zero) hysteresis compared with films with large tensile stress. Therefore, selection of a suitable substrate (with a larger thermal expansion coefficient than the TiNi film) could help generate a large compressive stress, thus a smaller hysteresis. An alternative is to use an external heat sink. TiNi based films can be deposited on a suitable substrate with good thermal conductivity, like Cu plate, thus significantly improving thermal dissipation and working frequency. However, this brings in more critical issues, such as integration and compatibility with the MEMS batch process, residual stress and adhesion.

1.2.6 Adhesion and interfacial analysis

When TiNi films are deposited on an Si substrate, there exist interfacial diffusion and chemical interactions at the interface whereby titanium and nickel silicides may form during high-temperature deposition or post-deposition annealing. These interfacial reaction products could be complex, heterogeneous and meta-stable [76,77]. Since the thickness of TiNi film required in MEMS applications is usually less than a few microns, a relatively thin reaction layer could have a significant adverse effect on adhesion and shape memory properties. TiNi film adheres well to silicon substrate provided it is clean and pre-chemically etched. TiNi films deposited on a glass substrate can be easily peeled off, which is quite useful for obtaining freestanding films. In MEMS processes, there is a need for an electrically and thermally insulating or sacrificial layer. Thermally grown SiO_2 is often used as this sacrificial layer. However, the adhesion of TiNi films on an SiO_2 layer (or on glass and polymer substrate) is poor owing to the formation of a thin intermixing layer and a fragile and brittle TiO_2 layer [78]. Upon a significant deformation or during a complex interaction involving scratch, this layer is easily broken, thus peeled off. Wolf *et al.* [7] proposed a two-step deposition method to solve this problem: pre-deposition of 0.1 μm TiNi film on SiO_2 at 700 °C to promote interdiffusion of elements, followed by bulk film deposition at room temperature. Reference [79] reported that the addition of an Si_3N_4 interlayer between film and Si substrate did not cause much change in phase transformation behavior or adhesion properties. There is significant interdiffusion of elements and formation of a Ti–N bond at the Si_3N_4/TiNi interlayer. If compared with poor adhesion of TiNi films on an SiO_2 interlayer, the Si_3N_4 interlayer seems to be a good choice for an electrically insulating and diffusion barrier layer in respect to its adhesion properties. Adhesion of TiNi films on polysilicon and amorphous silicon layers is also quite good.

1.2.7 Stability, degradation and fatigue

Stability and fatigue have always been concerns in the development of TiNi thin films for applications. The fatigue of TiNi films is referred to the non-durability and deterioration of the shape memory effect after many cycles. The repeated phase changes will alter the microstructure and hysteresis of the transformation and in turn will lead to changes in transformation temperatures, transformation stresses and strains. The performance degradation and fatigue of thin films are

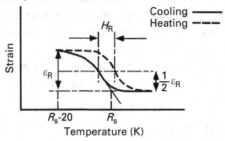

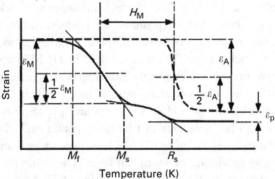

Figure 1.33   Schematic strain–temperature curves representing shape memory behavior associated with (a) R-phase transformation and (b) martensitic transformation, respectively [Miyazaki & Ishida, 1999 [6], with permission from Elsevier].

influenced by a complex combination of internal (alloy composition, lattice structure, precipitation, defects, film/substrate interface) and external parameters (thermomechanical treatment, applied maximum stress, stress and strain rate, the amplitude of temperature cycling frequency) after long-term thermal–mechanical cycles.

Figures 1.33(a) and (b) show schematic strain vs. temperature curves representing shape memory behavior associated with the R-phase transformation (from B2 to R phase) and the martensitic transformation (from R phase to monoclinic phase), respectively. R_s, M_s, M_f, are abbreviations of the temperatures for the start of R-phase transformation and the martensitic transformation start and finish, respectively. H_R and H_M represent the hysteresis of the R-phase and martensitic transformations, respectively. ε_R, ε_M, ε_A and ε_P indicate the strains due to R-phase transformation, martensitic transformation, reverse martensitic transformation and plastic deformation, respectively.

Figure 1.34 shows the strain vs. temperature curves as a function of the number of cyclic deformations associated with the R-phase transformation under 50 MPa in a Ti-43.9at%Ni solution-treated thin film. In the cooling process of the initial cycle ($N = 1$), the R-phase transformation starts at 333 K and finishes at 316 K, resulting

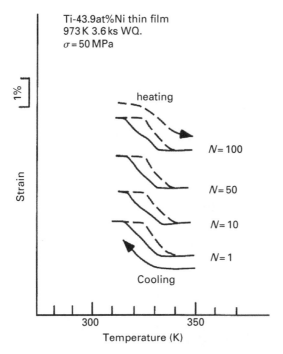

Ti-43.9at%Ni thin film
973 K 3.6 ks WQ.
$\sigma = 50$ MPa

Figure 1.34   The effect of thermal cyclic deformation on strain vs. temperature curves associated with the R-phase transformation for a solution-treated TiNi thin film, N indicating the number of cycles [Miyazaki & Ishida, 1999 [6], with permission from Elsevier].

in a shape change of 0.13% strain. In the heating process, the reverse R-phase transformation starts at 323 K and finishes at 336 K, resulting in a perfect shape recovery and small temperature hysteresis (H_R) of 4 K. Because of the small hysteresis, a quick response is expected in microactuators using such an R-phase transformation. After cycling 100 times, no significant change appears in the shape of the curves. The reason for the stability can be explained by the fact that the R-phase transformation strain is so small that slip deformation hardly occurs.

Figure 1.35 shows the effect of thermally induced cyclic deformation on the strain–temperature curve under a stress of 250 MPa in a solution-treated Ti-43.9at%Ni thin film. The curves of the cyclic deformation show a two-stage deformation; upon cooling a shape change appears at R_s due to the R-phase transformation and upon further cooling a second shape change occurs at M_s due to the martensitic transformation, as shown in the initial cyclic deformation curve. The strains ε_R and ε_M are 0.28% and 1.12%, respectively. Upon heating, the original shape of the specimen is almost recovered due to a two-stage deformation associated with the reverse-transformations occurring at A_s for the first stage and at RA_s for the second stage. The first stage is associated with the reverse martensitic transformation from the martensitic phase to the R phase, and the second with the reverse R-phase transformation from the

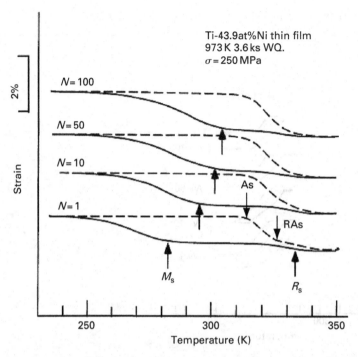

Figure 1.35   The effect of thermal cyclic deformation on strain vs. temperature curves associated with the martensitic transformation for a solution-treated TiNi thin film, N indicating the number of cycles [Miyazaki & Ishida, 1999 [6], with permission from Elsevier].

R phase to the parent phase. Although there is an unrecoverable strain which is caused by slip deformation, it is only 0.03 %.

In increasing the number of cycles, the R-phase transformation characteristics such as R_s, ε_R and H_R are kept almost constant, while the martensitic transformation characteristics apparently change. For example, M_s rises and hence the temperature difference between R_s and M_s decreases. Besides, the temperature hysteresis (H_M) decreases and martensitic transformation strain increases gradually. Such changes in the martensitic transformation can be considered to be caused by the internal stress field which is formed by the introduction of dislocations during cyclic deformation. The internal stress field overlaps with the external applied stress so that the martensitic transformation temperatures increase. However, the R-phase transformation characteristics show few changes during cyclic deformation, because the R-phase transformation involving a small strain is not so sensitive to applied stress.

Reference [80] studied the fatigue of the constrained TiNi films using the changes of recovery stress during cycling, and showed that the recovery stress of TiNi films from curvature measurement decreased dramatically in the first tens of cycles, and became stable after thousands of cycles. This reduction of the recovery stress is believed to result from the dislocation movement, grain boundary sliding, void formation, or partial de-bonding at the film/substrate interfaces, non-recoverable plastic deformation, changes in stress, etc. Transformation temperatures also changed dramatically during cycling. The repeated phase changes will alter the microstructure and hysteresis of the transformation and in turn lead to changes in transformation temperatures, stresses and strains. All these changes in the martensitic transformation behavior become insensitive to thermal cycling after the number of cycles exceeds 50, indicating that training is effective in stabilizing the shape memory behavior. Such a steady state has been achieved by the work hardening during cyclic deformation. Therefore, it is concluded that the stabilization of the shape memory characteristics against thermal cycles under stress can be improved by training.

1.2.8 Film thickness effect

Effects of film thickness on crystallization and shape memory effect have recently been investigated by several groups [81, 82, 83]. TiNi films usually undergo a high temperature (between 400 °C and 650 °C) during deposition or post-annealing. At such a temperature, the surface oxidation and interfacial diffusion between the film and substrate could significantly affect the phase transformation behavior if the film is too thin. It is important to know how thin the TiNi based film can go or how small the TiNi structure can be without losing the shape memory effect. In Ref. [84], the stress–temperature evolution in the TiNi films with different thickness was measured using the curvature method. The stress–temperature response of a 50 nm film is linear, i.e., this film experiences only a thermal effect (due to the difference in thermal expansion between film and substrate) with no apparent phase transformation. Thicker films (up to 4 µm thick) produce stress/temperature hysteresis loops upon thermal cycling, demonstrating the shape memory effect. With increase of film thickness, the residual stress decreases sharply and then remains at a low value, whereas the recovery stress increases significantly, and reaches a maximum at a film thickness of 820 nm before it gradually decreases with further increase in film thickness.

Results revealed that a minimum thickness (about 100 nm) is necessary to guarantee an apparent shape memory effect in TiNi films. A surface oxide and oxygen diffusion layer as well as an interfacial diffusion layer are dominant in the films with a thickness of tens of nanometers. The combined constraining effects from both surface oxide and interfacial diffusion layers in a very thin film will be detrimental to the phase transformation. As the film thickness increases to above a few hundred nanometers, the effects of the surface oxide, oxygen diffusion layer and inter-diffusion layer become relatively insignificant. Therefore, phase transformation becomes significant and the recovery stress increases as thickness increases. Due to the significant phase transformation effect, thermal and intrinsic stresses in the films are drastically relieved, resulting in a significant decrease in residual stress. With a further increase in film thickness, more and more grain boundaries form in the films. The grain boundaries are the weak points for

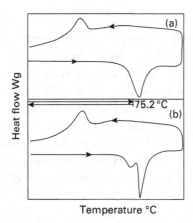

Figure 1.36   DSC results of (a) TiNiCu$_4$ film and (b) temperature memory effect of the same sample with a single incomplete cycle on heating at 75.2 °C [Fu *et al.*, 2009, [9] with permission from Inderscience].

generation of large distortion and twinning processes. Therefore, as the film thickness increases, the constraining effect from the neighboring grains becomes more and more significant, causing decreases in recovery stress. A detailed discussion can be found in Chapters 6 and 7.

1.2.9   Temperature memory effect

A new phenomenon, temperature memory effect (TME), has also been reported in TiNi based films [85]. An incomplete thermal cycle upon heating in a shape memory alloy (arrested at a temperature between austenite transformation start and finish temperatures, A_s and A_f) induced a kinetic stop in the next complete thermal cycle [86], and the kinetic stop temperature is a "memory" of the previous arrested temperature (see Fig. 1.36) [87]. If a number N of incomplete heating processes with different arrested temperatures are performed in decreasing order, N temperatures can be memorized. TME can be fully eliminated by a following complete transformation heat cycle to above A_f. During the partial reverse transformation, only part of the martensite transforms into the parent phase, with the rest of the martensite, M1, remaining. With further decrease in the temperature to below the martensitic transformation finish temperature, the parent phase transforms back to martensite, and the newly formed martensite is called M2. M1 and M2 are different martensite variants with different elastic energies, which cause different transformation temperatures of M1 and M2 during the next heating process. On the other hand, if a partial austenite to martensite transformation is performed by an incomplete cycle on cooling, the next complete austenite to martensite transformation does not show any evidence of kinetic interruptions. The temperature memory effect can be found in both thermally and external-stress induced transformations.

1.2.10   Nanoscale mechanical evaluation

The nano-indentation test is a promising method for characterization of nanoscale pseudoelasticity (PE) [88]. The pseudoelasticity behavior in TiNi based thin films demonstrates their intrinsic capacity to undergo large deformations without permanent surface damage, which is known as self-healing behavior. Nano-indentation testing with or without changes of temperature could reveal the different elastic and plastic deformation behaviors of austenite and martensite, which is promising for characterization of the shape memory effect and thermomechanical properties of the constrained thin films [89, 90]. During loading and unloading in nano-indentation, there is a large force hysteresis, i.e. large energy dissipation during loading/unloading in TiNi based thin films. During the reverse phase transformation, the martensite variants must overcome the internal stress generated during phase change in order to return back to the parent austenite matrix. Therefore, energy is dissipated as friction heat due to the imperfect austenite matrix and the martensite. The energy dissipation associated with the pseudoelastic behavior contributes to the high vibration damping capacity of the TiNi films.

Using a sharp tip, it is difficult to obtain pseudoelastic behavior, since the plastic deformation due to dislocation movement is dominant over the phase transformation [91]. Therefore, spherical-shaped tips have recently been widely used to characterize the nano-scale superelasticity behavior of TiNi and thin films [92]. Indentation using a spherical indenter could avoid large plastic deformation if the indentation force is not too high. During nano-indentation on shape memory alloy thin films using a spherical indenter, Yan *et al.* [93] found that there exist two characteristic points, the bifurcating point and the returning point, in one indentation loading/unloading curve, which are associated with the forward transformation stress and the reverse transformation stress. They proposed a method to determine the transformation stresses of SMA films based on the measured bifurcating and returning forces.

Nanometer-scale indentations in TiNi thin films (for example) less than 100 nm in depth can be fully recovered upon further heating due to thermally induced reverse martensitic transformation. Using the spherical indentation method, surface protrusions can be made on the surfaces, which will disappear upon heating. Reversible circular surface protrusions can be produced due to the two-way shape memory effect [94]. Shape memory alloys with shape relief ability can find more optical and mechanical applications in their greater load bearing capacity and/or better durability than normally used polymers. For example, information storage technology has undergone a revolution in the past years, and magnetic storage is reaching a fundamental limit of about 100 Gbit/in^2 (or 6500 nm^2 / bit) [95]. This is because with the shrinkage of the size of magnetic domains, the fluctuation in temperature could easily cause the random changing of the moments of the magnetic domains, thus the loss of the stored data. Recently the nano-indentation method has been proposed for high-density mechanical storage applications [96].

Storage devices with a capacity of $1\,\text{Tbit/in}^2$ are achievable. The write–read or erase–rewrite operations can be performed with a nano-indenter or atomic force microscope. Information is written into the martensite TiNi thin film by nano-indentation with probe tips. The indents are then scanned and the nano-indentation tip is heated for the shape memory effect, thus erasing the data recorded. Subsequent rewriting can again be performed by the indentation. For this application, the use of shaper probes rather than the relatively blunt or spherical diamond indenter could increase storage density. There are some drawbacks in this mechanical storage method: slow speed, strong dependence on film planarity (and roughness) and tip wear.

1.2.11   Functionally graded and composite TiNi based films

To further improve the properties of TiNi films, multi-layer, composite or functionally graded TiNi based films can be designed. So far, there are different design models for the functionally graded TiNi thin films. The first type is through the gradual change in composition (Ti/Ni ratio), crystalline structures, transformation temperatures, and/or residual stress through film thickness [97]. As the Ti or Ni content changes in the micron-thick film, the material properties could change from pseudo elastic (similar to rubber) to shape memory. The seamless integration of pseudo elastic with shape memory characteristics produces a two-way reversible actuation, because variations in the residual stress in thickness direction will enable a biasing force to be builtup inside the thin films. These functionally graded TiNi films can be easily prepared by slightly changing the target powers during deposition. Another novel way is to vary the target temperature during sputtering, and the films produced by hot targets will compositions similar to that of the target while films produced from a cold target are Ti deficient. In order to successfully develop functionally graded TiNi thin films for MEMS application, it is necessary to characterize, model and control the variations in composition, thermomechanical properties and residual stress in these films.

The second type of functionally graded films involves new materials and functions other than TiNi films. Recently there have been some reports [98, 99] on the deposition of a functionally graded TiN/TiNi layer to fulfill this purpose. The presence of an adherent and hard TiN layer (300 nm) on a TiNi film (3.5 µm) forms a good passivation layer (to eliminate the potential Ni release), and improves the overall hardness, load bearing capacity and tribological properties without sacrificing the shape memory effect of the TiNi films. Also, a TiN layer is able to restore elastic strain energy during heating and to provide a driving force for martensitic transformation on subsequent cooling, forming a two-way SMA effect. In order to improve biocompatibility and adhesion of TiNi films, a functionally graded Ti/TiNi/Ti/Si graded layer could be used. A thin covering layer of Ti can improve biocompatibility (to prevent potential Ni allergic reactions), while the Ti interlayer can be used to improve film adhesion. Using co-sputtering with multi-targets, or controlling the gases during sputtering, these graded film designs can be easily realized.

Some surface modification methods, such as irradiation of TiNi films by electrons, ions (Ar, N, He, Ni or O ions), laser beams or neutrals can be used (1) to modify the surface physical, mechanical, metallurgical, wear, corrosion and biological properties for application in hostile and wear environment; (2) to cause lattice damage and/or alter the phase transformation behaviors along the thickness of the film, forming novel two-way shape memory actuation [100, 101, 102]. The problems of these surface treatments are their high cost, possible ion-induced surface damage, amorphous phase formation, or degradation of shape memory effects. Surface oxidation in TiNi bulk materials has often been reported to prevent the Ni ion releasing and improve the biocompatibility [103, 104]. It is possible to have the same process for TiNi films with a slight sacrifice in the shape memory effect.

Other functionally graded or composite designs include the combination of TiNi films with piezoelectric, ferromagnetic, or magnetostrictive thin films [105, 106]. Response time of the piezoelectricity mechanisms (PZT films) is fast, but the displacement is relatively small. TiNi film, on the other hand, has a larger force and displacement, but with slower response frequency. By coupling TiNi and PZT films to fabricate a new hybrid heterostructure composite or functionally graded film, it is possible to tune or tailor the static and dynamic properties of TiNi thin films, which may generate a larger displacement than conventional piezoelectric or magnetorestrictive thin films and have an improved dynamic response compared with that of single layer TiNi films. Both PZT and TiNi films can be prepared by sputtering methods, or PZT film by the sol-gel method and TiNi film by sputtering. Either TiNi or PZT films can be the bottom layer. However, the complexity of the fabrication processing, the interfacial diffusion and adhesion, and dynamic coupling of dissimilar components remain tough issues for these types of composite thin film.

1.3   MEMS applications of TiNi thin films

1.3.1   Comparison of various microactuation mechanisms

Many types of material and methods have been proposed for fabricating microactuators. Their actuation capacity can be characterized by their work per unit volume and cycling frequency. These characteristics are summarized in Table 1.5 [5]. Figure 1.37 shows the work per unit volume as a function of cycling frequency for various actuators: i.e., (1) TiNi shape memory alloy (SMA), (2) solid–liquid phase change (SL), (3) thermopneumatic (TP), (4) thermal expansion (TE), (5) electromagnetic (EM), (6) electrostatic (ES), (7) piezoelectric (PE), (8) muscle (M), (9) microbubble (MB), which are also shown in Table 1.5. Among these actuators, only the first three can generate large forces over long displacements.

The work output per unit volume W can be defined as $W = Fu/v$, where F, u and v are force, displacement and volume, respectively. If an actuator material is

Table 1.5 Work per unit volume for various microactuators, after Krulevitch *et al.* [1996][5] [Miyazaki & Ishida, 1999[6] with permission from Elsevier].

Actuator type	W/v (J m^{-3})	Equation	Comments
1. Ni-Ti SMA(SMA)	2.5×10^7	$\sigma \cdot \varepsilon$	Maximum one time output: $a = 500$ MPa, $\varepsilon = 5\%$
	6.0×10^6	$\sigma \cdot \varepsilon$	Thousands of cycles: $\sigma = 300$ MPa, $\varepsilon = 2\%$
2. Solid liquid phase change (SL)	4.7×10^6	$\frac{1}{3}\left(\frac{\Delta v}{v}\right)^2 k$	k = bulk modulus = 2.2 GPa (H_2O) 8% volume change (acetamide)
3. Thermo-pneumatic (IP)	1.2×10^6	$\frac{F \cdot \delta}{v}$	Measured values: $F = 20$ N, $\delta = 50$ μm, $v = 4$ mm $\times$ 4 mm $\times$ 50 μm^3
4. Thermal expansion (TE)	4.6×10^5	$\frac{1}{2}\frac{(E_s + E_f)}{2}(\Delta\alpha \cdot \Delta T)^2$	Ideal, nickel on silicon, s = substrate, f = film, $\Delta T = 200$ K
5. Electro-magnetic (EM)	4.0×10^5	$\frac{F \cdot \delta}{v} \quad F = \frac{-M_s^2 A}{2\mu}$	Ideal, variable reluctance: v = total gap volume, $M_s = 1$ V s m^{-2}
	2.8×10^4	$\frac{F \cdot \delta'}{v}$	Measured values, variable reluctance: $F = 0.28$ mN, $\delta = 250$ μm, $v = 100 \times 100 \times 250$ μm^3
	1.6×10^3	$\frac{T}{v}$	Measured values, external field: torque = 0.185 nN m^{-1}, $v = 400 \times 40 \times 7$ μm^3
6. Electrostatic (ES)	1.8×10^5	$\frac{F \cdot \delta}{A \cdot gap} \quad F = \frac{\varepsilon V^2 A}{2\delta^2}$	Ideal: $V = 100$ V, δ = gap = 0.5 μm
	3.4×10^3	$\frac{F \cdot \delta}{v}$	Measured values, comb drive: $F = 0.2$ mN (60 V), $v = 2 \times 20 \times 3000$ μm^3 (total gap) $\delta = 2$ μm
	7.0×10^2	$\frac{F \cdot \delta}{v}$	Measured values, integrated force array: γ = device volume, 120 V
7. Piezoelectric (PE)	1.2×10^5	$\frac{1}{2}(d_{33}E)^2 E_f$	Calculated, PZT: $E_f = 60$ GPa(bulk), $d_{33} = 500$ (bulk), $E = 40$ kV cm^{-1}
	1.8×10^2	$\frac{1}{2}(d_{33}E)^2 E_f$	Calculated. ZnO: $E_f = 160$ GPa (bulk) $d_{33} = 12$ (bulk), $E = 40$ kV cm^{-1}
8. Muscle (M)	1.8×10^4	$\frac{1}{2}(\sigma \cdot \varepsilon)$	Measured values: $\sigma = 350$ kPa, $\varepsilon = 10\%$
9. Microbubble (MB)	3.4×10^2	$\frac{F \cdot \delta}{v \, b}$	Measured values: bubble diameter = 71 μm, F = 0.9 μN, $\delta = 71$ μm

deformed in the course of performing the work, the work output per unit volume equals the elastic strain energy, i.e., $W = \sigma\varepsilon/2$, where σ and ε are stress and strain, respectively. However, for SMA, $W = \sigma\varepsilon$, which means a constant force over the actuation. According to Fig. 1.37, SMA films can generate the greatest work per unit volume up to a reasonably high cycling frequency.

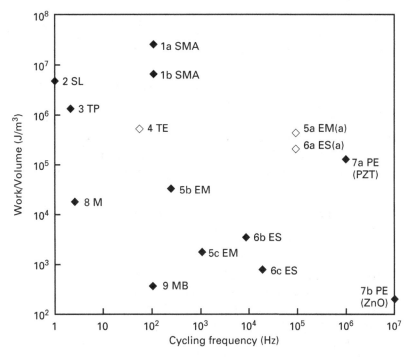

Figure 1.37   Work per volume vs. cycling frequency for various microactuators. Numbers refer to Table 1.5, which gives details on the calculations. Ideal values ($\diamond$) represent the energy available for actuation. Other values ($\blacklozenge$) are based on actual microactuator data [Miyazaki & Ishida, 1999 [6], with permission from Elsevier].

1.3.2   Modeling and optimal design of TiNi thin film microactuators

Numerical modeling and computational simulation of the behaviors of TiNi films and their microactuators, together with experimental characterization efforts, will lead to the optimization of technical factors, such as structural configuration, geometry and processing procedures and further improvement in the overall performance of TiNi thin film based microactuators [107]. There are two levels of simulation. The first level is the simulation and modeling of thermo-mechanical behaviors of TiNi films, and the second is the design of geometry and structures as well as performance of TiNi microactuators. There are many models describing the constitutive behaviors based on thermodynamics and continuum mechanics [108], but only a few have been used in engineering practice. It is difficult to obtain an accurate constitutive relationship for a particular TiNi film. The intrinsic hysteresis, non-linearity and history-dependent behaviors make it more difficult to accurately predict the response of a TiNi thin film microactuator. At present, only phenomenological models appears to be realistic for engineers [109], and the transformation can be assumed as either a linear or a sine/cosine function [110, 111, 112]. As compared with bulk TiNi materials, there are several special issues in the simulation of TiNi films: (1) smaller grain size in TiNi films and a constraint effect on substrates; (2) large film biaxial stress after deposition and stress evolution during the phase transformation process; (3) possible textured structure in the thin films. References [113, 114, 115] reported different thermo-dynamic simulations for TiNi thin films. Jin *et al.* [116] developed a relaxed self-consistent model to simulate the thermomechanical behavior of TiNi films, and it is confirmed that thermally induced phase transformation has a narrower range of transformation temperatures in the films, and the work hardening characteristics are lower than the bulk material due to geometrical relaxation. For TiNi thin film based micro-devices, the non-uniform stress and temperature distribution could affect the precision of deformation and lead to inaccurate position control.

1.3.3   Freestanding microactuators based on a two-way shape memory effect

Freestanding films usually show an intrinsic "two-way" shape memory effect, with large displacement, but relatively small force in actuation. This is applicable in microsensors, microswitches and micropositioners. The origin of the two-way shape memory effect observed in TiNiCu films can be attributed to the difference in sputtering yields of titanium and nickel, which produces a compositional gradient through the film thickness [117, 118]. The film layer near the substrate is normally nickel rich, and no shape memory effect is observed, but the material may possess superelasticity. As the Ti/Ni content changes through the film thickness, the material properties change from being superelastic to having a shape memory. A stress gradient is generated due to the changing microstructure and composition as a function of thickness, thus causing free standing structures to bend upward. When heated, the film layer returns to a flat position due to the shape memory effect. Figure 1.38 shows some examples of simple structures which can be actuated by heating/cooling through the two-way shape memory effect [119].

TiNi alloys with Ni-rich content can show a two-way shape memory effect if aged under elastic constraint [120]. Since this effect enables a TiNi thin film to deform spontaneously upon thermal cycling without any bias force, it is ideal for miniaturizing and simplifying actuators [16]. Since the two-way shape memory effect is related to the precipitation process and distribution of fine Ti_3Ni_4 plate precipitates in the matrix, this effect is sensitive to aging treatment conditions, such as temperature and time.

Ti-51.3at%Ni alloy thin films with a thickness of 8.5 μm were made on glass substrates by sputter deposition. They were peeled off the glass substrates mechanically. Samples of 20 mm length and 1 mm width were cut out of the thin films. They were first solution-treated at 973 K for 3.6 ks followed by aging treatment at various temperatures between 573 K and 773 K for three different times, i.e., 3.6, 36 and 360 ks, under elastic constraint and without constraint, respectively. These heat-treatments form a single parent phase of TiNi with a surplus of Ni atoms (solution-treatment) and fine Ti_3Ni_4 precipitates in the TiNi parent phase in the following age-treatment. The constraint during the aging-treatment produced preferentially oriented precipitates which have an intrinsic

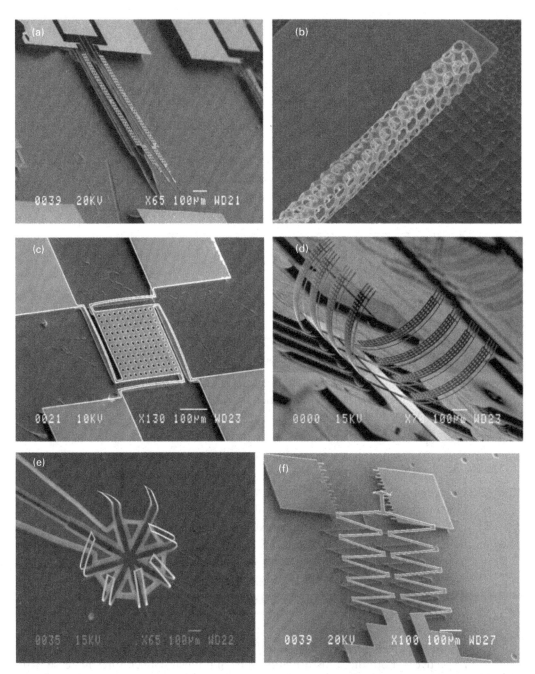

Figure 1.38 Freestanding TiNi based film structures; (a) a microtweezer structure which has both horizontal and vertical movement due to both shape memory and thermal effects; (b) a microstent which can be opened by heating; (c) a micromirror structure which can be actuated by four arms when electrically heated; (d) a microfinger which can operate both horizontally and laterally, and can be designed and integrated into a walking robot; (e) a microcage structure with fingers opening/closing by the two-way effect; (f) microspring structures [Fu *et al*, 2009 [9], with permission from Inderscience].

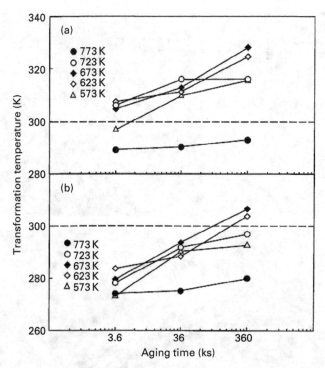

Figure 1.39   R-phase transformation (a) start and (b) finish temperatures of Ti-51.3at%Ni thin films aged at various temperatures as a function of aging time [Miyazaki & Ishida, 1999 [6], with permission from Elsevier].

stress field around them, causing the two-way shape memory effect to appear. The constraint was given by winding the thin films on to a stainless steel pipe with an outer diameter of 7.5 mm. The two-way shape memory effect was evaluated by measuring the curvature of the thin film in both boiling and iced water.

All the aging-treated thin films exhibited two-stage transformation behavior both upon cooling and heating, i.e., the R-phase transformation in a higher temperature region and the martensitic transformation in a lower temperature region. The major part of the two-way shape memory effect was associated with the R-phase transformation. Figure 1.39 shows the start and finish temperatures of the R-phase transformation as a function of aging temperature and aging time. The R-phase transformation start and finish temperatures are defined as temperatures at which the R-phase transformation starts and finishes upon cooling, respectively. Both the transformation start and finish temperatures increase with increasing aging time.

The two-way shape memory behavior is sensitive to aging conditions. For example, the film aged at 573 K for 3.6 ks and the film aged at 723 K for 3.6 ks bend in the same direction as the constrained direction in boiling water. However, in iced water, they show different behavior; the film aged at 573 K bends forward, but the film aged at 723 K bends backward and shows an opposite curvature. As described earlier, Ti$_3$Ni$_4$ plate precipitates are formed in the TiNi matrix during age-treatment. These precipitates form in a disk shape on {111} planes of the TiNi matrix and reduce the volume by 2.3 % along the ⟨111⟩ direction. When a film is aged under constraint, the precipitates form on one of the {111} planes selectively, relaxing the constraint stress by the volume change. If the film is aged at a low temperature for a short time, the precipitation of Ti$_3$Ni$_4$ is not sufficient for relaxing the constraint stress. In this case, when the constraint is removed, the film tends to go back to the original shape before the constraint-aging. However, even after the removal of the constraint, internal stress seems to remain locally between relaxed and unrelaxed regions in this film. This internal stress may determine a specific R-phase variant upon the transformation, so that the film shape approaches the constrained shape. However, if the constraint stress is fully relaxed after aging at a high temperature for a long time, such internal stress becomes small. Instead, the coherent strain around the precipitates becomes large. This stress field determines the specific R-phase variant so that the film shape returns to the original shape. This effect has been also found in bulk specimens by Nishida *et al.* [120]. They called it "all-round shape memory effect", since this effect is so prominent that the film curvature reverses. However, "two-way shape memory effect" is more appropriate to reflect the nature of the phenomenon.

Figure 1.40 shows the shapes of aged films which are put in boiling water and in iced water. The shape change is described by a parameter r_i/r_t, where r_i represents the radius of curvature of a film under constraint, while r_t is the radius of curvature of the film at a given temperature. For example, when the film is straight, the r_i/r_t is zero. If the film bends in the opposite direction to the constraint direction, this parameter becomes negative. The corresponding strain at the inner surface can be calculated by the equation $\varepsilon = -t/2r$ (ε: strain at the inner surface, t: film thickness, r: curvature radius of a film). In Fig. 1.40, the open circles represent the film shape in boiling water, where the film is in the parent phase. The closed circles represent the film shape in iced water, where the film is in the R phase. When a film is aged at 573 K for 3.6 ks, the film bends in the constrained direction upon cooling. However, as aging time increases, the bending direction changes and the film aged for 360 ks bends backward upon cooling and shows the so-called all-round shape memory effect. This effect becomes stronger with increasing aging temperature and aging time, since the stress fields around the Ti$_3$Ni$_4$ precipitates develop as the precipitates grow. However, at a higher aging temperature region between 723 K and 773 K, the effect gradually decreases with increasing aging time. These results clearly indicate that there are optimum aging conditions for a two-way shape memory effect: i.e., an aging temperature of 673 K and an aging time of 360 ks. It is to be noted that these aging conditions are also the optimum aging conditions for a high transformation temperature. In other words, it turns out that a strong two-way shape memory effect is obtained above room temperature when a Ti-51.3at%Ni thin film is aged at 673 K for 360 ks.

A simple thin film actuator aged under the optimum condition was produced: the shape is a double-beam cantilever and the ends of the two beams are connected

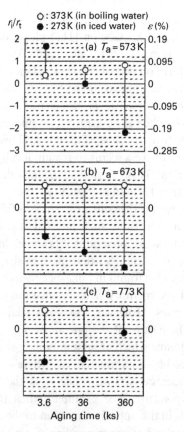

Figure 1.40   Change in curvature of Ti-51.3at%Ni thin films aged at (a) 573 K (b) 673 K and (c) 773 K for 3.6, 36 and 360 ks [Miyazaki & Ishida, 1999 [6], with permission from Elsevier].

to a battery. The cantilever made of the TiNi thin film was confirmed to be able to work with a battery of 1.5 V at room temperature [28]. It showed bending forward and backward reversibly by turning on and off the battery alternately.

1.3.4   TiNi diaphragms, micropump and microvalves

Several types of micromachining process have been proposed for fabricating microactuators or microstructures utilizing sputter-deposited thin films. Figure 1.41 shows the outline of the main processing steps for the deposition of TiNi thin film and the release of active micromechanical components. First, a polyimide layer 3 μm thick was spun on and cured to act as a spacer and sacrificial layer for the TiNi film. The polyimide was spun on at 3000 rpm for 30 s and soft baked at 423 K. Secondly, TiNi thin films were sputter-deposited on the polyimide layer. Typical thicknesses of the TiNi thin films ranged from 1 to 2 μm. Thirdly, photoresist was spun on over the TiNi film, baked, patterned and developed to expose specific portions of the TiNi film. Fourthly, the TiNi films were etched with a 1:1 mixture of

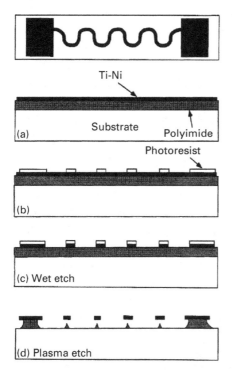

Thin-film shape memory alloy process outline: (a) deposition of polyimide spacer and
SMA film; (b) deposition and patterning of photoresist; (c) wet etch of SMA film, and
(d) plasma etch of polyimide and release of SMA structure, after Walker *et al.* [Miyazaki
& Ishida, 1999 [6], with permission from Elsevier].

commercially available titanium and nickel etchants (hydrofluoric and nitric acid-
based, respectively) to define the desired TiNi thin film components. Finally, the
TiNi film structures were released through a dry-etch procedure in order to minimize damage to the metal films, which can occur during wet etch and rinse
sequences. The dry-etch process consists of reactive ion etching the sacrificial
polyimide layer in a 1:1 SF_6-O_2 plasma at high pressure (100 mtorr) and moderate
power (150 W).

   TiNi diaphragms were fabricated in the following process, as shown in Fig. 1.42.
For the fabrication of TiNi diaphragms, substrate micromachining was done prior
to deposition, in order to minimize the exposure of the TiNi films to hot ethylene
diamine pyrocatechol (EDP) etchant. Si wafers with a thermal oxide (SiO_2) layer
on both sides were coated with photoresist and soft-baked. The photoresist was
then patterned on the back side to open EDP etch windows in the oxide. This
pattern contains 2 mm × 2 mm square diaphragms and division lines separating the
wafer into individual 10 mm × 10 mm diaphragm sections. EDP was then used to
etch a majority of the diaphragm cavities, as well as the division lines. The oxide
was removed from the front side of the wafer just prior to deposition using a
buffered oxide etchant. After the TiNi was deposited and annealed, the remainder

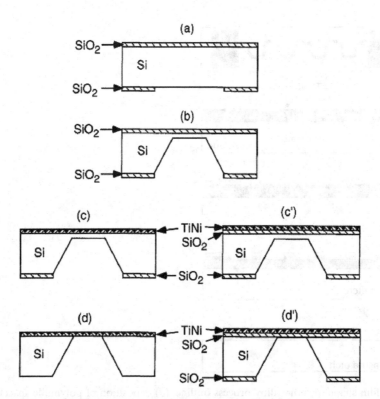

Figure 1.42 Microfabrication of TiNi diaphragms without a SiO_2 layer (a)–(b)–(c)–(d) and with a SiO_2 layer (a)–(b)–(c′)–(d′) [Miyazaki & Ishida, 1999 [6], with permission from Elsevier].

of the silicon diaphragm support was removed using EDP, thus releasing a suspended diaphragm of TiNi. In some other cases, the oxide layers were not
removed, so that TiNi was deposited on the oxide. In order to adhere TiNi films to
SiO_2, a very thin (0.1 μm) Ti adherence layer was deposited on substrates heated to
973 K. The bulk of the TiNi film was then deposited at room temperature and
annealed after deposition in the usual way.

Another type of microfabrication process was proposed, as shown in Fig. 1.43 [6].
The process is as follows. (1) A sacrificial Cr layer is deposited on an SiO_2
substrate and patterned, if necessary. An 11 μm thick polyimide layer is spun
coated and covered with another Cr layer for an etching mask. (2) The masking
Cr layer is patterned by photolithography and wet etched. The polyimide layer is
vertically etched by O_2 plasma using the Cr mask. (3) A TiNi layer of 7–10 μm
thickness is sputtered. (4) The polyimide layer is removed by wet etching.
Unnecessary parts of TiNi films are also removed with polyimide. Since this process
does not include etching TiNi layers but polyimide etching, which is typically very
good, the aspect ratio of TiNi patterns is not limited. Utilizing similar micromachining processes, prototypes of springs, double-beam cantilevers, switch-type
cantilevers, valves, mirror actuators, diaphragms, microgrippers, etc. have been
fabricated.

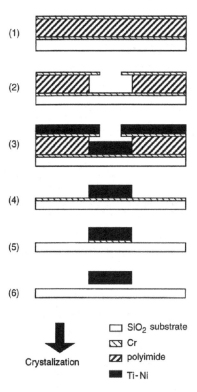

(1)

(2)

(3)

(4)

(5)

(6)

Crystalization

☐ SiO₂ substrate
◺ Cr
▨ polyimide
■ Ti-Ni

Figure 1.43 Fabrication process: (1) deposition of Cr:polyimide (11 mm): Cr layers; (2) upper Cr patterning and O₂ RIE of polyimide; (3) TiNi sputtering (7–10 mm); (4) lift-off by removing polyimide with KOH; (5, 6) Cr wet etching to release movable parts [Miyazaki & Ishida, 1999 [6], with permission from Elsevier].

MEMS based micropumps and microvalves are attractive for many applications such as implantable drug delivery, chemical analysis and analytical instruments, etc. TiNi thin films are suitable for building microvalves and pumps [121]. There are different designs for TiNi film-based micropumps and microvalves, and most of them use a TiNi membrane (or diaphragm, microbubble, etc.) for actuation with one example shown in Fig. 1.44 [122, 123, 124]. Both freestanding TiNi films and constrained TiNi films are used. Although freestanding TiNi film has an intrinsic two-way shape memory effect, in order to maximize this effect, extra process, such as 3-D hot-shaping of TiNi film (membrane) and externally biased structure (such as a polyimide layer, or a bonded Si cap or glass cap) are often applied. All of these extra processes could result in a complicated structure and difficulty in MEMS processing [125,126]. Another concern is the effective thermal isolation between the heated TiNi films and the fluid being pumped. TiNi/Si bimorph membrane-based micropumps and valves are more commonly reported. The advantages of using a TiNi/Si membrane as the driving diaphragm include [127]: (1) large actuation force; (2) simplicity in process and no special bias structure needed because the silicon

Figure 1.44   TiNi microvalve fabricated with a TiNi electrode on a silicon membrane structure: (a) top view of membrane and TiNi electrode; (b) bottom view. [Fu *et al.*, 2009 [9], with permission from Inderscience].

substrate can provide bias force; and (3) no isolating structure is needed because the silicon structure can separate the working liquid from the SMA film completely.

1.3.5   Microgrippers

The wireless capsule endoscope (WCE) is a new diagnostic tool in searching for the cause of obscure gastrointestinal bleeding. A WCE contains video imaging, self-illumination, image transmission modules and a battery [128, 129]. The indwelling camera takes images and uses wireless radio transmission to send the images to a receiving recorder device that the patient wears around the waist. However, there are two drawbacks in the current WCE: (1) lack of ability for biopsy; and (2) difficulty in identifying the precise location of pathology. Without tissue diagnosis, it is often difficult to differentiate inflammatory lesions from tumor infiltration. The former may require only medical treatment while the latter may need a surgical solution. Therefore, there are two potential microactuator applications in capsule endoscopy: (1) a microgripper for biopsy or tissue sampling [130]; (2) a microclipper or pin tagging device, to firmly attach to the tissue. Shape memory alloy thin-film based microactuators are promising for these applications.

Grasping and manipulating small or micro-objects with high accuracy is required for a wide range of important applications, such as the assembly in microsystems, endoscopes for microsurgery, and drug injection micromanipulators for cells [130, 131, 132, 133]. There are some basic requirements for microgrippers, for example, large gripping force, sufficient opening distance for assembling works, etc. TiNi films are promising in these applications. So far, two types of TiNi film based microgripper design are available.

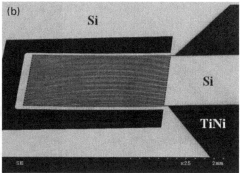

Figure 1.45   TiNi/Si microgripper with cantilever structure with out-of-plane bending mode, (a) microgripper [5]; (b) the patterned TiNi electrodes on silicon cantilevers [Fu *et al.*, 2004 [10], with permission from Elsevier].

The popular design is an out-of-plane bending mode, mostly with two integrated TiNi/Si cantilevers (or other substrate, such as SU-8 or polyimide, etc) with opposite actuation directions (see Fig. 1.45(a)) [135, 136]. Fig. 1.45(b) shows the patterned TiNi electrodes on silicon cantilevers. When the electrodes are electrically heated, the cantilever bends up due to the shape memory effects of TiNi films, thus generating a gripping force. These types of gripper design usually need a further bonding process to combine two cantilevers to form the gripping movement. The force and displacement generated can be very large. A novel microwrapper was fabricated using freestanding TiNi films with out-of-plane movement. The overall dimension of the microwrapper arms is 100 μm, approximately the diameter of a human hair. The microwrapper has a small current passing through it to maintain the flat shape. Upon removal of the current, the small arms close to form a cage. This micro-wrapper can be used to manipulate micro-organisms or possibly in minimally invasive surgery to remove anomalies such as tumors. Reference [132] reported a novel microelectrode with a TiNi clipping structure, which can be used for minimally invasive microelectrodes to clip a nerve cord or other living organisms. The TiNi film is actuated when a current is applied to the electrode. The clipping force of the electrode to the nerve is enhanced by a hook structure and two C-shaped probes, as shown in Figs. 1.46(a) and (b).

Another gripper design is in-plane mode, in which the deformation of two arms (using freestanding TiNi films or TiNi/Si beams) is within a plane realized by compliant structure design [137]. Reference [138] reported a microtweezer structure, in which residual stress in TiNi film is used as a bias force load. This can eliminate the need for providing a bias force for device operation. However, the force from the deformation of the free-standing films is not large enough to grasp large objects. The other problem in this type of design is how to prevent out-of-plane bending, beam deformation and fracture during operation caused by intrinsic film stress.

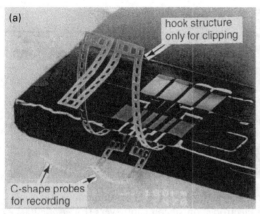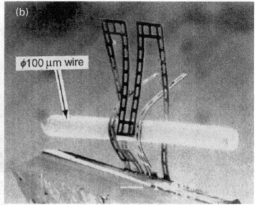

Figure 1.46   (a) A TiNi electrode with a hook structure is returned to its memorized shape when it is heated, while two C-shaped probes for recording are not heated. (b) The microelectrode clipping a wire (100 microns) after the hook structure is heated [132] [Fu *et al.*, 2004 [10], with permission from Elsevier].

1.3.6   Microsensors, microswitches and microrelays

TiNi thin films are sensitive to environmental changes such as thermal, stress, magnetic and electrical fields, and thus should be ideal for applications in microsensors. However, only a few studies and applications are reported in this respect. Possible reasons for the lack of studies include: (1) the sensing function is limited only within the appropriate temperature range; (2) low response speed and frequency are inherent to the SMA. TiNi film was reported as a gate of a metal-on-silicon (MOS) capacity sensor, for detecting the increase in capacitance of TiNi films during heating and cooling [139]. Other potential applications such as switches or microrelays include on-chip circuit breakers against overheating caused by a short circuit or overload, probe tips for automatic test equipment, fiber optics switching, automotive fuel injectors, microlens positioners, etc. [26]. However, the environmental stimulus must reach a critical temperature value to trigger the operation. Also, the exact relationship between external heat or electrical current and movement is difficult to control.

When TiNi film undergoes a phase transformation, both its surface roughness (corresponding to light scattering) and its refractive index change [140, 141]. The reflection coefficient of the austenite phase is higher than that of the martensite phase by more than 45 %, thus it is possible to use TiNi films as a light valve or on–off optical switch for spatial light modulators. TiNi film can also be used as a lever to move an optical lens up or down (instead of left or right), thus forming an out-of-plane microactuator for optical switches. The potential areas of application include field emission flat panel displays technology, in which TiNi film based microactuators can erect a large number of micromachined spacers between the pixels. Reference [142] reported an infrared radiation (IR) imaging sensor based on a TiNi film cantilever. One side of the cantilever is coated with an IR absorbing layer, and the other side is coated with a reflecting layer of gold. Upon adsorption of IR, the temperature of the TiNi cantilever changes, causing a large tilting effect. The detection of the mechanical movements in the cantilever is realized by illuminating the reflective gold side with a laser beam.

1.3.7   Other applications

High passive and active damping capacity is considered as one of the important functional properties of SMAs [143]. Since the process involves the hysteresis movement of interfaces (martensite variant interfaces, like twin boundaries, phase boundaries, lattice defects), thus a large amount of energy is dissipitated upon cycling. The changes of damping capacity of TiNi films have been used to study their phase transformation behavior. Damping capacity is temperature and frequency dependent and peaks near the martensitic transformation temperature. These factors can be explored for the design of anti-vibration damping structures. In hard disk drives, the positioning accuracy of the read/write heads strongly depends on the inherent dynamic characteristics of the head actuator assembly (i.e., vibrations of the head actuator assembly system). TiNi film with a large damping property is promising in minimizing the vibrations during the operations of the hard disk drive. However, fabrication of the actual damping devices must take into account factors of economics, reliability, versatility and construction needs. Also important to be considered are film stress and potential distortion of structures.

Good wear resistance is an important property required in some MEMS and biomedical applications, such as pumps, grippers, valves, etc. It was reported that bulk TiNi alloys in austenite exhibit good wear resistance (due to its rapid work hardening and pseudoelastic properties) [144]. However, poor wear resistance of B19′ (martensite) and a high coefficient of friction are potential problems. In the case of SMA films, interfacial adhesion, a large coefficient of friction and stress are other major concerns for their tribological application [145, 146, 147]. Functionally graded layer design (TiN/TiNi) or surface modifications could provide viable solutions.

Bulk TiNi is a common and well-known material for the medical industry [148]. At present, increasing attention has also been paid to the use of TiNi thin film in minimally invasive surgery, microstents and bioMEMS applications [149]. Some microactuators made from TiNi thin films may be used to infuse drugs, or placed in strategic locations in the body to assist circulation. Glass, silicon and polymers are the commonly used substrates for biological applications. However, high deposition or annealing temperatures for preparation of TiNi films and the poor adhesion on these substrates pose potential problems. Superelasticity of TiNi, a non-linear peudoelasticity as much as 7–10% strain, has already found many applications for bulk materials, but few explorations have been carried out in MEMS applications so far using thin films. TiNi thin film SMA in its superelastic state is promising for some compliant elements in MEMS devices.

Since thin films with nanometer-sized grains (about tens or hundreds of nanometers) still show a shape memory effect, it is promising to fabricate nanoscale SMA thin film structures with the aid of precision tools (such as focused ion milling or FIB). These structures may be able to perform physical actuation (push, pull, etc.) at nanoscale. Possible difficulties of TiNi films in nanoscale structures include: (1) a large amount of oxygen and carbon adsorption on TiNi surface due to the extremely reactive nature of Ti elements, and the oxide and oxygen diffusion depth could be as large as tens of nanometers; (2) the difficulty in fabrication and manipulation of these nanostructures. Laser beam and FIB are promising techniques for this application.

1.4   Summary

Sputter-deposited TiNi and TiNiX (X = Cu, Pd) thin films have been successfully produced and show stable shape memory effect and thermomechanical properties which are equivalent to those of bulk materials. The development of TiNi based SMA thin films and microactuators has seen considerable progress in recent years. Some important issues pertaining to the preparation of high performance shape memory TiNi films using sputtering methods and their MEMS applications were reviewed in this chapter. Successful application of TiNi thin films in MEMS requires consideration of the following issues: preparation and characterization, residual stress and adhesion, frequency improvement, fatigue and stability, patterning and modeling of behavior. Development of a variety of characterization methods is needed to evaluate the actuator characteristics of the TiNi thin films in service in micromechanical systems. Systematic investigation of the interaction between TiNi films and substrates is also demanded to achieve tight bonding of the films onto the substrates without much chemical reaction between them. With further development of fabrication and characterization techniques, it is clear that a variety of micromechanical systems utilizing microactuators of TiNi thin film will be developed and present important technical impacts in the quite near future. At microscale, TiNi actuators out-perform other actuation mechanisms in work/volume (or power/weight) ratio, large deflection and force, but with a relatively low frequency (less than 100 Hz) and efficiency, as well as non-linear behavior. More functional and complex designs based on TiNi film devices are needed with multi-degree-of-freedom and compact structure. TiNi film based microactuators will find potential applications in medicine, aerospace, automotive and consumer products. Miniature TiNi actuated devices based on sputtered TiNi films are ready for the huge commercial market, especially for medical microdevices and implantable applications.

References

[1] S. Miyazaki and K. Otsuka, Development of shape memory alloys, *ISIJ International*, **29** (1989) 353–377.

[2] J. V. Humbeeck, Non-medical applications of shape-memory alloys, *Mater. Sci. Engng*, **A273–275** (1999) 134–148.

[3] R. D. James and K. F. Hane, Martensitic transformations and shape-memory materials, *Acta Mater.*, **48** (2000) 197–222.

[4] K. Otsuka and X. Ren, Physical metallurgy of Ti-Ni-based shape memory alloys, *Prog. Mater. Sci.*, **50** (2005) 511–678.

[5] P. Krulevitch, A. P. Lee, P. B. Ramsey, J. C. Trevino, J. Hamilton and M. A. M. A. Northrup, Mixed-sputter deposition of Ni-Ti-Cu shape memory films, *J. MEMS*, **5** (1996) 270–282.

[6] S. Miyazaki and A. Ishida, Martensitic transformation and shape memory behavior in sputter-deposited TiNi-base thin films, *Mater. Sci. Engng.*, **A 273–275** (1999) 106–133.

[7] R. H. Wolf and A. H. Heuer, TiNi shape-memory alloy and its applications for MEMS, *J. MEMS*, **4** (1995) 206–212.

[8] H. Kahn, M. A. Huff and A. H. Heuer, The TiNi shape-memory alloy and its applications for MEMS, *J. Micromech. Microeng.*, **8** (1998) 213–221.

[9] Y. Q. Fu, J. K. Luo, A. J. Flewitt, *et al.*, Thin film shape memory alloys and micro-actuators, *Inter. J. Nanomanufacturing*, in press (2009).

[10] Y. Q. Fu, Hejun Du, Weimin Huang, Sam Zhang and Min Hu, TiNi-based thin films in MEMS applications: a review, *Sensors & Actuators: A. Physical*, **112** (2004) 395–408.

[11] B. Winzek, S. Schmitz, H. Rumpf, *et al*. Recent developments in shape memory thin film technology, *Mater. Sci. Engng*, **A378** (2004) 40–46.

[12] S. Miyazaki, *Engineering Aspects of Shape Memory Alloys*, ed. T.W. Duerig *et al.*, London: Butterworth-Heinemann Ltd. (1990) 394.

[13] J.A. Walker, K. J. Gabriel and M. Mehregany, Thin film processing of shape memory alloy, *Sensors and Actuators*, **A21–23** (1990) 243.

[14] A. Ishida, M. Sato, A. Takei and S. Miyazaki, Effect of heat treatment on shape memory behavior of Ti-rich Ti-Ni thin films, *Materials Trans., JIM*, **36** (1995) 1349.

[15] S. Kajiwara, T. Kikuchi, K. Ogawa, T. Matsunaga and S. Miyazaki, Strengthening of Ti-Ni shape-memory films by coherent subnanometric plate precipitates, *Phil. Mag. Letters*, **74** (1996) 137.

[16] M. Sato, A. Ishida and S. Miyazaki, Two-way shape memory effect of sputter-deposited thin films of Ti 51.3 at% Ni, *Thin Solid Films*, **315** (1998) 305–309.

[17] Li Hou and D. S. Grummon, Progress on sputter-deposited thermotractive titanium-nickel films, *Scripta Metall. et Materialia*, **33** (1995) 989.

[18] A. Ishida, K. Ogawa, M. Sato and S. Miyazaki, Microstructure of Ti-48.2 at. pct Ni shape memory thin films, *Metall. and Mater. Trans. A*, **28** (1997) 1985–1991.

[19] S. Miyazaki, T. Hashinaga and A. Ishida, Martensitic transformations in sputter-deposited Ti-Ni-Cu shape memory alloy thin films, *Thin Solid Films*, **281–282** (1996) 364.

[20] X. L. Meng, M. Sato and A. Ishida, Structure of martensite in Ti-rich Ti-Ni-Cu thin films annealed at different temperatures, *Acta Mater.* **56** (2008) 3394–3402.

[21] X. L. Meng, M. Sato and A. Ishida, Transmission electron microscopy study of the microstructure of B19 martensite in sputter-deposited Ti50.2Ni30Cu19.8 thin films, *Scripta Mater.* **59** (2008) 451–454.

[22] A. Ishida, M. Sato and K. Ogawa, Microstructure and shape memory behavior of annealed Ti-36.8 at.% Ni-11.6 at.% Cu thin film, *Mater. Sci. Engng., A*, **481** (2008) 91–94.

[23] E. Quandt, C. Halene, H. Holleck, *et al.*, *Sensors and Actuators A*, **53** (1996) 434.

[24] E. Baldwin, B. Thomas, J. W. Lee and A. Rabiei, Processing TiPdNi base thin-film shape memory alloys using ion beam assisted deposition, *Surf. Coat. Technol.* **200** (2005) 2571–2579.

[25] J. W. Lee, B. Thomas and A. Rabiei, Microstructural study of titanium-palladium-nickel base thin film shape memory alloys, *Thin Solid Film*, **500** (2006) 309–315.

[26] A. D. Johnson, NiTinal thin film three dimensional devices-fabrication and 3-D TiNi shape memory alloy actuators, *Micromachine Devices*, **4** (1999) 1.

[27] S. M. Tan and S. Miyazaki, Ti-content and annealing temperature dependence of deformation characteristics of TiXNi(92-X)Cu-8 shape memory alloys, *Acta Materialia*, **46** (1997) 2729.

[28] A. Ishida, M. Sato, A. Takei, K. Nomura and S. Miyazaki, Effect of aging on shape memory behavior of Ti-51.3 at pct Ni thin films, *Metall. Mater. Trans. A*, **27A** (1996) 3753–3759.

[29] C. L. Shih, B. K. Lai, H. Kahn, S. M. Philips and A. H. Heuer, A robust co-sputtering fabrication procedure for TiNi shape memory alloys for MEMS, *J. MEMS*, **10** (2001) 69–79.

[30] A. Ohta, S. Bhansali, I. Kishimoto and A. Umeda, Novel fabrication technique of TiNi shape memory alloy film using separate Ti and Ni targets, *Sensors and Actuators A*, **86** (2000) 165–170.

[31] K. K. Ho, K. P. Mohanchandra and G. P. Carman, Examination of the sputtering profile of TiNi under target heating conditions, *Thin Solid Films*, **413** (2002) 1–7.

[32] K. K. Ho and G. P. Carman, Sputter deposition of TiNi thin film shape memory alloy using a heated target, *Thin Solid Films*, **370** (2000) 18–29.

[33] H. Cho, H. Y. Kim and S. Miyazaki, Alloying process of sputter-deposited Ti/Ni multilayer thin films. *Mater. Sci. Engng., A*, **438** (2006) 699–702.

[34] T. Lehnert, S. Crevoiserat and R. Gotthardt, Transformation properties and microstructure of sputter-deposited Ni-Ti shape memory alloy thin films, *J. Mater. Sci.*, **37** (2002) 1523–1533.

[35] P. Surbled, C. Clerc, B. L. Pioufle, M. Afaka and H. Fujita, Effect of the composition and thermal annealing on the transformation temperatures of sputtered TiNi shape memory alloy thin films, *Thin Solid Films*, **401** (2001) 52–59.

[36] A. Isalgue, V. Torra, J. -L. Seguin, M. Bendahan, J. M. Amigo and V. Esteve-Cano, Shape memory TiNi thin films deposited at low temperature, *Mater. Sci. Engng*, **A 273–275** (1999) 717–721.

[37] W. M. Huang, Q. He, M. H. Hong, Q. Xie, Y. Q. Fu and H. J. Du, On the fabrication of TiNi shape memory alloy micro devices using laser, *Photonics Asia 2002*, 14–18 October 2002, Shanghai, China, *SPIE* **4915**, 2002 234–240.

[38] Y. Bellouard, T. Lehnert, J. E. Bidaux, *et al.*, Local annealing of complex mechanical devices: a new approach for developing monolithic micro-devices, *Mater. Sci. Engng.*, **A 273–275** (1999) 795–798.

[39] A. Ishida, M. Sato and S. Miyazaki, Mechanical properties of Ti-Ni shape memory thin films formed by sputtering, *Mater. Sci. Engng.*, **A 273–275** (1999) 754–757.

[40] A. Ishida, A. Takei, M. Sato and S. Miyazaki, Stress–strain curves of sputtered thin films of Ti-Ni, *Thin Solid Films*, **281–282** (1996) 337–339.

[41] Y. Q. Fu, H. J. Du, S. Gao and S. Yi, Mechanical properties of sputtered TiNiCu shape memory alloy thin films, *Mater. Sci. Forum*, **437** (2003) 37–40.

[42] T. Matsunaga, S. Kajiwara, K. Ogawa, T. Kikuchi and S. Miyazaki, High strength Ti-Ni-based shape memory thin films, *Mater. Sci. Engng.*, **A273–275** (1999) 745–748.

[43] D. S. Grummon, J. P. Zhang and T. J. Pence, Relaxation and recovery of extrinsic stress in sputtered titanium-nickel thin films on (100)-Si, *Mater. Sci. Engng.*, **A273–275** (1999) 722–726.

[44] H. D. Espinosa, B. C. Prorok and M. Fischer, A methodology for determining mechanical properties of freestanding thin films and MEMS materials, *J. Mech. Phys. Solids*, **51** (2003) 46–67.

[45] S. Moyne, C. Poilane, K. Kitamura, S. Miyazaki, P. Delobelle and C. Lexcellent, Analysis of the thermo mechanical behavior of Ti-Ni shape memory alloy thin films by bulging and nanoindentation procedures, *Mater. Sci. Engng*, **A273–275** (1999) 727–732.

[46] E. Makino, T. Shibata and K. Kato, Dynamic thermo-mechanical properties of evaporated TiNi shape memory thin film, *Sens. Acutator A*, **78** (1999) 163–167.

[47] E. Makino, T. Mitsuya and T. Shibata, Micromachining of TiNi shape memory thin film for fabrication of micropumps, *Sens. Actuator A*, **79** (2000) 251–259.

[48] C. Craciunescu and M. Wuttig, Extraordinary damping of Ni-Ti double layer films, *Thin Solid Films*, **378** (2000) 173–175.

[49] W. Ni, Y. T. Cheng and D. S. Grummon, Recovery of microindents in a nickel-titanium shape-memory alloy: a "self-healing" effect, *Appl. Phys. Lett.*, **80** (2002) 3310–3312.

[50] G. A. Shaw, D. D. Stone, A. D. Johnson, A. B. Ellis and W. C. Crone, Shape memory effect in nanoindentation of nickel-titanium thin films, *Appl. Phys. Lett.*, **83** (2003) 257–259.

[51] X. G. Ma and K. Komvopoulos, Nanoscale pseudoelastic behavior of indented titanium-nickel films, *Appl. Phys. Lett.*, **83** (2003)3773–3775.

[52] Y. Q. Fu and H. J. Du, RF magnetron sputtered TiNiCu shape memory alloy thin film, *Mater. Sci. Engng.*, **A 339** (2003) 10.

[53] S. Miyazaki, S. Kimura and K. Otsuka, Shape memory effect and psedoelectricity associated with the R-phase transition in Ti-50.5 at in single crystals, *Phil. Mag. A*, **57** (1988) 467.

[54] S. Miyazaki and C. M. Wayman, The R-phase transition and associated shape memory mechanism in TiNi single crystals, *Acta Metall.*, **36** (1988) 181.

[55] Y. C. Shu and K. Bhattacharya, The influence of texture on the shape-memory effect in polycrystals, *Acta Materialia*, **46** (1998) 5457–5473.

[56] K. Gall and H. Sehitoglu, The role of texture in tension–compression asymmetry in polycrystalline NiTi, *Int. J. Plasticity*, **15** (1999) 69–92.

[57] Y. S. Liu, D. XU, B. H. Jiang, Z. Y. Yuang and P. Van Houtte, The effect of crystallizing procedure on microstructure and characteristics of sputter-deposited TiNi shape memory thin films, *J. Micromech. Microeng.*, **15** (2005) 575–579.

[58] H. Inoue, N. Miwa and N. Inakazu, Texture and shape memory strain in TiNi alloy sheets, *Acta Mater.*, **44** (1996) 4825.

[59] S. Miyazaki, V. H. No, K. Kitamura, A. Khantachawwana and H. Hosoda, Texture of Ti-Ni rolled thin plates and sputter-deposited thin films, *Int. J. Plasticity*, **16** (2000) 1135–1154.

[60] R. Hassdorf, J. Feydt, P. Pascal, *et al.*, Phase formation and structural sequence of highly-oriented MBE-grown NiTiCu shape memory thin films, *Mater. Trans.* **43** (2002) 933–938.

[61] S. Miyazaki, S. Kimura, K. Otsuka and Y. Suzuki, Shape memory effect and psedoelectricity in a TiNi single crystal, *Scri. Mat.* **18** (1984) 883.

[62] J. X. Zhang, M. Sato and A. Ishida, Deformation mechanism of martensite in Ti-rich Ti-Ni shape memory alloy thin films, *Acta Mater.*, **54** (2006) 1185–1198.

[63] H. J. Du and Y. Q. Fu, Deposition and characterization of Ti1-x(Ni,Cu)x shape memory alloy thin films, *Surf. Coat. Technol.*, **176** (2004) 182–187.

[64] L. Chang and D. S. Grummon, Structure evolution in sputtered thin films of Ti-x(Ni, Cu)(1-x) 1. Diffusive transformations, *Phil. Mag. A*, **76** (1997) 163, 191.

[65] T. Sawaguchi, M. Sato and A. Ishida, Microstructure and shape memory behavior of Ti-51.2 (Pd27.0Ni21.8) and Ti-49.5(Pd28.5Ni22.0) thin films, *Mater. Sci. Engng.*, **A332** (2002) 47–55.

[66] D. S. Grummon, Thin-film shape-memory materials for high-temperature applications, *JOM*, **55** (2003) 24–32.

[67] Y. Q. Fu and H. J. Du, Magnetron sputtered Ti50Ni40Pt10 shape memory alloy thin films, *J. Mater. Sci. Lett.*, **22** (2003) 531–533.

[68] C. Craciunescu, J. Li and M. Wuttig, Thermoelastic stress-induced thin film martensites, *Scri. Mat.*, **48** (2003) 65–70.

[69] Y. Q. Fu, H. J. Du and S. Zhang, Sputtering deposited TiNi films: relationship among processing, stress evolution and phase transformation behaviors, *Surf. Coat. Technol.* **167** (2003)120–128.

[70] Y. Q. Fu and H. J. Du, Effects of film composition and annealing on residual stress evolution for shape memory TiNi film, *Mater. Sci. Engng.*, **A342** (2003) 236–244.

[71] Q. He, W. M. Huang, M. H. Hong, *et al.*, Characterization of sputtering deposited NiTi shape memory thin films using a temperature controllable atomic force microscope, *Smart Mater. Struct.*, **13** (2004) 977.

[72] Y. Q. Fu, S. Sanjabi, Z. H. Barber, *et al.*, Evolution of surface morphology in TiNiCu shape memory thin films, *App. Phys. Lett.*, **89** (2006) 171922.

[73] M. J. Wu and W. M. Huang, In situ characterization of NiTi based shape memory thin films by optical measurement, *Smart Mater. Struct.*, **15** (2006) N29–N35.

[74] B. Winzek and E. Quandt, Shape-memory Ti-Ni-X-films (X = Cu, Pd) under constraint, *Z. Metallkd.*, **90** (1999) 796–802.

[75] M. Tomozawa, H. Y. Kim and S. Miyazaki, Microactuators using R-phase transformation of sputter-deposited Ti-47.3Ni shape memory alloy thin films, *J. Intel. Mater. System. Struct.*, **17** (2006) 1049–1058.

[76] S. Stemmer, G. Duscher, C. Scheu, A. H. Heuer and M. Ruhle, The reaction between a TiNi shape memory thin film and silicon, *J. Mater. Res.*, **12** (1997) 1734–1740.

[77] S. K. Wu, J. Z. Chen, Y. J. Wu, *et al.*, Interfacial microstructures of rf-sputtered TiNi shape memory alloy thin films on (100) silicon, *Phil. Mag. A*, **81** (2001) 1939–1949.

[78] Y. Q. Fu, H. J. Du and S. Zhang, Adhesion and interfacial structure of magnetron sputtered TiNi films on Si/SiO$_2$ substrate, *Thin Solid Films*, **444** (2003) 85–90.

[79] Y. Q. Fu, H. J. Du, S. Zhang and S. E. Ong, Effects of silicon nitride interlayer on phase transformation and adhesion of TiNi films, *Thin Solid Films*, **476** (2004) 352–357.

[80] Y. Q. Fu, H. J. Du and S. Zhang, Curvature method as a tool for shape memory effect, *Surface Engineering: Science and Technology II Symposium at TMS 2002*

Annual Meeting, ed. A. Kumar, Y.W. Chung, J.J. Moore, G.L. Doll, K. Yahi and D.S. Misra, TMS, pp. 293–303, Feb. 17–21, 2002, Seattle, Washington, USA.

[81] A. Ishida and M. Sato, Thickness effect on shape memory behavior of Ti-50.0at.%Ni thin film, *Acta Mater.*, **51** (2003) 5571–5578.

[82] D. Wan and K. Komvopoulos, Thickness effect on thermally induced phase transformations in sputtered titanium-nickel shape-memory films, *J. Mat. Res.* **20** (2005) 1606.

[83] X. Wang, M. Rein and J.J. Vlassak, Crystallization kinetics of amorphous equiatomic NiTi thin films: effect of film thickness, *J. Appl. Phys.* **103** (2008) 023501.

[84] Y.Q. Fu, Sam Zhang, M.J. Wu, *et al.*, On the lower thickness boundary of sputtered TiNi films for shape memory application, *Thin Solid Films*, **515** (2006) 80–86.

[85] Z.G. Wang, X.T. Zu, Y.Q. Fu and L.M. Wang, Temperature memory effect in TiNi-based shape memory alloys, *Thermochimica Acta*, **428** (2005) 199.

[86] Y.J. Zeng, L.S. Cui and J. Schrooten, Temperature memory effect of a nickel-titanium shape memory alloy, *Appl. Phy. Lett.*, **84** (2004) 31.

[87] Z.G. Wang and X.T. Zu, Incomplete transformation induced multiple-step transformation in TiNi shape memory alloys, *Scripta Mater.* **53** (2005) 335.

[88] X.G. Ma and K. Komvopoulos, Nanoscale pseudoelastic behavior of indented titanium-nickel films, *Appl. Phys. Lett.*, **83** (2003) 3773–3775.

[89] G.A. Shaw, D.D. Stone, A.D. Johnson, A.B. Ellis and W.C. Crone, Shape memory effect in nanoindentation of nickel-titanium thin films, *Appl. Phys. Lett.*, **83** (2003) 257–259.

[90] D.P. Cole, H.A. Bruck and A.L. Roytburd, Nanoindentation studies of graded shape memory alloy thin films processed using diffusion modification, *J. Appl. Phys.* **103** (2008) 064315.

[91] Y.J. Zhang, Y.T. Cheng and D.S. Grummon, Finite element modeling of indentation-induced superelastic effect using a three-dimensional constitutive model for shape memory materials with plasticity, *J. Appl. Phys.*, **98** (2005) 033505.

[92] A.J.M. Wood, S. Sanjabi and Y.Q. Fu, Nanoindentation of binary and ternary Ni-Ti-based shape memory alloy thin films, *Surf. Coat. Technol.*, **202** (2008) 3115–3120.

[93] W. Yan, Q. Sun, X.Q. Feng, L. Qian, Determination of transformation stresses of shape memory alloy thin films: A method based on spherical indentation, *Appl. Phys. Lett.*, **88** (2006) 241912.

[94] Y. Zhang, Y.T. Cheng and D.S. Grummon, Shape memory surfaces, *Appl. Phys. Lett.*, **89** (2006) 041912.

[95] S. Chikazume, *Physics of Ferromagnetism*, Oxford, UK, Oxford University Press, 1997.

[96] G.A. Shaw, J.S. Trethewey and A.D. Johnson, Thermomechanical high-density data storage in a metallic material via the shape-memory effect, *Adv. Mater.*, **17** (2005) 1123.

[97] S. Takabayashi, E. Tanino, S. Fukumoto, Y. Mimatsu, S. Yamashita Y. Ichikawa, Functionally gradient TiNi films fabricated by sputtering, *Japn. J. Appl. Phys.*, **35** (1996) 200–204.

[98] Y.Q. Fu, H.J. Du and S. Zhang, Deposition of TiN layer on TiNi thin films to improve surface properties, *Surf. Coat. Technol.* **167** (2003) 129–136.

[99] Y.Q. Fu, H.J. Du and S. Zhang, Functionally graded TiN/TiNi shape memory alloy films, *Mater. Lett.*, **57** (2003) 2995–2999.

[100] F. Goldberg and E. J. Knystautas, The effects of ion irradiation on TiNi shape memory alloy thin films, *Thin Solid Films*, **342** (1999) 67–73.

[101] D. S. Grummon and R. Gotthardt, Latent strain in titanium-nickel thin films modified by irradiation of the plastically-deformed martensite phase with 5 MeV Ni2+, *Acta Mater.*, **48** (2000) 635–646.

[102] T. B. Lagrange and R. Gotthard, Microstructrual evolution and thermo-mechanical response of Ni ion irradiated TiNiSMA thin films, *J. Optoelectr. Adv. Mater.*, **5** (2003) 313–318.

[103] G. S. Firstov, R. G. Vitchev, H. Kumar, B. Blanpain and J. Van Humbeeck, Surface oxidation of TiNi shape memory alloy, *Biomaterials*, **23** (2002) 4863–4871.

[104] L. Tan and W. C. Crone, Surface characterization of TiNi modified by plasma source ion implantation, *Acta Mater.* **50** (2002)4449–4460.

[105] C. M. Craciunescu and M. Wuttig, New ferromagnetic and functionally graded shape memory alloys, *J. Optoelectr. Adv. Mater.*, **5** (2003) 139–146.

[106] T. J. Zhu, X. B. Zhao and L. Lu, Pb(Zr0.52Ti0.48)O-3/TiNi multilayered heterostructures on Si substrates for smart systems, *Thin Solid Films*, **515** (2006) 1445–1449.

[107] F. Auricchio, R. L. Taylor and J. Lubliner, Shape-memory alloys: macromodelling and numerical simulations of the superelastic behavior, *Comp. Meth. Appl. Mech. Engng.*, **146** (1997)281–312.

[108] A. Bhattacharyya, M. G. Faulkner and J. J. Amalraj, Finite element modeling of cyclic thermal response of shape memory alloy wires with variable material properties, *Comput. Mater. Sci.* **17** (2000) 93–104.

[109] A. Ishida, M. Sato, W. Yoshikawa, *et al.*, Graphical design for thin-film SMA microactuators, *Smart Mater. Strucut.*, **16** (2007) 1672–1677.

[110] W. M. Huang, Modified shape memory alloy (SMA) model for SMA wire based actuator design, *J. Intell. Mater. Sys. Struct.*, **10** (1999) 221–231.

[111] X. Y. Gao and W. M. Huang, Transformation start stress in non-textured shape memory alloys, *Smart Mater. Struct.*, **11** (2002) 256–268.

[112] W. M. Huang and J. J. Zhu, To predict the behavior of shape memory alloys under proportional load, *Mech. Mater.*, **34** (2002) 547–561.

[113] B. Gabry, C. Lexcellent, V. H. No and S. Miyazaki, Thermodynamic modeling of the recovery strains of sputter-deposited shape memory alloys Ti-Ni and Ti-Ni-Cu thin films, *Thin Solid Films*, **372** (2000) 118–133.

[114] C. Lexcellent, S. Moyne, A. Ishida and S. Miyazaki, Deformation behaviour associated with the stress-induced martensitic transformation in Ti-Ni thin films and their thermodynamical modeling, *Thin Solid Films*, **324** (1998) 184–189.

[115] K. Bhattacharya and R. D. James, A theory of thin films of martensitic materials with applications to microactuators, *J. Mech. Phys. Solids*, **47** (1999) 531–576.

[116] Y. M. Jin and G. J. Weng, Micromechanics study of thermo mechanical characteristics of polycrystal shape-memory alloy films, *Thin Solid Films*, **376** (2000) 198–207.

[117] A. Gyobu, Y. Kawamura, T. Saburi and M. Asai, Two-way shape memory effect of sputter-deposited Ti-rich Ti-Ni alloy films, *Mater. Sci. Engng.*, A, **312** (2001) 227–231.

[118] J. J. Gill, K. Ho and G. P. Carman Three-dimensional thin-film shape memory alloy microactuator with two-way effect, *J. MEMS*, **11** (2002) 68–77.

[119] Y. Q. Fu, J. K. Luo, A. J. Flewitt, *et al.*, Shape memory microcage of TiNi/DLC films for biological applications, *J. Micromech. Microeng.* **18** (2008) 035026.

[120] M. Nishida and T. Honma, All round shape memory effect in Ni rich TiNi alloys generated by constrained aging, *Scripta Metall.* **18** (1984) 1293.

[121] D. D. Shin, D. G. Lee, K. P. Mohanchandra *et al.* Thin film NiTi microthermostat array, *Sens. Actuat. A*, **130** (2006) 37–41.

[122] D. Reynaerts, J. Peirs and H. Van Brussel, An implantable drug-delivery system based on shape memory alloy micro-actuation, *Sens. Actuat. A*, **61** (1997) 455–462.

[123] W. L. Benard, H. Kahn, A. H. Heuer and M. Huff, Thin-film shape-memory alloy actuated micropumps, *J. MEMS*, **7** (1998) 245–251.

[124] M. Kohl, K. D. Skrobanek and S. Miyazaki, Development of stress-optimised shape memory microvalves, *Sens. Actuat.*, **A72** (1999) 243–250.

[125] M. Kohl, D. Dittmann, E. Quandt, B. Winzek, S. Miyazaki and D. M. Allen, Shape memory microvalves based on thin films or rolled sheets, *Mater. Sci. Engng*, **A 273–275** (1999) 784–788.

[126] E. Makino, T. Mitsuya and T. Shibata, Fabrication of TiNi shape memory micro-pump, *Sens. Actuat.* **A 88** (2001) 256–262.

[127] D. Xu, L. Wang, G. F. Ding, Y. Zhou, A. B. Yu and B. C. Cai, Characteristics and fabrication of TiNi/Si diaphragm micropump, *Sens. Actuat. A*, **93** (2001) 87–92.

[128] G. Iddan, G. Meron, A. Glukhovsky and P. Swain, Wireless capsule endoscopy, *Nature*, **405** (2000) 417.

[129] J. D. Waye, Small-bowel endoscopy, *Endoscopy*, **35** (2003) 15.

[130] T. Sugawara, K. Hirota, M. Watanabe, *et al.*, Shape memory thin film actuator for holding a fine blood vessel, *Sens. Actuat.*, **130** (2006) 461–467.

[131] J. J. Gill, D. T. Chang, L. A. Momoda and G. P. Carman, Manufacturing issues of thin film TiNi microwrapper, *Sens. Actuat.*, **A 93** (2001) 148.

[132] S. Takeuchi and I. Shimoyama, A three-dimensional shape memory alloy microelectrode with clipping structure for insect neural recording, *J. MEMS*, **9** (2000) 24–31.

[133] W. M. Huang, Q. Y. Liu and L. M. He, Micro NiTi-Si cantilever with three stable positions, *Sens. Actuat.*, **A 114** (2004) 118–122.

[134] V. Seidemann, S. Butefisch and S. Buttgenbach, Fabrication and investigation of in-plane compliant SU8 structures for MEMS and their application to micro valves and micro grippers, *Sens. Actuat. A*, **97–98** (2002) 457–461.

[135] A. P. Lee, D. R. Ciarlo, P. A. Krulevitch, S. Lehew, J. Trevino and M. A. Northrup, A practical microgripper by fine alignment, eutectic bonding and SMA actuation, *Sens. Actuato.*, **A54** (1996) 755–759.

[136] Y. Q. Fu and H. J. Du, Fabrication of micromachined TiNi-based microgripper with complaint structure, *SPIE* **5116**, (2003), 38–47. *Smart Sensors, Actuators, and MEMS*, 19–21 May, 2003, Gran Canaria, Spain, SPIE-Int. Soc. Opt. Eng., USA.

[137] R. X. Wang, Y. Zohar M. Wong, Residual stress-loaded titanium-nickel shape-memory alloy thin-film micro-actuators, *J. Micromech. Microeng.* **12** (2002) 323–327.

[138] M. Bendahan, K. Aguir, J. L. Seguin and H. Carchano, TiNi thin films as a gate of MOS capacity sensors, *Sens. Actuat.*, **74** (1999) 242–245.

[139] M. Tabib-Azar, B. Sutapun and M. Huff, Applications of TiNi thin film shape memory alloys in micro-opto-electro-mechanical systems, *Sens. Actuat.*, **77** (1999) 34–38.

[140] B. Sutapun, M. Tabib-Azar and M. Huff, Applications of shape memory alloys in optics, *Appl. Opt.* **37** (1998) 6811–6815.

[141] P. M. Chan, C. Y. Chung and K. C. Ng, NiTi shape memory alloy thin film sensor micro-array for detection, *J. Alloys Comp.*, **449** (2008) 148–151.

[142] J. Van Humbeeck, Damping capacity of thermoelastic martensite in shape memory alloys, *J. Alloys Comp.*, **355** (2003) 58–64.

[143] D. Y. Li, A new type of wear-resistant material: pseudo-elastic TiNi alloy, *Wear*, **221** (1998) 116–123.

[144] T. Girardeau, K. Bouslykhane, J. Mimault, J. P. Villain and P. Chartier, Wear improvement and local structure in nickel-titanium coatings produced by reactive ion sputtering, *Thin Solid Films*, **283** (1996) 67–74.

[145] J. L. He, K. W. Won, C. T. Chang, K. C. Chen and H. C. Lin, Cavitation-resistant TiNi films deposited by using cathodic are plasma ion plating, *Wear*, **233–235** (1999) 104–110.

[146] Y. J. Zhang, Y. T. Cheng and D. S. Grummon, Novel tribological systems using shape memory alloys and thin films, *Surf. Coat. Technol.*, **202** (2007) 998–1002.

[147] T. Duerig, A. Pelton and D. Stockel, An overview of NiTinol medical applications, *Mater. Sci. Engng.*, **A273–275** (1999) 149–160.

[148] D. S. Levi, R. J. Williams, J. Liu, *et al.*, Thin film NiTinol covered stents: design and animal testing, *ASAIO J.* **54** (2008) 221–226.

2 Martensitic transformation in TiNi alloys

Shuichi Miyazaki

Abstract

The basic characteristics of the martensitic transformation of TiNi shape memory alloys are described. They include the crystal structures of the parent and martensite phases, the recoverable strain associated with the martensitic transformation, the transformation temperatures, the temperature and orientation dependence of deformation behaviour, etc. Shape memory and superelasticity related to the martensitic transformation are also explained.

2.1 Introduction

The shape memory effect (SME) and superelasticity (SE) are associated with the crystallographically reversible nature of the martensitic transformation which appears in shape memory alloys (SMAs). Such a crystallographically reversible martensitic transformation has been named "thermoelastic martensitic transformation". The name originates from the characteristic of the martensitic transformation in shape memory alloys, i.e., the total free energy change associated with the thermoelastic martensitic transformation mainly consists of two thermoelastic terms, chemical free energy and elastic energy, while the total free energy change associated with the conventional martensitic transformation, which appears in steels for instance, consists of the energy of interfaces and plastic deformation in addition to the two thermoelastic terms. Therefore, the interface between transformed and untransformed regions moves smoothly according to the temperature variation so that the transformation temperature hysteresis is small, from several to several tens of degrees K, compared with those of steels that are several hundreds of degrees K. The characteristic that plastic deformation does not occur in the thermoelastic martensitic transformation is one of the necessary factors for the perfect shape recovery upon the reverse transformation in shape memory alloys.

Thin Film Shape Memory Alloys: Fundamentals and Device Applications, eds. Shuichi Miyazaki, Yong Qing Fu and Wei Min Huang. Published by Cambridge University Press. © Cambridge University Press 2009.

The martensitic transformation itself is not a new phenomenon. It was first found a long time ago in a steel [1] which was heat-treated at a high temperature followed by rapid quenching: the martensitic transformation in most steels is not thermoelastic, hence the SME does not appear. It has been found that several tens of alloys show the shape memory and superelastic behaviour [2]. Of these, the TiNi alloys have been successfully developed as practical materials for many applications.

The TiNi alloys have been investigated since the first report on SME in a TiNi alloy in 1963 [3]. However, TiNi alloys had presented many difficult problems with many puzzling phenomena for about 20 years until 1982, when the basic understanding was established on the relationship between the microstructure and the corresponding deformation behavior such as SME and SE [4, 5]. Since then, many puzzling phenomena have been clarified: e.g., the microstructures which cause the rhombohedral phase (R-phase) transformation to appear [6, 7], the orientation dependence of shape memory and superelastic behavior observed in single crystals [8, 9, 10, 11], the temperature dependence of deformation and fatigue behavior [5, 7, 12], the shape memory mechanism[13, 14], etc.

In this chapter, the basic characteristics such as the martensitic transformation and shape-memory/superelastic properties of TiNi alloys are reviewed.

2.2   TiNi phase diagram

An equilibrium phase diagram of the TiNi system is shown in Fig. 2.1, which describes a middle composition region including an equiatomic composition TiNi. Full information of the equilibrium phase diagram can be found in Ref. [15]. In this chapter, TiNi includes both nearly equiatomic and exactly equiatomic compositions. TiNi locates around the equiatomic composition region, while Ti_2Ni and $TiNi_3$ intermetallic compounds locate at 33.3at%Ni and 75at%Ni, respectively. These three alloys are equilibrium phases. There is another phase, Ti_3Ni_4, which is not in equilibrium but is important for affecting both the transformation temperatures and shape memory behaviour [5]. The TiNi single phase region terminates at 903 K in Fig. 2.1, however, the region seems to extend to around room temperature in a narrow Ni-content width according to empirical information.

2.3   Crystallography of martensitic transformation

The parent phase of the TiNi has a CsCl-type B2 superlattice, while the martensite phase is three-dimensionally close packed (monoclinic or B19′), as shown in Fig. 2.2. The TiNi alloy also shows another phase transformation prior to the martensitic transformation according to heat-treatment and alloy composition. This transformation (rhombohedral phase or R-phase transformation) can be formed by

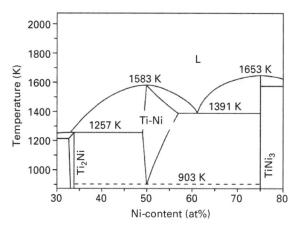

Figure 2.1   An equilibrium phase diagram of the TiNi system.

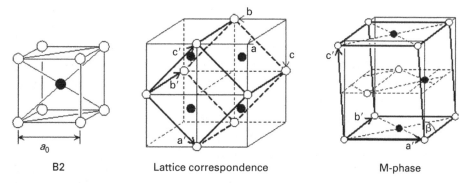

Figure 2.2   Crystal structures of the parent (B2) and martensite (B19') phases and the lattice correspondence between the two phases.

elongating along any one of the $\langle 111 \rangle$ directions of the B2 structure as shown in Fig. 2.3, and is characterized by a small lattice distortion when compared with that of the martensitic transformation. The R-phase transformation usually appears prior to the martensitic transformation when the martensitic transformation start temperature M_s is lowered by some means below the R-phase transformation start temperature T_R. There are many factors to effectively depress M_s as follows [7]:

(1) increasing the Ni-content;
(2) aging at intermediate temperatures;
(3) annealing at temperatures below the recrystallization temperature after cold working;
(4) thermal cycling;
(5) substitution of a third element.

Among these factors, factors (2)–(5) are effective in realizing the R-phase transformation.

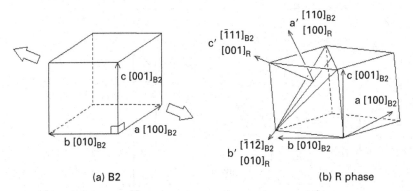

Figure 2.3   Crystal structure of the R phase which is formed by elongation along one of the $\langle 111 \rangle$ directions of the B2 lattice.

The martensitic transformation occurs in such a way that the interface between the martensite variant and parent phase becomes an undistorted and unrotated plane (invariant plane or habit plane) in order to minimize the strain energy. In order to form such a martensite variant (habit-plane variant), it is necessary to introduce a lattice invariant shear such as twins, dislocations or stacking faults. The lattice invariant shear is generally twinning, which is reversible, in the shape memory alloys.

The crystallographic characteristics of the martensitic transformation are now well understood by phenomenological crystallographic theory [16, 17, 18]. This theory states that the transformation consists of the following three operational processes: (1) a lattice deformation B creating the martensite structure from the parent phase, (2) a lattice invariant shear P_2 (twinning, slip or faulting) and (3) a lattice rotation R. Thus, the total strain (or the shape strain) associated with the transformation is written in the following matrix form:

$$P_1 = RP_2B. \tag{2.1}$$

This theory requires that the shape strain produced by the martensitic transformation is described by an invariant plane strain, i.e., a plane of no distortion and no rotation, which is macroscopically homogeneous and consists of a shear strain parallel to the habit plane and a volume change (an expansion or contraction normal to the habit plane). Thus, the shape strain can also be represented in the following way:

$$P_1 = I + m_1 d_1 p'_1, \tag{2.2}$$

where I is the (3×3) identity matrix, m_1 the magnitude of the shape strain, d_1 a unit column vector in the direction of the shape strain, and p'_1 a unit row vector in the direction normal to the invariant plane. If we know the lattice parameters of the parent and martensite phases, a lattice correspondence between the two phases and a lattice invariant shear, the matrix P_1 can be determined by solving Eq. (2.1) under an invariant plane strain condition. Then, all crystallographic parameters such as P_1, m_1, d_1 and orientation relationship are determined. The lattice invariant shear of the TiNi is the $\langle 011 \rangle_M$ Type II twinning [19, 20].

There are generally 6, 12 or 24 martensite variants with each shape strain P_1. Each variant requires the formation of other variants to minimize the net strain of the group of variants. This is called self-accommodation, hence the whole specimen shows no macroscopic shape change except surface relief corresponding to each variant by the martensitic transformation upon cooling.

2.4   Transformation strain

The strain induced by the martensitic transformation shows strong orientation dependence in TiNi alloys [9, 11]. It is conventionally assumed that the most favorable martensite variant grows to induce the maximum recoverable transformation strain ε_M^i in each grain: ε_M^i can be calculated by using the lattice constants of the parent phase and martensite phase. The lattice constants of the parent and martensite phases of a TiNi alloy are as follows: $a_0 = 0.3013$ nm for the parent phase and $a = 0.2889$ nm, $b = 0.4150$ nm, $c = 0.4619$ nm and $\beta = 96.923$ for the martensite phase, respectively.

The following is a calculation process for transformation strain. Using the lattice constants of the parent phase and martensite phase, the transformation strain produced by lattice distortion due to the martensitic transformation can be calculated. If it is assumed that the most favorable martensite variants grow to induce the maximum transformation strain in each grain, the lattice distortion matrix T' is expressed in the coordinates of the martensite as follows using the lattice constants of the parent phase (a_0) and those of the martensite phase (a, b, c, β):

$$T' = \begin{bmatrix} \dfrac{a}{a_0} & 0 & \dfrac{c'\gamma}{\sqrt{2}a_0} \\ 0 & \dfrac{b}{\sqrt{2}a_0} & 0 \\ 0 & 0 & \dfrac{c'}{\sqrt{2}a_0} \end{bmatrix} \qquad (2.3)$$

where $c' = c \sin\beta$ and $\gamma = 1/\tan\beta$.

Then, the lattice distortion matrix T, which is expressed in the coordinates of the parent phase, can be obtained as follows:

$$T = RT'R^t, \qquad (2.4)$$

where R is the coordinate transformation matrix from the martensite to the parent phase and R_t is the transpose of R. R corresponding to the most favorable martensite variant is expressed as follows:

$$R = \begin{bmatrix} -1 & 0 & 0 \\ 0 & \dfrac{1}{\sqrt{2}} & -\dfrac{1}{\sqrt{2}} \\ 0 & -\dfrac{1}{\sqrt{2}} & \dfrac{1}{\sqrt{2}} \end{bmatrix}. \qquad (2.5)$$

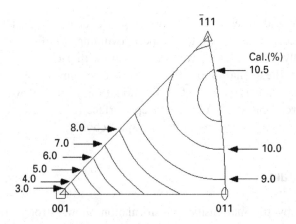

Figure 2.4   Orientation dependence of calculated strain induced by the martensitic transformation.

Since any vector x in the coordinates of the parent phase is transformed to x' due to the martensite transformation using the following equation

$$x' = Tx, \qquad (2.6)$$

the maximum transformation strain ε_M^i in each grain can be calculated as follows:

$$\varepsilon_M^i = \frac{|x'| - |x|}{|x|}. \qquad (2.7)$$

Figure 2.4 shows the calculated result of the transformation strain ε_M^i expressed by contour lines for each direction in a $[001] - [011] - [\bar{1}11]$ standard stereographic triangle. For example, the transformation strains along $[001]$, $[011]$, $[\bar{1}11]$ and $[\bar{3}11]$ are 3.0%, 8.4%, 9.9% and 10.7%, respectively. By applying similar calculation for the R-phase transformation, the transformation strain ε_R^i at a temperature 35 K lower than T_R can be found as shown in Fig. 2.5. The result indicates that the strain is a maximum along $[\bar{1}11]$ and that along $[001]$ the minimum is nearly equal to zero. The strain decreases with decreasing temperature from T_R, because the rhombohedral angle of the R-phase lattice shows temperature dependence.

By averaging ε_M^i for a representative 36 orientations which locate periodically in a stereographic standard triangle, the transformation strain for a polycrystal can be estimated as follows if there is no specific texture and the axis density distributes uniformly [21]:

$$\bar{\varepsilon}_M^0 = \left(\sum_{i=1}^{36} \varepsilon_M^i \right) \Big/ 36. \qquad (2.8)$$

If there is texture, the axis density I^i is not uniform in each inverse pole figure so that it is necessary to consider I^i in the calculation of the transformation

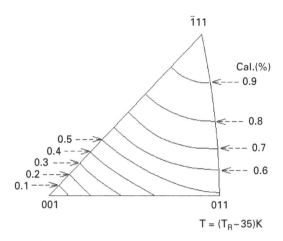

Figure 2.5   Orientation dependence of calculated strain induced by the R-phase transformation.

strain as follows [22]:

$$\bar{\varepsilon}_M = \left(\sum_{i=1}^{36} \varepsilon_M^i \Gamma^i \right) \Big/ 36. \qquad (2.9)$$

2.5   Transformation temperatures

The martensitic transformation temperatures are conventionally determined by the electrical resistivity measurement or by the differential scanning calorimetry (DSC). Figures 2.6 and 2.7 show example results of such measurements applied to an equiatomic TiNi alloy which was solution-treated at 1273 K for 3.6 ks. When the specimen is cooled from the parent B2 phase, the martensitic transformation starts at M_s (the martensitic transformation start temperature) by evolving transformation heat, i.e., the change in chemical enthalpy ΔH is negative and the reaction is exothermic, as shown in the DSC curve upon cooling. The electrical resistivity shows a normal decrease upon cooling in the B2 phase region and increase of the decreasing rate at the onset of the transformation at M_s because of the crystal structural change. Upon further cooling, the martensitic transformation finishes at M_f (the martensitic transformation finish temperature).

Upon heating the specimen from the martensite phase, the martensite phase starts to reverse transform to the B2 phase at A_s (the reverse martensitic transformation start temperature) and finish at A_f (the reverse martensitic transformation finish temperature). The DSC curve upon heating shows an endothermic reaction, i.e., ΔH is positive and heat is absorbed by the transformation.

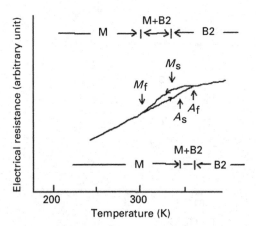

Figure 2.6   Electrical resistance vs. temperature curve showing the transformation temperatures of the Ti-50.0at%Ni alloy.

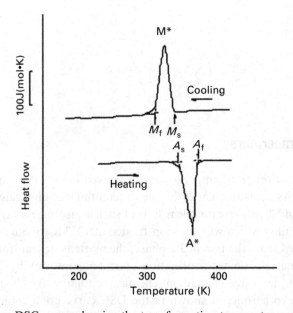

Figure 2.7   DSC curves showing the transformation temperatures of the Ti-50.0at%Ni alloy.

The M_s is shown in Fig. 2.8 as a function of Ni-content. In the composition range of the TiNi, M_s decreases with increasing Ni-content above 49.7at%Ni, while they are constant below 49.7at%Ni. The A_f is about 30 K higher than M_s in all of the composition region. The reason for the constant M_s in the Ni-content region less than 49.7 at% can be ascribed to the constant Ni-content in the TiNi phase, because the Ti_2Ni appears in the Ni-content region less than 49.7 at% as shown in Fig. 2.1, keeping the Ni-content of the TiNi to be 49.7 at%.

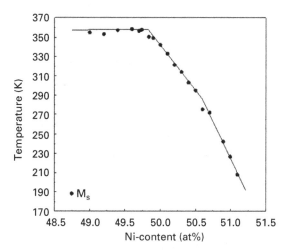

Figure 2.8   Ni-content dependence of M_s temperature.

2.6   Shape memory and superelasticity based on martensitic transformation

The mechanisms of the shape memory effect and superelasticity are explained in the following using a two-dimensional crystal model shown in Fig. 2.9. The crystal structure of the parent phase is shown in Fig. 2.9(a). It perfectly transforms to the martensite upon cooling below M_f, as shown in Fig. 2.9(b), where two martensite variants labelled A and B with the same crystal structure but different orientations are shown. Thermally induced martensite should be a habit-plane variant, since it will be formed in the parent phase and needs to contact the parent phase along the habit plane. However, the lattice invariant shear is not shown in the A and B variants for simplicity. The lattices of both martensite variants are made by distorting the parent phase lattice upon the transformation, creating the same shear strain with senses opposite to each other. Therefore, the martensite morphology in Fig. 2.9(b) is self-accommodated to minimize the macroscopic net strain. In three-dimensional real crystal structures, the number of martensite variants is generally 24 with different shear systems.

By applying stress to the shape memory alloy below M_f, the A variant grows by the movement of the interface between A and B variants, which is usually a twinning plane. The interface moves easily under an extremely low stress so that the shape of the alloy changes to any shape. The selection of martensite variants is such that the most preferential variant which creates the maximum strain along the applied stress grows. In this case, the variant A grows: after replacing B with A, the martensite morphology consisting of a single variant A is obtained, generating a macroscopic shear deformation, as shown in Fig. 2.9(c). The same specimen shape is maintained after unloading except for an elastic recovery. When the alloy is heated to the A_s point, the alloy starts to recover the original shape by

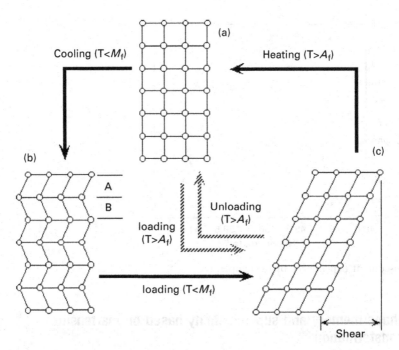

Figure 2.9   Schematic figure showing the specimen shapes and crystal structures upon cooling, heating, loading and unloading during the shape memory and superelastic behavior.

the onset of the reverse martensitic transformation. Upon further heating above the A_f point, the reverse martensitic transformation finishes, resulting in a perfect shape recovery. This is the process of the shape memory effect.

The martensitic transformation generally occurs by cooling the specimen. However, an applied stress also induces the martensitic transformation even above the M_s point. The reason why the applied stress assists the transformation is that the martensitic transformation is achieved by distorting the lattice of the parent phase. Therefore, the shape of the alloy changes from Fig. 2.9(a) directly to Fig. 2.9(c) by the lattice distortion associated with the stress-induced martensitic transformation. If the deformation temperature is below the A_f point, the shape recovery is not perfect upon unloading. However, if it is above the A_f point, the alloy recovers the original shape upon unloading without the following heating. This is the process of superelasticity. Since both the shape memory effect and superelasticity are associated with the same martensitic transformation, they show the same amount of shape recovery. The driving forces for shape recovery in both phenomena originate from the recovery stress associated with the reverse martensitic transformation.

2.7   Deformation behavior

The deformation behavior of SMAs is strongly temperature sensitive, because the deformation is associated with the martensitic transformation: this is different

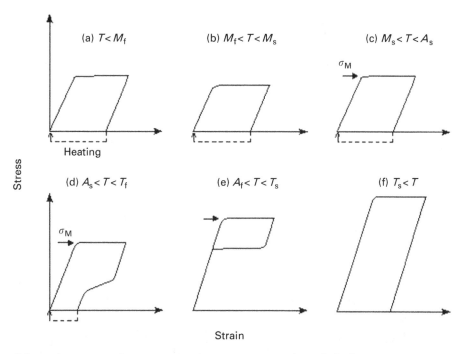

Figure 2.10   Schematic stress–strain curves at various temperatures in a TiNi alloy.

from plastic deformation by slip which occurs in conventional metals and alloys. Schematic stress–strain curves of a TiNi alloy obtained at various temperatures (T) are shown in Fig. 2.10. In the temperature range of $T < M_f$, the specimen is fully transformed before applying stress so that the elastic deformation takes place in the martensite phase at first, as shown in Fig. 2.10(a), where many martensite variants self-accommodate each other before loading. Upon further loading, twin planes in the martensite phase move to create an apparent plastic deformation. Therefore, the yield stress in Fig. 2.10(a) corresponds to the critical stress for twinning deformation in the martensite phase. In the temperature range $M_f < T < M_s$, the parent and martensite phases coexist so that yielding occurs due to twinning in the martensite phase and/or stress-induced martensitic transformation in the parent phase. Both the yield stresses by twinning and stress-induced transformation in Fig. 2.10(b) are lowest in this temperature range, because the former decreases with increasing temperature and the latter decreases with decreasing temperature until reaching this temperature region. The stress–strain curves in Figs. 2.10(a) and (b) are essentially the same, except the yield stress is a little lower in Fig. 2.10(b) than that in Fig. 2.10(a). In the temperature range of $M_s < T < A_s$, the parent phase elastically deforms at first and yielding occurs due to the stress-induced martensitic transformation. Therefore, the yield stress linearly increases with increasing temperature satisfying the Clausius–Clapeyron relationship. The stress-induced martensite phase remains after unloading, because the

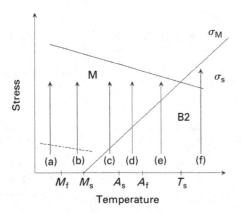

Figure 2.11   Critical stresses for inducing the martensitic transformation (σ_M) and for slip deformation (σ_s) shown as a function of test temperature in the specimen of Fig. 2.10.

temperature is below A_s. The shape of the stress–strain curve of Fig. 2.10(c) is similar to those of Figs. 2.10(a) and (b). In the temperature range $A_s < T < A_f$, the deformation induced by the stress-induced martensitic transformation recovers partially upon unloading, as shown in Fig. 2.10(d), resulting in partial super-elasticity and partial shape memory effect by the following heating. In the temperature range $A_f < T < T_s$, perfect superelasticity appears, as shown in Fig. 2.10(e), where T_s stands for the critical temperature above which the martensitic transformation does not take place and deformation occurs by slip. If T is above T_s, plastic deformation occurs as in conventional metals and alloys, as shown in Fig. 2.10(f).

According to Fig. 2.10, the effect of temperature on the critical stress for inducing martensite (σ_M) and the critical stress for slip (σ_s) are shown by two solid lines in Fig. 2.11. The former line shows positive temperature dependence, while the latter line shows negative temperature dependence, resulting in intersecting at T_s. The stress for the rearrangement of martensite variants due to the movement of twin planes is shown by a dashed line in the temperature range below M_s. The slope of the dashed line shows negative temperature dependence as well as the solid line for slip deformation, because both the deformation modes, slip and twinning, are thermal activation processes. Deformation paths corresponding to those shown in Figs. 2.10(a)–(f) are shown in Fig. 2.11.

TiNi alloys show successive stages of transformation in the stress–strain curve. The deformation is associated with both the R phase and the martensite in TiNi alloys which include a high density of dislocations and/or fine Ti_3Ni_4 precipitates. Therefore, the deformation behavior is sensitive to test temperature; it is classified into six categories according to the relative relationship between test temperature and transformation temperatures, as schematically shown in Fig. 2.12.

In range 1 ($T < M_f$), only one stage associated with the rearrangement of martensite variants appears, as shown in Fig. 2.12(a). In range 2 ($M_f < T < M_s$), both the R phase and the martensite coexist, revealing two stages associated with the rearrangement of the R-phase and martensite variants, as shown by the two

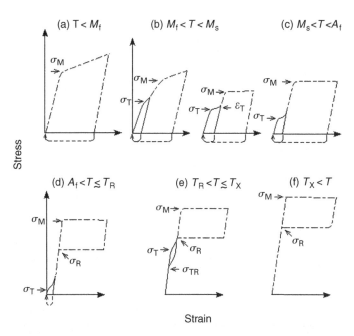

Figure 2.12 Schematic typical stress–strain curves at specific temperatures in a TiNi alloy which exhibits both R-phase (rhombohedral phase) and martensite phase (monoclinic phase) transformations.

stress–strain curves in Fig. 2.12(b). In the figure, the stress–strain curve associated with the R phase is drawn by a solid line, while that associated with the martensite is by a broken line; the dashed line shows the shape recovery associated with the two reverse transformations upon heating. In range 3 ($M_s < T < A_f$), the specimen is in a full R-phase state prior to loading, and hence deformation proceeds by the rearrangement of the R-phase variants to a favourable one, as shown in Fig. 2.12(c). Upon further loading the martensite is stress-induced in the second stage. In range 4 ($A_f < T < T_R$), the superelasticity associated with the forward and reverse martensitic transformations appears, although a part of the deformation is still associated with the rearrangement of the R-phase variants. In range 5 ($T_R < T < T_x$), the R-phase is also stress-induced, exhibiting two-stage superelasticity. The critical stresses for inducing both the R phase and martensite phase satisfy the Causius–Clapeyron relationship, as shown in Fig. 2.13, where the critical stresses are plotted against test temperature. Since the slope of the stress–temperature relation for the R phase is steeper than that for the martensite, both lines cross each other at a temperature T_x. Thus, the deformation associated with the R phase does not appear in range 6 ($T_x < T$), as shown in Fig. 2.13(f). The steepness of the Clausius–Clapeyron relationship for the R-phase transformation mainly originates from the small transformation strain associated with the R-phase transformation, i.e., only a tenth of that associated with the martensitic transformation [10, 11].

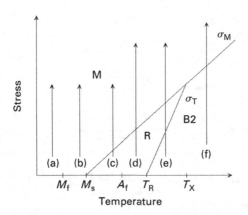

Figure 2.13   Schematic phase diagram of TiNi alloy in a stress–temperature plot.

2.8   Summary

The equilibrium phase diagram of the TiNi system was explained, followed by the Ni-content dependence of the martensitic transformation temperature. The crystallography of the martensitic transformation of the TiNi was also explained, then the orientation dependence of the transformation strain was derived both for the R-phase and martensitic transformations. It was shown that the transformation temperatures can be determined by measuring electrical resistance or heat flow as a function of temperature. The mechanisms of the shape memory effect and superelasticity were explained by using a two-dimensional model. Finally, the deformation behavior associated with a single martensitic transformation and with both the R-phase and martensitic transformations. Readers can refer to Refs. [23, 24, 25, 26, 27, 28, 29, 30, 31, 32, 33, 34, 35, 36, 37] for more details of the martensitic transformation, deformation behavior and other properties of TiNi alloys.

References

[1] Z. Nishiyama, *Martensitic Transformation*, Burlington, MA: Academic Press, Inc. (1978).

[2] S. Miyazaki and K. Otsuka, *ISIJ Int.*, **29** (1989) 353.

[3] W. J. Buehler, J. V. Gilfrich and K. C. Weiley, *J. Appl. Phys.*, **34** (1963) 1467.

[4] S. Miyazaki, Y. Ohmi, K. Otsuka and Y. Suzuki, *J. de Phys.*, **43** Suppl. **12** (1982) C4–255.

[5] S. Miyazaki, *Engineering Aspects of Shape Memory Alloys* (eds. T. W. Duerig *et al.*), London: Butterworth-Heinemann (1990) 394.

[6] S. Miyazaki and K. Otsuka, *Phil. Mag.*, **A 50** (1984) 393.

[7] S. Miyazaki and K. Otsuka, *Met. Trans.*, **A 17** (1986) 53.

[8] F. Takei T. Miura, S. Miyazaki, S. Kimura, K. Otsuka and Y. Suzuki, *Scripta Met.*, **17** (1983) 987.

[9] S. Miyazaki, S. Kimura, K. Otsuka and Y. Suzuki, *Scripta Met.*, **18** (1984) 883.

[10] S. Miyazaki and C. M. Wayman, *Acta Met.*, **36** (1988) 181.

[11] S. Miyazaki, S. Kimura and K. Otsuka, *Phil. Mag.*, A **57** (1988) 467.

[12] S. Miyazaki, Y. Igo and K. Otsuka, *Acta Met.*, **34** (1986) 2045.

[13] S. Miyazaki, K. Otsuka and C. M. Wayman, *Acta Met.*, **37** (1989) 1873.

[14] S. Miyazaki, K. Otsuka and C. M. Wayman, *Acta Met.*, **37** (1989) 1885.

[15] J. L. Murray (ed.), *Phase Diagrams of Binary Titanium Alloys*, Monograph Series on Alloy Phase Diagrams, Materials Park, Ohio ASM International, (1987).

[16] M. S. Wechsler, D. S. Lieberman and T. A. Read, *Trans. AIME*, **197** (1953) 1503.

[17] J. S. Bowles and J. K. Mackenzie, *Acta Met.*, **2** (1954) 129, 138, 224.

[18] D. S. Lieberman, M. S. Wechsler and T. A. Read, *J. Appl. Phys.*, **26** (1955) 473.

[19] K. M. Knowles and D. A. Smith, *Acta Met.*, **29** (1981) 101.

[20] O. Matsumoto, S. Miyazaki K. Otsuka and H. Tamura, *Acta Met.*, **35** (1987) 2137.

[21] S. M. Tan and S. Miyazaki, *Acta Mat.*, **46** (1997) 2729.

[22] S. Miyazaki, V. H. No, K. Kitamura, K. Anak and H. Hosoda, *Int. J. Plasticity*, **16** (2000) 1135.

[23] A. Bansiddhi, T. D. Sargeant, S. I. Stupp, D. C. Dunand, *Acta Biomaterialia*, **4** (2000) 773.

[24] Y. Bellouard, *Mater. Sci. Eng. A*, **481** (2008) 582.

[25] S. A. Wilson, R. P. J. Jourdain, Q. Zhang, *et al.*, *Mater. Sci. Eng. R*, **56** (2007) 1.

[26] L. Ponsonnet, D. Treheux, M. Lissac, *et al.*, *Int. J. Appl. Electromag. Mech.*, **23**, Issue 3–4 (2006) 147.

[27] A. Paiva and M. A. Savi, *Mathematical Problems in Engineering*, Article Number: 56876 (2006).

[28] L. Janke, C. Czaderski, M. Motavalli, *et al.*, *Mat Struct.*, **38** (2005) 578.

[29] G. S. Firstov, J. Van Humbeeck, and Y. N. Koval, *J. Intelligent Systems and Structures*, **17** (2006) 1041.

[30] G. Song, N. Ma and H. N. Li, *Eng. Struct.*, **28** (2006) 1266.

[31] K. Otsuka and X. Ren, *Prog. Mat. Sci.*, **50** (2005) 511.

[32] N. B. Morgan, *Mater. Sci. Eng. A*, **378** (2004)16.

[33] Y. Q. Fu, H. J. Du, W. M. Huang, *et al.*, *Sensors and Actuators A – Physical*, **112** (2004) 395.

[34] S. A. Shabalovskaya, *Biomed. Mat. Eng.*, **12** (2002) 69.

[35] F. El Feninat, G. Laroche, M. Fiset, *et al.*, *Adv Eng Mat*, **4** (2002) 91.

[36] S. A. Shabalovskaya, *Inter. Mat. Revi.*, **46** (2001) 233.

[37] R. D. James and K. F. Hane, *Acta Mat.*, **48** (2000) 197.

3 Deposition techniques for TiNi thin film

A. David Johnson

Abstract

Direct current vacuum sputter deposition is the commonly used method of creating TiNi thin film. Polished silicon wafers are a preferred substrate. Limitations on composition and impurities are similar to those for bulk material. These limitations impose severe constraints on sputtering conditions for obtaining optimal performance of the resulting material. Obtaining material with desirable shape memory properties, uniform composition and uniform thickness requires understanding and control of the processes used. With sputter deposition it is possible to produce thin films with a range of transition temperatures from 173 K to 373 K. Superelastic thin film can be made without cold work. After deposition, photolithography and chemical etching are used to create shapes and combine thin film with other materials to produce microdevices. Producing thin film with shape memory properties is not difficult. But, to obtain uniformity and high yield requires specialized equipment and great care in process control. This chapter introduces some specific recommendations for fabrication of TiNi thin film and incorporation in useful devices. Applications of TiNi thin film are described elsewhere (see Chapter 10).

3.1 Introduction to methods of making TiNi thin film

Titanium nickel shape memory alloy (TiNi SMA) in the form of thin film has been available in limited quantities for nearly two decades [1, 2]. This chapter describes the technology and methods used in forming TiNi thin film and combining it with other materials to create useful microdevices from the perspective of TiNi Alloy Company personnel who have been involved in the development from some of the earliest efforts [3, 4, 5].

Thin Film Shape Memory Alloys: Fundamentals and Device Applications, eds. Shuichi Miyazaki, Yong Qing Fu and Wei Min Huang. Published by Cambridge University Press. © Cambridge University Press 2009.

Nickel titanium thin film fabricated by vacuum sputter deposition, the most common form of physical vapor deposition, has intrinsic thermomechanical properties at least equal to those of bulk material created from melt if several critical process parameters are controlled. The quality of vacuum, smoothness of substrate, purity of gases and targets, strength and uniformity of magnetic field, temperature and target-to-substrate distance are all important to successful results.

TiNi thin film was developed during the same period as microelectro mechanical systems (MEMS) from 1989 to 2000. It was natural for thin film to be exploited for microactuators. TiNi produces large force and displacement; it can be shaped by microlithography; and it has appropriate electrical properties for actuation by joule heating [6]. Microlithography and chemical etching, the MEMS processes, have been used to fabricate microactuators by combining titanium nickel (TiNi) with silicon and to make patterned freestanding super-elastic film.

3.2   Sputter deposition

Shape memory alloy thin films with thicknesses ranging from 0.5 to 2 μm are produced by magnetron sputtering, a process which may include simultaneous deposition from several targets. In a sputtering process, individual atoms are knocked from a TiNi target by atoms of inert argon gas that are accelerated in an electric field [7].

The titanium and nickel atoms condense onto a substrate such as a silicon wafer, polyimide film or glass plates to build up a thin film. Film deposited onto a cold substrate is amorphous and requires a heat treatment to become crystalline and to exhibit a phase change.

A direct current (DC) magnetron is favored for sputtering because the deposition rate is greater for DC than for radio frequency (RF) sputtering. In DC magnetron sputtering, a magnetic field confines electron orbits to maintain an intense plasma and to increase the collision rate, thereby enhancing the deposition rate.

Other methods of depositing thin films have been introduced. E-beam evaporation has been demonstrated [8]. In this method an energetic electron beam vaporizes very local regions of the target, thus preserving the stoichiometry in the film. Uniformity of film thickness is not easily controlled, and the rate of deposition is much smaller than for sputtering. Pulsed laser deposition, in which electromagnetic radiation incident on a target vaporizes its surface, has been shown to be feasible [9]. Chemical vapor deposition (CVD) has been suggested for large-scale production, but has not yet been put into common practice. CVD processes generally consist of deposition of a solid material from a gaseous phase. Chemical precursors are exposed to a substrate under pressure and temperature controlled conditions such that the precursors react to form a solid film on the substrate and a gas that is removed. Suggested precursors include titanium chloride ($TiCl_4$) and nickel carbon monoxide Ni $(CO)_4$ [10].

The TiNi Alloy Company's experience is confined to planar sputtering. The first sputtering machine used was a diffusion-pumped NRC vacuum system equipped with a home-built DC magnetron. Since 1992, nearly all deposition at the TiNi Alloy Company has been performed using Perkin-Elmer 4400 and 4450 machines equipped with three sputtering sources, a rotating platen, and turbo and cryogenic pumps. The source used is nearly always a single target of titanium-nickel.

From the start, silicon wafers have been the preferred substrates because of their high quality surfaces and availability. But the versatility of TiNi thin film was not realized until sputter deposition was combined with micromachining processes including photolithography, chemical machining by selective etching, and wafer processing. This step, which required significant investment in equipment and training, replaced conventional machining and enabled fabrication of true microdevices [6].

3.3   Description of the sputtering process

For DC sputter deposition, a vacuum system capable of being pumped to high vacuum is required. It is convenient if the system has more than one target for co-sputtering (for example TiNi plus nickel or titanium) and for incorporating a sacrificial layer such as chromium. During sputtering, atoms from the target are transferred to a suitable substrate such as polished silicon. Target to substrate distance is 2–5 cm. The base pressure used is below 5×10^{-7} torr, achieved with assistance from liquid nitrogen Meissner traps.

After pumping, the chamber is filled with pure argon to a pressure of 2–5 millitorr and an electric field of 500 volts is applied between the target and the substrate to generate a plasma. A magnetic field of about 200–500 gauss is supplied, which confines the electrons to orbits within the plasma (an 'electron racetrack') and enables higher deposition rates (up to 10 microns of deposited material per hour). Ions of argon impinge on the target, knocking out individual atoms of titanium and nickel, which build up a layer on the substrate. Typical deposition rates are 2–10 microns per hour with 500 volts and 1–10 amperes direct current.

The most critical consideration in making a shape memory alloy, whether in bulk or in thin film, is chemical composition. Equiatomic TiNi – the composition having the best ductility, largest amount of shape recovery, and best work output – consists of 50 atomic percent nickel and 50 atomic percent titanium. Changing the titanium-to-nickel ratio by as little as 0.1 atomic percent results in a noticeable shift of the phase transformation temperature. Deviation from the ideal 50/50 ratio by more than one or two percent results in material without a shape memory [11].

At the start of the use of a new target, because the sputtering rate for nickel is nearly three times that for titanium, the film is nickel rich. After an hour or more of sputtering the composition stabilizes. The composition gradually varies as the target 'wears' an erosion ring.

The phase transition temperature of TiNi is mainly determined by the relative atomic percentages of nickel and titanium. The thermo-mechanical behavior is sensitive to small deviations in composition. Control of transition temperature for production of shape-memory or superelastic characteristics requires fine-tuning of the target composition, target to substrate distance, power, and deposition rate. A target of equiatomic titanium and nickel does not produce equiatomic thin film because titanium is selectively lost to residual oxygen in the chamber. This loss due to residual oxygen can be offset by using a target that is rich in titanium. A recommended target composition is 46.75 wt percent titanium (51.82 atomic percent titanium) and 53.25 wt percent nickel (48.18 atomic percent nickel).

The transition temperature can be lowered to form superelastic film by adding nickel, and possibly chromium, from a separate target. One must then calibrate the deposition rate versus power to determine the appropriate power to use on each target.

Since oxygen contamination causes brittleness and composition shift in deposited films by forming titanium-rich oxide, a high vacuum is created to minimize oxygen. Incorporation of about 0.1 atomic percent of oxygen reduces the transition temperature by about 100 K. To avoid oxygen contamination, a high vacuum system having base pressure less than 1×10^{-6} torr is necessary. Sputtering from a titanium target with a closed shutter prior to sputtering (pre-sputtering) helps reduce the oxygen partial pressure by the gettering action of titanium.

In the sputtering process the surface finish of the substrate is critical to the quality of deposited film. Films deposited on optically smooth surfaces, such as silicon wafers, display a crystal structure with random orientation of crystal boundaries and very strong cohesion between the individual grains, as shown in Fig. 3.1. Such a film is capable of being deformed plastically in tension up to 15% before it necks and breaks. Surface features as small as a fraction of a micron can produce flawed material. In particular, sharp edges such as 'steps' may result in discontinuities like those shown in Fig. 3.2. TiNi film sputter-deposited on ground and polished metal surfaces reveals a related phenomenon shown in Fig. 3.3. Here the imperfect surface results in a columnar structure perpendicular to the film surface. Irregularities on the substrate surface interfere with the merger of nanometer-scale 'islands' of initially-deposited material to form a continuous surface. Instead, some of these islands grow vertically to form separate columns. The resulting film is subject to brittle

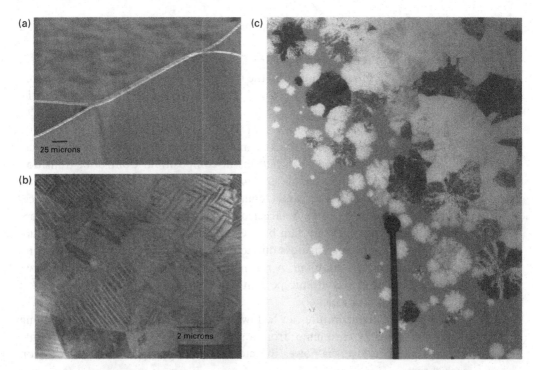

Figure 3.1   Electron microscope images of sputter-deposited thin film TiNi after crystallization by heat treatment. Crystallization is initiated at multiple sites forming individual domains a few microns apart. These crystals coalesce and form strong bonds. (a) SEM image of one edge of a multiple-layer TiNi film showing strong cohesion between crystal domains and between layers. (b) TEM image. Crystal domains are 1–3 microns in size. Twinning is evident, usually at right angles. (c) TEM image. Crystals of TiNi in the process of being formed, showing how they nucleate and collide.

fracture because of poor adhesion between columns comprising adjacent crystal grains.

Even when deposition is on a smooth substrate, brittle material may result from formation of columnar crystal structure if the substrate is heated above some threshold temperature during the sputter deposition process. Probably the increased mobility of atoms allows them to aggregate into clumps, and these clumps act as surface irregularities to produce a columnar structure similar to that seen in film deposited onto substrates that are not smooth.

It is difficult to obtain films that are uniform in composition throughout a large area or from run to run. Variations in film composition arise from differences in sputtering yield and angular flux distribution of atoms knocked from the target. The target itself is not of absolutely uniform composition, and as the target is depleted and wear rings form, subtle geometry changes occur that affect the composition differently in different areas of the substrate. As a result

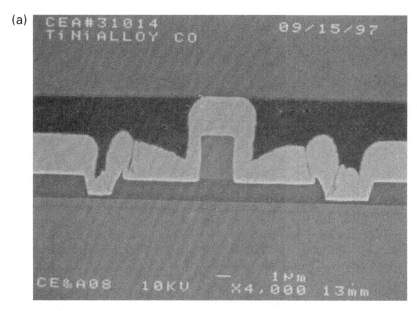

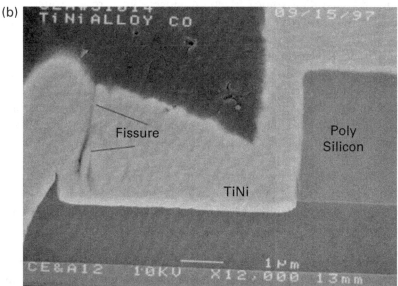

Figure 3.2 Sputter deposition requires a substrate with maximum surface roughness of about one micron. Surface defects may propagate through the deposited layer and form flaws in the film, including discontinuities. These are not visible in surface inspection by optical microscope but cause failure during the micromachining processes. (a) SEM image showing a cross-section through polysilicon deposited on silicon wafer and patterned to form steps. TiNi film deposited on this surface is discontinuous due to the propagation of surface features through the film. (b) Higher magnification reveals a fissure through the film that will cause failure during subsequent processing.

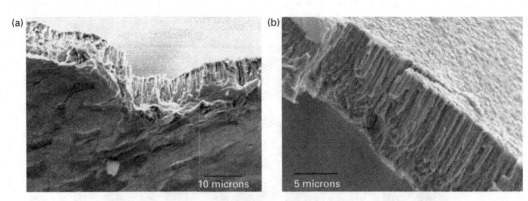

Figure 3.3   SEM images of thin film TiNi deposited on a ground steel rod. Surface irregularities act as crystallization sites that propagate through the film layer as columns that are weakly bound to each other. This film tears apart easily. (a) Substrate side and edge of torn film showing how columns form. (b) Edge and top side of film showing that the columns, once formed, propagate through the thickness of the film.

the deposited film is not perfectly homogeneous across a wafer. Homogeneity is greatly improved if the substrate rests on a rotating platen so that the substrate moves across the face of the target during deposition. The tradeoff for this improvement in uniformity is a decrease in the effective deposition rate (Fig. 3.4(a) and (c)).

Further improvement in uniformity of deposition can be made by replacing the conventional fixed DC magnet array with a rotating magnet system (Fig. 3.4(b) and (d)). This distributes the erosion of the target over a larger area, and reduces the variation in angular distribution with time. A system developed for the Perkin-Elmer 4450 is shown in Fig. 3.4(e).

Consistent sputtering results depend on several other parameters including the intrinsic properties of each vacuum system. Increasing the distance from substrate to target depletes titanium because of increased interaction with residual oxygen. Low argon pressure (less than about 2 millitorr) is associated with low deposition rates, while high pressure (greater than 20 millitorr) results in increased scatter of the titanium and nickel atoms. Running at high power heats the substrate, which affects the mobility of atoms arriving at the substrate and the deposition rate. The target temperature affects the diffusion at the surface and the consequent relative sputter rates for titanium and nickel. Substrate bias affects the mobility of atoms arriving at the surface, and can be used to manipulate the effects of substrate surface irregularities.

Base pressure of about 5×10^{-7} torr is sufficient. Little or no improvement in properties is seen from decreased pressure. This suggests that oxidation becomes the rate-limiting process at about 1×10^{-6} torr. At this pressure each atom composing the surface will on average be struck by one gas molecule per second [12].

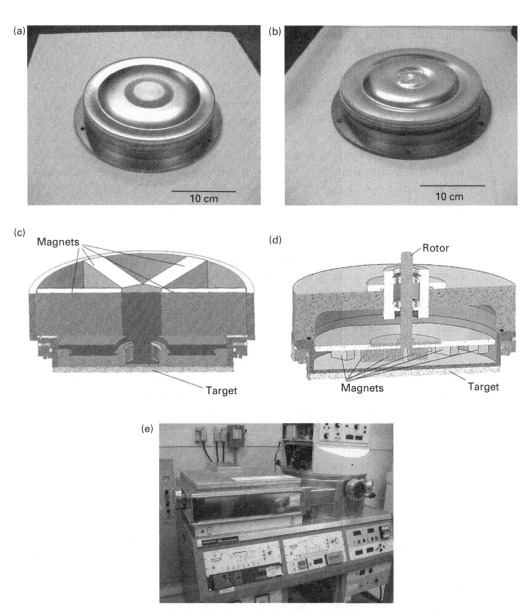

Figure 3.4   Photographs of targets and drawings of conventional and rotating magnet sputtering target
assemblies. (a) Fixed-magnet source produces one erosion ring, resulting in non-uniform
deposition rate and composition. The dark ring at center is re-deposited Ti 'whiskers.'
(b) A rotating magnet source produces concentric erosion rings, and more uniform
deposition across the substrate. (c) Conventional fixed-magnet DC sputtering source.
Radially-disposed magnets produce a toroid shaped electron racetrack to enhance plasma
density. (d) In this fixture, an array of magnets is arranged so that a closed electron orbit
is created, and the assembly is rotated to spread the deposition more evenly across the
substrate surface. (e) Perkin-Elmer 4450 sputtering system used for deposition of TiNi
thin film. The load lock at the left permits the use of the process chamber (on the right)
without loss of vacuum between runs. The process chamber head contains RF and DC
magnetron sources.

3.4   Characterization of thin film by electrical resistivity and stress–strain measurement

Understanding the behavior of SMAs depends on knowledge of the phase transition temperatures: austenite start, austenite finish, martensite start and martensite finish. Differential scanning calorimetry (DSC) measures these quantities directly, while the change of curvature of a substrate with varying temperatures allows them to be estimated. Since the thermodynamics of SMAs are primarily determined by elemental composition, a direct measurement of the percentage of Ti and Ni would be a desirable tool for characterizing TiNi thin film.

Unfortunately the chemical composition of film cannot be determined by conventional techniques (such as energy dispersive X-ray spectroscopy, wavelength dispersive X-ray spectroscopy, and other methods using incident beams) with sufficient accuracy to make quantitative, reliable prediction of the thermomechanical characteristics. Variation in composition of 0.1 percent will affect the behavior, and accuracy of composition measurement is limited to about 1 percent by several factors: dead-time in counters, finite volume of the interactive region, noise background, beam intensity and stability of the electronics of the measuring apparatus which limits the counting statistics attainable in a sample [13]. Inductively coupled plasma mass spectroscopy (ICPMS), used to measure composition of thin film, has been shown to give repeatable results accurate to about 0.5 percent.

Rather than measuring composition, a simple and reliable method of predicting the phase transformation temperature is measurement of resistivity versus temperature. The actual phase transition shows up as a change in the slope of the temperature versus resistivity curve. A thin film sample is heated and cooled while changes in voltage are measured and recorded using a four-probe constant-current technique to produce temperature versus resistivity data, as shown in Fig. 3.5.

Binary TiNi alloys have an upper limit to the transition temperature of approximately 373 K. Ternary alloys, in particular TiNi with 20 atomic percent of hafnium added, have transition temperatures above 423 K. The mechanical properties of TiNi thin film have been characterized using stress–strain measurement to determine the mechanical strength. A device resembling an Instron machine elongates a strip of thin film while simultaneously measuring force and displacement (Fig. 3.6).

3.5   Methods of joining thin film

TiNi thin film is perceived as a component of other devices, not as a product in itself, so methods of joining it to other materials are a critical step in making it usable. For example, construction of actuators and intravascular devices usually involves joining thin film to itself or to other materials. A TiNi thin film sheet that

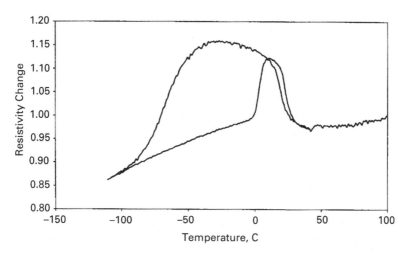

Figure 3.5 Resistivity measurement is used to study the phase transformation temperature and thermal hysteresis characteristics of SMA thin film. Four-point electrical contacts are made to the sample, and the varying voltage is measured as a constant current run through the sample while it is heated and cooled. The large broad excursion shows the temperature at which martensite transformation occurs during cooling. The small rise between 0 and 30 degrees centigrade is due to the R-phase transformation. (Reprinted with permission of ASM International®. All rights reserved. Gupta *et al.* [14], figure 1.)

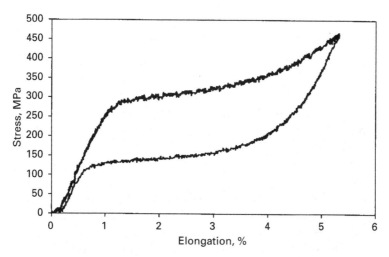

Figure 3.6 Stress versus strain measurement characterizing thin film with austenite finish temperature slightly above ambient. The stress–strain curve has a stress plateau typical of superelastic TiNi thin film. Force and elongation are measured while the thin film sample, clamped at the ends, is forced to elongate and allowed to contract. Stress-induced martensite formation permits strain to increase at near constant stress. (Reprinted with permission of ASM International®. All rights reserved. Gupta *et al.* [14], figure 2.)

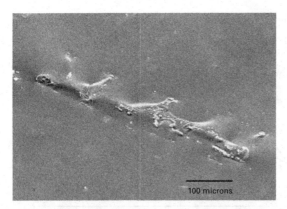

100 microns

Figure 3.7   Various methods have been used to bond TiNi film to itself or to other materials. Some results appear to be successful until viewed at high magnification. This SEM image shows the surface of a welded thin film TiNi locally damaged by heat. In resistance welding, the two layers of TiNi thin film were pressed together by two electrodes, and a current was run through the film between the electrodes to melt the two layers and form a union. The film is thinner than the irregularities on the surface of the electrodes, so the current delivery is uneven, resulting in hot and cold spots.

has been released from its substrate can be rolled up to make a conical or cylindrical shape. It would be desirable to join such a sheet at the edge to make three-dimensional devices. Three different technologies, resistance welding, laser welding, and ultrasonic bonding, were tested but did not produce satisfactory results.

3.5.1   Resistance welding

For resistance welding trials, TiNi Alloy Company used a conventional resistance welding system. The energy delivered per unit time was limited by installing a 0.15-ohm TiNi wire shunt. Attempts were made to spot-weld different materials – sputter-deposited samples of bare TiNi film, TiNi film that was sputtered with titanium-copper-silver (a widely used silver-based brazing material), and samples of TiNi film sputtered with gold.

For lack of previous work on welding of thin film SMA, ranges of welding profiles were tested by trial and error. These profiles included variations in electrical current, force, squeeze time, up-slope time, weld time, down-slope time and cooling time. A qualitative assessment of the attributes and properties of the welds was made by observation under stereo optical and scanning electron microscopes.

Even at the lowest power setting available with the welder, the heating was concentrated in small areas, tending to burn holes in the films rather than melt them together (Fig. 3.7). The welds produced were strong enough to tear the film. The film was locally damaged so that it was no longer strong and supple. A brazing material layer was deposited on the film surface in an attempt to provide a union of the two films without melting. Burn-through holes to the TiNi film persistently occurred.

3.5.2   Laser welding

A small set of sample welds from Nd:YAG laser welding was made, using varying power and pulse duration (Tim Weber of Rofin-Sinar Inc., suppliers of industrial welders, personal communication, 2007). These welds were examined in the scanning electron microscope (SEM) with the finding that the pulses used were not consistently controlled with enough precision: some spots were burned through, while adjacent spots were clearly not melted.

3.5.3   Ultrasonic bonding

Ultrasonic bonding was expected to produce better results than resistive welding because this technology simultaneously binds and cleans the metal oxide surface layer during the welding operation. Conventional apparatus available in the market is not designed to weld TiNi thin film of the order of 3–5 microns thick. Therefore, the welding systems were modified to adapt to the requirements imposed by the nature of this material. For ultrasonic welding, TiNi Alloy Company used a Nippon Avionics Ribbon Bonder NAW-1087 equipped with a UTHE model 20 G ultrasonic generator-transducer. The equipment produced a maximum output power of 20 watts and high frequency vibratory motion of about 62 kilohertz. At all power levels tested, the welds were irregular, with holes burned through in places. Welding profiles at power levels below 5 watts and 4 seconds did not produce welds.

No successful method of joining TiNi thin film was found. Holes were very likely to be created. It is hypothesized that the surface oxide layer (native) on the TiNi film may have prevented formation of metal-to-metal bonds. The combinations of force, time and energy were either insufficient to overcome the oxide layer (producing no welds) or were strong enough to overcome the oxide layer but too strong to preserve the integrity of the TiNi film.

3.6   TiNi thin film and MEMS processes

Thin film SMA has a natural connection with MEMS. Evaporative and sputter deposition, single crystal silicon substrates, microlithography, chemical etching and use of sacrificial layers for separation from substrate are MEMS processes adapted from microelectronics manufacturing processes. Integration of sputter-deposited TiNi film into a MEMS fabrication operation required: compatibility with the high annealing temperature (~773 K) of deposited film, compatibility of patterning methods for TiNi films, and development of techniques for selective release of TiNi film from the substrate.

3.6.1   Heat treatment

Sputtered TiNi must generally be heat-treated to crystallize while it is still on a substrate, which imposes critical temperature limitations on materials that may be used for substrates. Silicon and silicon oxide have worked very well, and polished metal has been used successfully. Some polyimides are capable of surviving at 773 K, but most polymers are not usable as substrates.

3.6.2   Patterning TiNi thin film

Whereas nickel and especially titanium are chemically active elements, TiNi alloy is highly resistant to attack by most chemicals. It is also highly biocompatible and has received approval by the Food and Drug Administration (United States) for several intravascular applications. However, TiNi is readily etched by nitric-hydrofluoric acid combinations – a characteristic that is exploited in chemical etching of microfabricated TiNi components. A satisfactory TiNi etchant solution is composed of 1 part buffered oxide etch (BOE) plus 12 parts nitric acid (HNO_3) and 13 parts water (H_2O).

3.6.3   Sacrificial layer

TiNi thin film can be separated from the silicon substrate using a sequence of processes that involve putting down an intermediate sacrificial layer between TiNi and the silicon substrate, and then chemically etching that layer away after the TiNi film has been annealed.

A thin chromium layer (about 50 nanometers thick) is sputter-deposited prior to the deposition of TiNi. After the TiNi has been deposited and annealed, the underlying chromium layer is chemically etched away in a commercially available chromium etchant. Chromium makes an excellent sacrificial material for several reasons – chromium does not diffuse into TiNi during the high temperature annealing process resulting in no change in composition of TiNi alloy, the etchant for chromium does not effect TiNi or the silicon substrate and, similarly, the etchant for TiNi does not affect chromium, and very thin layers of chromium can be deposited by RF sputtering.

Successfully implementing selective etching requires particular chemicals and specific sequences of operations. Only a few etchants are really completely selective. As an example, TiNi absorbs hydrogen especially at elevated temperatures. Buffered oxide etch (BOE) used to etch silicon oxide contains hydrofluoric acid which affects TiNi very drastically. Hydrogen atoms generated in the chemical reaction are absorbed in the film crystal boundaries and make the film brittle. Extended exposure to hydrofluoric acid must be avoided.

Using sputtered TiNi thin film in micromachining processes introduces additional variations. The etch rate of silicon as well as metals depends on the temperature and age of the etchant. In etching thin film metal, the rate varies across a wafer so that one region of the surface may be undercut while another is not fully processed. In multiple-step processes, incomplete etching may not be apparent until later in the MEMS sequence.

3.7   Fabrication of miniature actuators

Miniature actuators of two types have been made with thin film shape memory alloy – valve poppet actuators and bending-beam actuators. The TiNi Alloy Company has developed the poppet actuator, which is a millimeter size device, for pneumatic valves and pump applications. Bending-beam actuators were developed as a means of moving sub-millimeter diameter mirrors.

3.7.1   Valve poppet actuators

Sputtering and chemical photolithography processes are in some ways easier than casting, rolling and drawing operations performed on bulk NiTinol. MEMS technology is suitable for making submillimeter-sized features as small as a few microns. The TiNi microvalve actuator is made in a series of steps using contact masks from AutoCAD drawings (Fig. 3.8). Bare silicon wafers are oxidized to produce a layer of silicon oxide 1–2 microns thick. Silicon oxide acts as a barrier preventing inter-diffusion of TiNi with silicon during the crystallization heat treatment. The wafer is patterned and etched to remove most of the silicon where thin film will be freestanding. This is necessary to avoid long exposure of TiNi film to silicon etchants. A protective layer of chromium is deposited and patterned on the front surface to protect TiNi from the final oxide etch. Areas are left bare to act as anchors for TiNi microribbons.

TiNi film from 1 to 10 microns thick is deposited and heat-treated. A thin layer of gold is applied and patterned to act as electrical contacts. The valve actuator is finished by completely etching away the remaining silicon under the microribbons.

3.7.2   Bending-beam actuators

The difference in coefficients of thermal expansion for silicon (7.6×10^{-6} per °C) and TiNi (10×10^{-6} per °C) has been used to make bending-beam actuators. TiNi is deposited on a thin silicon beam, then crystallized and annealed at 773 K. During cooling, the TiNi contracts more than the silicon, leaving about 0.5 percent strain at 373 K. This constitutes a stress sufficient to bend the silicon beam. As the TiNi continues to cool and undergoes a transformation to martensite, this stress is relieved and the beam becomes flat. However, when heated, the stress is induced and the beam again bends (Fig. 3.9).

3.8   Fabrication of intravascular medical devices

Today's TiNi stents are normally made using bulk nitinol tubes, even though these drawn tubes are relatively large and too inflexible to be delivered by intravascular microcatheters to brain lesions. Nitinol tubes with wall thickness as small as 50 microns have been achieved by etching and electropolishing

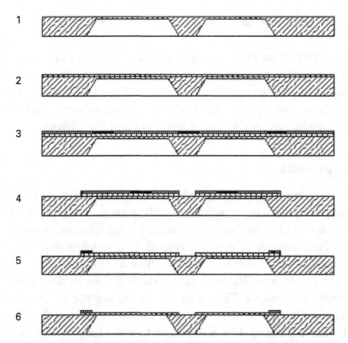

Figure 3.8   This figure illustrates the microfabrication steps in making valve actuators. (1) Back-side anisotropic wet etch, creating thin membranes of silicon (~40 microns thick) for supporting an island of silicon which eventually becomes the poppet of the valve. (2) Sputter deposition of layer of TiNi on the front side and subsequent crystallization heat treatment. (3) Sputtering gold on top of TiNi for facilitating masking in wet etching and to provide electrical contact to the conductive path of the actuator. (4) Patterning and wet etching of the gold and TiNi film. (5) Removing the gold from the active parts of the actuators. Gold remains only on the contact tabs and interconnecting parts of TiNi. (6) Final backside etch to release the poppet and active parts of the actuator. (Adapted from Ishida & Martynov, (2002) [19], p. 114, figure 6.)

techniques. Recently a NiTinol neurovascular stent was developed at SMART Therapeutics (now an entity of Boston Scientific Corp.) and is being used in an occlusion system for aneurysms in the brain. However, such techniques aren't suitable to achieve wall thickness much smaller than 50 microns. Sputter deposition techniques can produce TiNi film as thin as one micron. The medical industry represents a growing demand for smaller and thinner stents that can be surgically implanted or delivered via catheter into small diameter, highly tortuous blood vessels. Specially designed medical devices need to be developed to serve as blood clot retrievers or filters that can be delivered through a very small-diameter blood vessel in the brain.

Several variations of the sputtering process have been used in making three-dimensional objects for intravascular applications. Sputtering is generally done on a plane surface, but TiNi has also been successfully sputtered onto conical surfaces rotated along the cone axis and in systems with cylindrical targets.

(a)

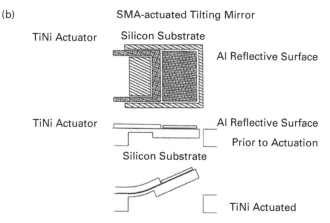

(b)   SMA-actuated Tilting Mirror

    TiNi Actuator   Silicon Substrate

                                  Al Reflective Surface

    TiNi Actuator   Al Reflective Surface

                                  Prior to Actuation

            Silicon Substrate

                                  TiNi Actuated

Figure 3.9   Differential thermal contraction during cooling of a TiNi-silicon bimorph from crystallization temperature creates a stress that causes the bimorph to bend. Below the transformation temperature of TiNi this stress is relieved by the formation of martensite. Heating (for example by electrical current) re-creates the bending stress. Thus a two-way actuator is formed. (a) Photograph of bending beam actuated mirror devices. (b) Mirror actuator concept drawing.

3.8.1   Planar sputtering on a three-dimensional substrate

TiNi film was sputtered from a planar sputtering source onto three-dimensional substrates. The Perkin Elmer 4450 sputtering system was used. High melting temperature glass rods were found to be reasonably suitable substrates. These were heat-softened to create other shapes such as cones and hemispheres. A liquid polyimide PI 2611 (manufactured by HD Microsystems) layer was spin-coated on to the rod to fill up the pits and to smooth the sharp edges on the glass rod surface. The polyimide was cured at 773 K in an air oven. The coated glass rod was loaded into the sputtering chamber via a small load lock chamber isolated from the process chamber by a gate valve. The glass rod substrate was rotated on its axis while at a fixed distance of about 2 inches from the TiNi target by an external electrical motor at a speed of 0.4 revolutions per minute.

A thin layer of chromium was sputter-deposited on the polyimide-coated surface of the rotating glass rod in the sputtering chamber after which the substrate was moved to the TiNi source. A thin film of TiNi was then sputter-deposited on top of the chromium layer. A second layer of chromium was deposited on top of the TiNi layer and used as a mask for patterning the underlying TiNi layer. The glass rod with sputter-deposited layers of TiNi and chromium was then heat treated at 773 K in a vacuum in order to crystallize the TiNi film.

A modified photolithography process was used to pattern film on three-dimensional substrates. Photoresist OCG 825 was spin-coated using a custom designed chuck, baked in an oven, exposed to ultraviolet light through a mask, and developed. Chromium and TiNi layers were etched with appropriate chemicals and the photoresist layer was dissolved away in a solvent. TiNi cylinders and cones were released from the substrate surface by chemically etching the underlying chromium layer.

TiNi film formed by planar sputtering on a rotating substrate was shown to have good shape memory and superelastic properties. However, this method produced only one device at a time, so other methods were developed.

3.8.2   Sputtering using the multiple-layering method

The multiple-layering method for making three-dimensional shaped objects has been developed to enable formation of a wide variety of TiNi film structures without the need to weld or bond. The process also enables fabrication of many devices simultaneously. In this method, successive layers of TiNi film are sputtered onto a plane surface, which eliminates the problems associated with sputtering on three-dimensional substrates [14]. Each of the multiple TiNi layers exhibits excellent shape memory and superelastic properties.

Fabrication of three-dimensional TiNi thin film structures is accomplished by the deposition of two (or more) layers of TiNi interspersed with layers of sacrificial material (chromium), part of which is removed to form bonds between the two TiNi layers at selected locations, producing structures that can be opened to produce cones, cylinders and other shapes.

The multiple-layered TiNi thin film parts are transformed into integral three-dimensional shapes by a "shape-setting" annealing process in which steel mandrels are inserted between TiNi layers (Fig. 3.10). The processing steps to fabricate three-dimensional TiNi film structures using the multiple-layer method are shown in Fig. 3.11. Two or more layers of TiNi thin film interspersed with layers of sacrificial material are sputter-deposited sequentially on polished and oxidized silicon wafer substrates. Chromium is used as a sacrificial material. The thickness of each deposited chromium film is about 50 nanometers. The top chromium layer serves two purposes: (1) as a masking layer for patterning the underlying TiNi layer in subsequent process and (2) also as an intermediate sacrificial layer which in the final steps of fabrication is chemically etched away.

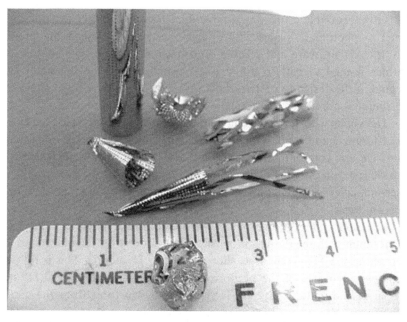

Figure 3.10 Photograph of cylinders, cones and other shapes made by the multiple-layer deposition method.

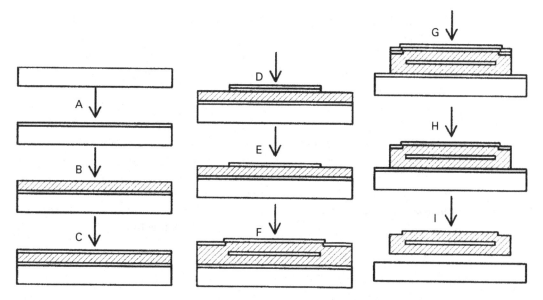

Figure 3.11 The drawing illustrates the processing steps (described in the text) in the fabrication of multiple-layer thin film TiNi devices. A pocket is formed between successive layers of TiNi wherever a sacrificial layer of chromium is provided. This pocket is opened by insertion of a mandrel, and a subsequent heat anneal provides shape-setting to produce the final desired form. (Reprinted with permission of ASM International®. All rights reserved. Gupta *et al.* [14], figure 6.)

In this example of making a three-dimensional thin film structure, a minimum of two photomask plates with appropriate pattern designs are prepared using AutoCAD software and printed onto mask plates. Mask 1 contains the design which is used to pattern the second chromium layer on the wafer and mask 2 contains the design to pattern the TiNi layers in the final stage. Standard MEMS techniques are used to pattern the TiNi and chromium layers.

The wafer with the selectively etched chromium layer is loaded back into the sputtering chamber. In a high vacuum, the substrate surface is sputter-etched to remove any contamination layer and also to remove the undesired thin native oxide layer on the TiNi film surface that forms during the lithography process. A second layer of TiNi film is sputter-deposited followed by a third layer of chromium.

The same photolithography steps are followed to pattern the top layer of chromium with the designs in mask 2. The underlying TiNi layers are chemically etched in the TiNi etchant as shown in Fig. 3.11, step D. This is followed by the complete removal of the photoresist layer in a solvent (Fig. 3.11, step E). To separate the multiple-layered TiNi thin film structures from the surface of the substrate, the wafer is immersed in the chromium etchant to completely dissolve all the chromium layers, including the first, second and third layers. The final etching of chromium not only separates the multiple-layered TiNi thin film structures from the silicon wafer but also creates a pocket between the first and second TiNi layers.

The multiple-layer method can be used for making TiNi structures with more than two layers without much change to the fabrication steps described above. The method of producing three-dimensional TiNi thin film structures using multiple layers has opened a window of opportunity for making a wide range of small and implantable medical devices.

Medical applications require materials that are superelastic at 310 K. As is well known, increasing the atomic percentage of nickel lowers the temperature at which the transition to martensite begins. In efforts to make superelastic thin film with an austenite finish (A_f) temperature below body temperature, it was found that in thin film the addition of nickel, which continues to decrease the martensite start (M_s) temperature, does not lower the austenite finish temperature below about 308 K. A solution was found – adding a fraction of less than one percent chromium lowers both phase transition temperatures together.

3.9   Literature of TiNi thin film

The earliest publication describing deposition of TiNi thin film was a paper by Ken Gabriel, Jim Walker and Mehran Mehregany at ATT Bell Laboratories presented at a Swiss Sensors and Actuators conference in the summer of 1989 [2]. This paper, and a subsequent patent issued to the authors, describes sputter deposition and processing, but it is not clear that the effects that they observed were due to phase change – small deformations may have been caused by differential thermal expansion.

TiNi Alloy's involvement began with a Small Business Innovation Research contract from the National Aeronautics and Space Administration for a thin film memory device. The resulting film was demonstrated at a poster session at the Lansing conference in 1989 [11]. The results on TiNi Alloy Company's first microvalve actuator were presented by John Busch at Napa Valley [3]. TiNi Alloy Company first published in a refereed journal in December 1990 [1].

During the past years, discussion of thin films appeared in numerous publications [4, 6, 15, 16, 17, 18, 19, 20, 21, 22, 23, 24, 25, 26]. Papers covering various aspects of thin films were presented at Shape Memory and Superelastic Technologies meetings at Asilomar and elsewhere as thin film TiNi has progressed from a subject for academic research to development of products protected by intellectual property. Internet searches now produce dozens of references from conference proceedings to US and foreign patents as well as new patent applications.

3.10   Summary

TiNi thin film can be fabricated using various physical deposition processes, the most common being vacuum sputter deposition from near equiatomic TiNi targets. Factors affecting successful sputter deposition are quality of vacuum, smoothness of substrate, purity of gases and targets, substrate–target distances, magnetic field, and temperature.

Chemical composition of sputtering targets is the primary determinant of phase transition temperatures. Titanium-rich targets produce the best results. Heat treatment is used to modify transition temperatures of the finished film. Joining or welding TiNi thin film to itself or other materials continues to represent a challenge for TiNi thin film applications. Incorporating MEMS processes, photolithographic patterning thin film and sacrificial layering in the sputter deposition production of TiNi thin film allows the creation of actuators that can exert large forces and recover large displacements, and possess desirable electrical properties. TiNi thin film combined with silicon by MEMS has been used to create a great variety of products of which small actuators and patterned freestanding superelastic film have received the most development effort.

Currently, fabrication of medical devices is most successful using planar sputtering on a three-dimensional substrate or using the multiple layer method, both of which allow manufacture of devices without welding or bonding steps.

Acknowledgements

Support for research and development leading to the results reported here have come from a variety of sources, but mainly from United States government agencies through the Small Business Innovation Research program including National Institutes of Health, National Aeronautics and Space Administration, United States Air Force, United States Navy, Defense Advanced Research Projects Agency, Ballistic Missile Defense Organization and United States Department of Education.

References

[1] J. D. Busch, A. D. Johnson, C. H. Lee and D. A. Stevenson. Shape-memory properties in Ni-Ti sputter-deposited film. *J. App. Phy.*, **68** (1990) 6224.

[2] J. A. Walker, K. J. Gabriel and M. Mehregany. Thin film processing of shape memory alloys. *Sensors and Actuators*, **A21–A23** (1990) 243.

[3] J. D. Busch and A. D. Johnson. Prototype micro-valve actuator. In *Micro Electro Mechanical Systems: 1990 Proceedings, An Investigation of Micro Structures, Sensors, Actuators and Robots* Piscatawny NJ: IEEE, (1990).

[4] A. D. Johnson. Vacuum-deposited TiNi shape memory film: characterization and applications in microdevices. *J. Micromech. Microeng.*, **1** (1991) 34–41.

[5] A. D. Johnson and V. Martynov. Applications of shape-memory alloy thin film. In *SMST 1997: Proceedings of the 2nd International Conference on Shape Memory and Superelastic Technologies*, Asilomar Conference Center, Pacific Grove, USA, (1997) 149.

[6] A. D. Johnson. Thin film shape-memory technology: a tool for MEMS. *Micromachine Devices*, **4** (1999) 12.

[7] K. Wasa and S. Hayakawa. *Handbook of Sputter Deposition Technology*, Pork Ridge, NJ: Noyes Publications (1992).

[8] T. M. Adams, S. R. Kirkpatrick, Z. Wang and A. Siahmakoun. NiTi shape memory alloy thin films deposited by co-evaporation. *Mat. Lett.*, **59**:10 (2005) 1161–1164.

[9] F. Ciabattari, F. Fuso and E. Arimondo. Pulsed laser deposition and characterization of NiTi-based MEMS prototypes. *App. Phys. A: Mat. Sci. Proc.*, **79**:4–6 (2004) 623–626.

[10] R. L. Villhard and R. J. Atmur. *Shape memory alloy MEMS component deposited by chemical vapor deposition*. United State Patent 20040252005 (2004). Available from www.freepatentsonline.com/20040252005.html

[11] T. W. Duerig, K. N. Melton, D. Stockel and C. M. Wayman (eds.) *Engineering Aspects of Shape Memory Alloys*, Vol. 1, London: Butterworth-Heinemann (1990).

[12] I. Brodie and J. J. Muray. *Physics of Microfabrication*, New York: Plenum Press (1982) 229.

[13] J. I. Goldstein. *Scanning electron microscopy and x-ray microanalysis*, second edn, New York: Plenum Press (1992).

[14] V. Gupta, A. D. Johnson, V. Martynov and L. Menchaca. Nitinol thin film three-dimensional devices-fabrication and applications. In *SMST 2003: Proceedings of the 4th International Conference on Shape Memory and Superelastic Technologies*, eds. S. M. Russell & A. R. Pelton, Asilomar Conference Center, Pacific Grove, USA (2003).

[15] W. J. Moberly, J. D. Busch and A. D. Johnson. In situ high voltage electron microscopy of the crystallization of amorphous TiNi thin films. *Materials Research Society Proceedings* (Spring 1991).

[16] K. R. C. Gisser, J. D. Busch, A. D. Johnson and A. B. Ellis. Oriented nickel-titanium shape memory alloy films prepared by annealing during deposition. *App. Phys. Lett.*, **61** (1992) 1632.

[17] P. Krulevitch, A. P. Lee, P. B. Ramsey, J. C. Trevino, J. Hamilton and M. A. Northrup. Thin film shape memory alloy microactuators. *J. MEMS*, **5**:4 (1996) 270–282.

[18] R. H. Wolf and A. H. Heuer. TiNi (shape memory) films for MEMS applications. *J. Micromech. Microeng.*, **4** (1995) 206–212.

[19] A. Ishida and V. Martynov. Sputter-deposited shape-memory alloy thin films: properties and applications. *MRS Bulletin*, **27**:2 (2002) 111–114.

[20] J. D. Busch, M. H. Berkson and A. D. Johnson. Phase transformation in sputtered Ni-Ti film: effects of heat treatment and precipitates. In *Phase Transformation Kinetics in Thin Films*, eds. M. Chen, M. O. Thompson, R. B. Schwarz and M. Libera, Pittsburgh: Materials Research Society, **230** (1992) 91–96.

[21] W. G. Moberly, J. D. Busch, A. D. Johnson and M. H. Berkson. In situ HVEM crystallization of amorphous TiNi thin films. In *Phase Transformation Kinetics in Thin Films*, eds. M. Chen, M. O. Thompson, R. B. Schwarz and M. Libera, Pittsburgh: Materials Research Society, **230** (1992) 85–90.

[22] D. S. Grummon, T. LaGrange and J. Zhang. Processing and deployment of sputtered thin films of NiTi and Nitix alloys for biomedical and MEMS applications. In *SMST 2003: Proceedings of the 4th International Conference on Shape Memory and Superelastic Technologies*, eds. S. M. Russell and A. R. Pelton, Asilomar Conference Center, Pacific Grove, USA (2003).

[23] S. Miyazaki and A. Ishida. Martensitic transformation and shape memory behavior in sputter-deposited TiNi-base thin films. *Mat. Sci. Eng. A*, **275** (1999) 106–133.

[24] Y. Q. Fu, H. J. Du, W. M. Huang, S. Zhang and M. Hu. TiNi-based thin films in MEMS applications: a review. *Sensors & Actuators A: Physical*, **112**: 2–3 (2004) 395–408.

[25] B. Winzek, S. Schmitz, H. Rumpf, *et al.*, Recent developments in shape memory thin film technology. *Mat. Sci. Eng. A*, **378**:1–2 (2004) 40–46.

[26] A. Ishida, M. Sato, O. Tabata W. Yoshikawa. Shape memory thin films formed with carrousel-type magnetron sputtering apparatus. *Smart Mat. Struct.*, **14** (2005) S216–S222.

4   TiNi multilayer thin films

H. Cho and S. Miyazaki

Abstract

The martensitic transformation temperatures of TiNi shape memory alloys are strongly affected by composition. However, the composition of the TiNi thin film fabricated by a conventional sputtering method using a TiNi alloy target is not easy to adjust precisely. In order to control the composition of thin films, a dual-source sputtering method using pure Ti and Ni targets can be an alternative candidate for fabricating TiNi multilayer thin films, which can be converted to TiNi alloy thin films by heat-treatment for alloying. This chapter explains the fabrication method and alloying process of the TiNi multilayer thin films in addition to the shape memory properties of the TiNi thin films made by alloying the TiNi multilayer thin films.

4.1   Introduction

TiNi shape memory alloys are considered as one of the most promising candidates for microactuators in MEMS (micro electro mechanical systems) because of their large recovery strain and recovery force [1]. In order to apply the TiNi alloys to MEMS, it is required to make them thin down to micron sizes. Rolling and melt-spinning methods are available for making thin plates with thickness larger than 15 μm. However, sputter-deposition methods are available for making thin films with thicknesses less than 10 μm. TiNi thin films fabricated by the sputter-deposition method are expected to be applied to microdevices such as microvalves [2], micropumps [3] and cantilevers [4], since they exhibit an excellent shape memory effect and good mechanical properties.

The martensitic transformation start temperature (M_s) of TiNi shape memory thin films is strongly affected by the composition [5, 6, 7, 8, 9]. Therefore, in order to apply it for various purposes, it is very important to control the composition of a TiNi thin film. A TiNi alloy target has been used in conventional sputter-deposition methods. The composition of the film is adjusted by placing pure Ti or

Thin Film Shape Memory Alloys: Fundamentals and Device Applications, eds. Shuichi Miyazaki, Yong Qing Fu and Wei Min Huang. Published by Cambridge University Press. © Cambridge University Press 2009.

Ni chips on the TiNi alloy target. This method is also applicable to adding a third element, if we use pure third element chips such as Cu or Pd. These ternary alloy thin films can indicate unique characteristics such as a low transformation temperature hysteresis and a high transformation temperature, respectively [10, 11, 12, 13].

However, the composition of the thin film fabricated by this conventional method is difficult to adjust precisely, because the composition of the deposited thin film is affected by the number and dimensions of the chips, the surface state including geometry and composition of the target, etc. To apply TiNi thin film for various applications, it is required to control the composition of the thin film more exactly.

As an alternative method which controls the composition more accurately than the conventional alloy target method, a dual-source sputtering system is applicable to make TiNi multilayer thin films. This sputtering system has parallel pure Ti and Ni targets in the same plane: a TiNi multilayer thin film can be fabricated by sputtering the Ti and Ni targets alternately by turning the substrate to face the either of the Ti and Ni targets periodically. By using this method, controlling the composition of the thin film seems to be easy in comparison with the conventional alloy target method, because the average composition of the multilayer thin film can be adjusted by the ratio of the thicknesses of the Ti and Ni layers. Specifially the composition of the thin film is controlled by setting the sputtering time and sputtering power for each target.

It has been reported that solid state amorphization (SSA) occurs to produce amorphous phase at the interface of Ti and Ni layers of a TiNi multilayer thin film by annealing [14, 15, 16, 17, 18, 19, 20, 21, 22, 23, 24, 25]. The TiNi multilayer thin film can be finally converted to a shape memory alloy thin film through SSA by annealing [26, 27]. In this chapter, the fabrication method of TiNi multilayer thin films, their alloying process to shape memory alloy thin films and the transformation/deformation behavior of shape memory alloy thin films are explained [28, 29].

4.2   Fabrication of multi-layer TiNi thin films

Figure 4.1 illustrates the dual-source DC sputtering system. The conventional alloy target sputtering system has one sputtering cathode and uses a TiNi alloy target. However, this dual-source sputtering system has two sputtering cathodes and two metal targets. It is possible to regulate the angle of each target in this system. The Ti (purity 99.9 wt%) and Ni (purity 99.99 wt%) targets are arranged in parallel. An SiO_2/Si substrate was placed on a turn-table. A TiNi multi-layer thin film can be made by repeating the following process many times, i.e. rotating the turn-table to make the substrate face the Ti target and stopping it for a fixed time followed by rotating the turn-table to make the substrate face the Ni target for another fixed time.

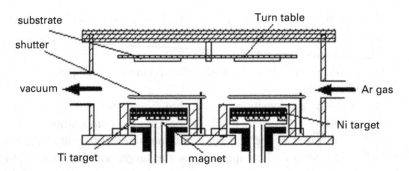

substrate
shutter
vacuum
Ti target
Turn table
Ar gas
Ni target
magnet

Figure 4.1   A schematic figure of the dual-source sputtering system [29], with permission from Elsevier.

The sputtering gas was pure Ar at a pressure of 2.5×10^{-4} Pa. The d.c. powers for the Ti and Ni targets were fixed as 280 W and 60 W, respectively. The sputtering time (the stopping time on each target) was coordinated in order to obtain an objective composition. The total number of Ti and Ni layers in the thin film was 200 (Ti: 100, Ni: 100), and the whole thickness of thin film was fixed at 3 μm. In this chapter, the thicknesses of Ti and Ni layers were adjusted to 18 nm and 12 nm, respectively, in order to fabricate Ti-51at%Ni multilayer thin films.

In a conventional method (using a TiNi alloy target), near-equiatomic TiNi thin films were fabricated in an amorphous state if the substrate was not heated [30, 31, 32]. Therefore, when we use TiNi amorphous thin film for making a shape memory thin film, it is necessary to make it crystallized by heat-treatment. However, the component metal layers of the TiNi multilayer thin films were already crystallized, as shown in the following section, when they are deposited using the dual-source sputtering method without substrate heating. In this case, heat-treatment is necessary only for alloying the multi-layer TiNi thin films to make shape memory alloy thin films.

In the next section, the alloying process of the TiNi multilayer film is revealed using differential scanning calorimetry (DSC), X-ray diffractometry (XRD) and transmission electron microscopy (TEM). A Ti-51at%Ni amorphous thin film was also fabricated by the conventional sputtering method utilizing an alloy target as a comparison.

4.3   Alloying process of TiNi multilayer thin films

4.3.1   DSC measurements

Figure 4.2 shows the DSC curve of an as-sputtered TiNi amorphous thin film and an as-sputtered TiNi multilayer thin film upon heating with a heating rate of 10 K/min. A single exothermic peak due to the crystallization reaction was observed at 750 K in the DSC curve of the TiNi amorphous thin film. It was reported that the crystallization process of a TiNi amorphous thin film was a single step from amorphous to crystalline state [30].

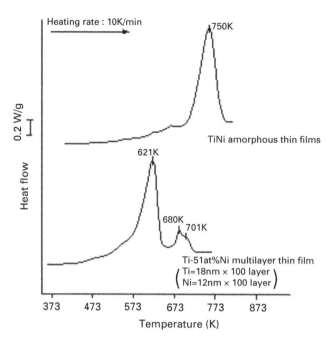

Figure 4.2 DSC curves of the as-sputtered TiNi amorphous thin film and the as-sputtered TiNi multilayer thin film measured with a heating rate of 10 K/min [29, 30], with permission from Elsevier and MRS.

On the other hand, three exothermic peaks were observed for the multilayer thin film at 621 K, 680 K and 701 K. This indicates that alloying of the TiNi multilayer thin film proceeded in multi-steps and the alloying reaction occurred at lower temperatures than that of the crystallization of amorphous thin films.

4.3.2 XRD measurements

To investigate the microstructural change upon heating in detail, XRD profiles were obtained for the Ti-51at%Ni multilayer thin film heated up to various temperatures with the same heating rate as the DSC measurement, followed by quenching into water. Figure 4.3 shows XRD profiles of the as-sputtered Ti-51at%Ni multilayer thin film and the Ti-51at%Ni multilayer thin films heated up to various temperatures with a heating rate of 10 K/min.

The diffraction peaks corresponding to Si were obtained from reference sample powders. For the as-sputtered thin film, $\{002\}_{Ti}$ and $\{111\}_{Ni}$ peaks were observed. This indicates that Ti and Ni have been crystallized during sputtering.

For the specimen heated up to 620 K, the $\{002\}_{Ti}$ and $\{111\}_{Ni}$ peaks became broader. It can also be seen that the intensities of the $\{002\}_{Ti}$ and $\{111\}_{Ni}$ peaks decreased. The intensity of the $\{111\}_{Ni}$ peak drastically decreased and the position shifted to a lower diffraction angle by heating up to 640 K. For this reason, the first exothermic peak in Fig. 4.2 corresponds to a certain reaction occurring in the Ni layers.

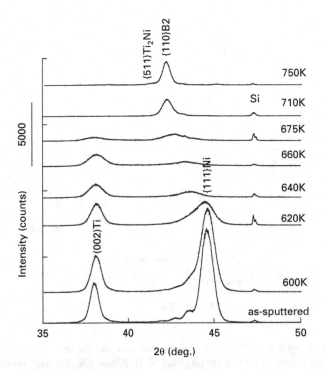

Figure 4.3   XRD profiles of the as-sputtered Ti-51at%Ni multilayer thin film and Ti-51at%Ni multilayer thin films heated up to various temperatures with a heating rate of 10 K/min [30], with permission from MRS.

On the other hand, the $\{002\}_{Ti}$ peak became broader and the intensity decreased gradually by heating up to 675 K, where the peak position was unchanged. A broad diffraction peak corresponding to the TiNi B2 phase appeared and the peak from the Ni layers disappeared after heating up to 675 K. From this result, it seems that the second exothermic peak corresponds to the formation of the TiNi B2 phase.

In the multilayer thin film heated up to 710 K, the intensity of the peak from the TiNi B2 increased and the peak from the Ti layers disappeared, indicating that the Ti layers were completely absorbed by the Ni layers to make the TiNi B2. Furthermore, for the specimen heated up to 750 K, a diffraction peak corresponding to the Ti_2Ni appeared. From this result, the third exothermic peak seems to indicate the formation of the Ti_2Ni phase.

4.3.3   TEM observation

To investigate the alloying process to the TiNi B2 and the formation process of the Ti_2Ni in more detail, microstructures in cross-sections were observed by the TEM for the as-sputtered specimen and the specimens heated up to various temperatures. The TEM specimens for cross-sectional observation were prepared by a focused ion beam (FIB) system.

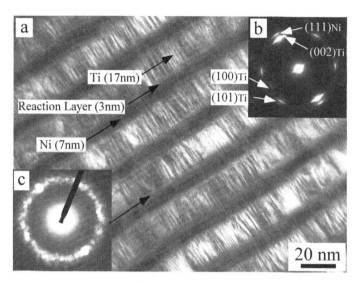

Figure 4.4   A bright field TEM image (a), the corresponding selected area diffraction pattern (b), and a NBD pattern of the cross-section including the Ti and Ni layers (c) in the as-sputtered TiNi multilayer thin film [30], with permission from MRS.

Figures 4.4(a) and 4.4(b) show a cross-sectional bright field image and the corresponding selected area diffraction pattern, respectively, of the as-sputtered multilayer specimen. The multilayer structure, which is composed of bright and dark layers, was clearly observed. Diffraction spots from $(100)_{Ti}$, $(101)_{Ti}$, $(002)_{Ti}$ and $(111)_{Ni}$ of Ti and Ni polycrystals were observed in the selected area diffraction pattern. The bright and dark layers were determined as Ti and Ni by dark field imaging. The average thicknesses of the Ti and Ni layers were 17 nm and 7 nm, respectively. It is also noted in the diffraction pattern that the Ti and Ni layers have strong fiber textures of $\{002\}_{Ti}$ and $\{111\}_{Ni}$ types, respectively.

In addition, gray layers were observed at every interface between the Ti and Ni layers. The thickness of the gray layer was about 3 nm. The expected thicknesses of the Ti and Ni layers were 18 nm and 12 nm, respectively, according to a calculation considering the sputtering conditions, but the observed values, 17 nm and 7 nm, were different from them. The difference is due to the formation of the additional gray layers.

A nano-beam diffraction (NBD) analysis was carried out for characterizing the gray layers. The result is shown in Fig. 4.4 (c). It can be seen that a diffused halo-ring was obtained from the gray layer. This indicates that the gray layer was amorphous, and these amorphous layers were formed at each interface between the Ti and Ni layers during sputtering. Each amorphous layer seems to be formed by collision of sputtered Ni atoms with the Ti layer or sputtered Ti atoms with the Ni layer during sputtering. As a conclusion, the as-sputtered TiNi multilayer thin film was composed of pure Ti, pure Ni and amorphous layers.

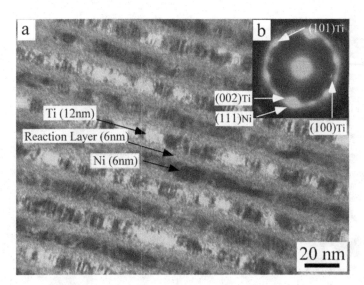

Figure 4.5   A bright field TEM image (a) and the corresponding selected area diffraction pattern (b) of the TiNi multilayer thin film heated up to 640 K [30], with permission from MRS.

Figure 4.5 shows (a) a cross-sectional bright field image and (b) the corresponding selected area diffraction pattern of the specimen heated up to 640 K, which corresponds to the end of the first exothermic peak.

It can be clearly seen that the thickness of the amorphous layer increased in comparison with the as-sputtered multilayer thin film. The average thicknesses of the Ti, Ni and amorphous layers were measured to be 12 nm, 6 nm and 6 nm, respectively. The selected area diffraction pattern clearly reveals not only diffraction spots from Ti and Ni but also a diffused halo-ring, indicating that the amorphous layers grew. It is also noticed that the amorphous phase was formed in part of the Ti layer, because the thickness of the Ti layer mostly decreased.

A similar solid state amorphization (SSA) phenomenon in TiNi multilayer films has been also reported [14, 15, 16, 17, 18, 19, 20, 21, 22, 23, 24, 25]. From the above result, it is considered that the first exothermic peak of the DSC curve is due to an SSA phenomenon consisting of the expansion of the amorphous layer into the Ti layer.

Figure 4.6 shows a cross-sectional bright field image (a) and the corresponding selected area diffraction pattern (b) of the specimen heated up to 675 K, which corresponds to the early stage of the second peak in the DSC curve. The halo-ring and the diffraction spots corresponding to the Ni and amorphous layers disappeared in the selected area diffraction pattern, whereas the diffraction ring corresponding to the TiNi B2 appeared in addition to the diffraction spots from the Ti layers. Two layers of thickness 18 nm and 12 nm were observed in the bright image. The wide and narrow layers were determined as the TiNi B2 and Ti layers, respectively, by dark field imaging and NBD analysis. From this

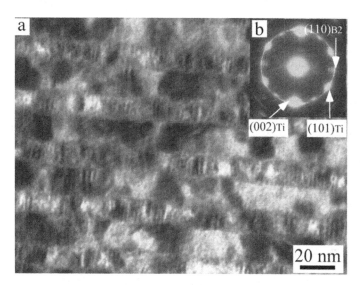

Figure 4.6   A bright field TEM image (a) and the corresponding selected area diffraction pattern (b) of the TiNi multilayer thin film heated up to 675 K [30], with permission from MRS.

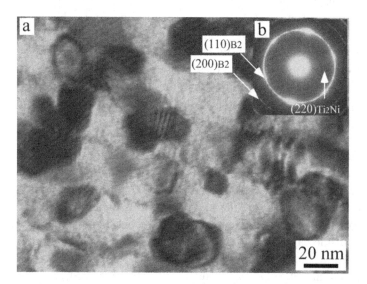

Figure 4.7   A bright field TEM image (a) and the corresponding selected area diffraction pattern (b) of the TiNi multilayer thin film heated up to 750 K [30], with permission from MRS.

result and the XRD profile, the second exothermic peak in the DSC curve is mainly due to the crystallization to the TiNi B2 from the amorphous layers.

Figure 4.7 shows a cross-sectional bright field image (a) and the corresponding diffraction pattern (b) of the specimen heated up to 750 K, which is higher than the end of the third exothermic peak.

Each layer of thickness 26–30 nm and the grains in the layer were observed in the bright field image. The diffraction rings corresponding to the TiNi B2 were clearly observed in comparison with those of the specimen heated up to 675 K, and diffraction rings corresponding to Ti_2Ni were also observed. By the NBD analysis, some grains near the interfaces were determined to be Ti_2Ni, and the main component of the layer was the TiNi B2 phase. Grains with Moiré patterns correspond to Ti_2Ni. These results indicate that the TiNi multilayer thin film was alloyed completely to TiNi B2 after heating up 750 K, except for the remaining Ti_2Ni. It is suggested that the Ti_2Ni was formed by the reaction between the TiNi B2 and the residual Ti. Hence, it is considered that the third exothermic DSC peak corresponds to the formation of Ti_2Ni due to this reaction.

4.4   Shape memory properties and mechanical properties

In order to investigate shape memory properties and mechanical properties of heat-treated TiNi multilayer thin films and to compare them with heat-treated TiNi amorphous thin films, thermal cycling tests under various constant stresses were carried out for the TiNi multilayer thin films and TiNi amorphous thin films, which were heat-treated at various temperatures between 673 K and 973 K for 3.6 ks for alloying and crystallizing.

Figure 4.8 shows an example of these results: i.e., the strain vs. temperature (S–T) relationship of the TiNi amorphous thin film (a) and TiNi multilayer thin film (b) heat-treated at 973 K for 3.6 ks. The solid and dashed lines correspond to cooling and heating curves, respectively.

The shape recovery strain induced upon heating due to the reverse martensitic transformation is denoted by ε_A. The permanent plastic strain is denoted by ε_p. Both ε_A and ε_p increase with increasing stress. The multilayer thin film fractured during the thermal cycling tests under 400 MPa. On the other hand the amorphous thin film fractured during the thermal cycling tests under 700 MPa. It seems that the amorphous thin film has a tensile strength higher than the multilayer thin film. However, in the case of ε_A, the multilayer thin film is almost equivalent to the amorphous thin film.

The maximum value of ε_A is denoted by ε_A^{max}. The critical stress for slip, which is denoted by σ_s, is the stress when ε_p reached 0.2%. Figure 4.9 shows the heat-treatment temperature dependence of ε_A^{max} of both the TiNi multilayer thin film and TiNi amorphous thin film.

The amorphous thin film heat-treated at 673 K for 3.6 ks did not exhibit a shape memory effect. This is because the amorphous thin film was not fully crystallized by the heat-treatment [33, 34]. When the heat-treatment temperature was higher than 723 K, the amorphous thin film was fully crystallized and exhibited the shape memory effect. The recovery strain increased slightly with increasing heat-treatment temperature, and reached 5.6% when heat-treated at 873 K.

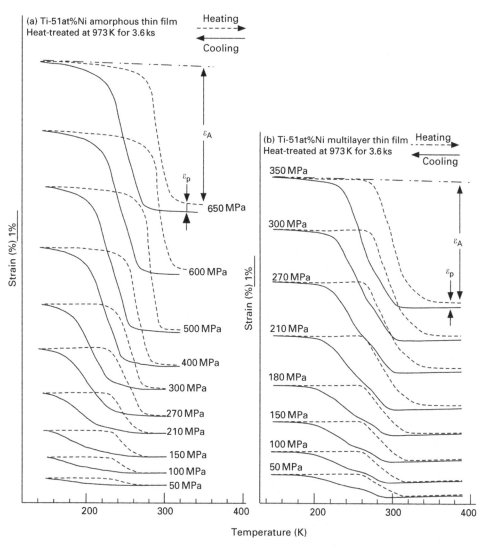

Figure 4.8 Strain vs. temperature curves under various constant stresses for TiNi amorphous thin film (a) and TiNi multilayer thin film (b) heat-treated at 973 K for 3.6 ks [29], with permission from Elsevier.

On the other hand, the shape memory effect was observed in the multilayer thin film heat-treated at 673 K for 3.6 ks. This indicates that the multilayer thin film exhibited the shape memory effect even if heat-treated at a lower temperature where the amorphous thin film was not crystallized. This result is consistent with the DSC curves in Figure 4.2.

The ε_A^{max} of the multilayer increased with increasing heat-treatment temperature, and it reached 4.6% when heat-treated to 973 K. In this result, the multilayer thin film exhibits not only an equivalent performance to amorphous thin film, but

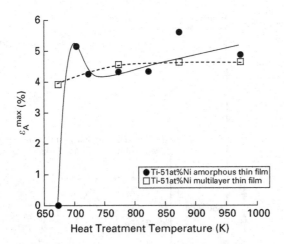

Figure 4.9   Heat-treatment temperature dependence of ε_A^{max} of TiNi multilayer thin film and TiNi amorphous thin film. [29, 30, 34], with permission from Elsevier and MRS.

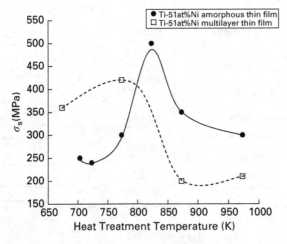

Figure 4.10   Heat-treatment temperature dependence of σ_s of TiNi multilayer thin film and TiNi amorphous thin film [29, 30, 34], with permission from Elsevier and MRS.

also the shape memory effect even if heat-treated to a lower temperature where the amorphous thin film did not exhibit it.

Figure 4.10 shows the heat-treatment temperature dependence of σ_s of TiNi multilayer thin film and TiNi amorphous thin film. The stress σ_s of amorphous thin film increases with increasing heat-treatment temperature, and it reached 500 MPa when heat-treated at 823 K. The increase of σ_s was due to the precipitation of Ti_3Ni_4. However, it decreased with further heating, since the Ti_3Ni_4 precipitates grew [33, 34]. On the other hand, σ_s of the multilayer thin film markedly decreased by heat-treatment above 873 K. This may be attributed to the fact that Ti_3Ni_4 precipitates were not formed in the multilayer thin films.

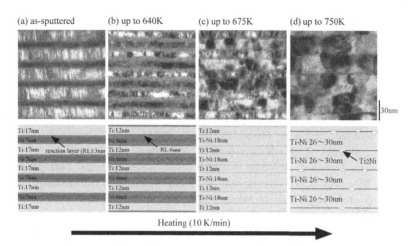

Figure 4.11   Cross-sectional bright field image and schematic figures corresponding to each
TEM image of the as-sputtered TiNi multilayer thin film (a) and TiNi multilayer
thin films heated up to 640 K (b), 675 K (c) and 750 K (d) [30], with permission
from MRS.

4.5   Summary

The alloying process of the Ti-51at%Ni multilayer thin film is summarized in
Figure 4.11, which shows the cross-sectional bright field image and schematic
figure corresponding to each of the TEM images of the as-sputtered TiNi
multilayer thin film (a) and the TiNi multilayer thin films heated up to 640 K (b),
675 K (c) and 750 K (d).

The as-sputtered multilayer thin film was composed of Ti, Ni and amorphous
layers. The amorphous layers were formed by the collision of sputtered Ni atoms
with the Ti layer or sputtered Ti atoms with the Ni layer during sputtering. When it
was heated up to 640 K, the amorphous layer was expanded by heating (SSA
phenomenon), i.e., parts of Ti and Ni layers reacted and became an amorphous
phase by the diffusion of Ti and Ni atoms. When it was heated up to 675 K, a TiNi
B2 phase was formed by the crystallization from the amorphous layer. But a part of
the Ti layer remained without reaction. Finally, when it was heated up to 750 K, it
became a multilayer structure with each layer consisting of about 30 nm TiNi B2
phase. Ti_2Ni grains remained near the interface of the TiNi B2 layers. Ti_3Ni_4 precipitates were not formed in the alloying process of the multilayer thin films,
although they were formed in the crystallization process of the amorphous thin films.

References

[1] S. Miyazaki and A. Ishida, Martensitic transformation and shape memory behaviour in
    sputter-deposited TiNi-base thin films, *Mater. Sci. Eng. A*, **273–275** (1999) 106–133.
[2] M. Kohl, K. D. Skrobanek and S. Miyazaki, Development of stress-optimised shape
    memory microvalves, *Sensors and Actuators A*, **72** (1999) 243–250.

[3] S. Miyazaki, M. Hirano and V. H. No, Dynamic characteristics of diaphragm microactuators utilizing sputter-deposited TiNi shape-memory alloy thin film, *Mater. Sci. Forum*, **349–395** (2002) 467–474.

[4] Y. Fu, W. Huang, H. Du, X. Huang, J. Tan and X. Gao, Characterization of TiNi shape-memory alloy thin films for MEMS applications, *Surface Coating and Tec.*, **145** (2001) 107–112.

[5] A. Ishida, A. Takei and S. Miyazaki, Shape memory thin film of TiNi formed by sputtering, *Thin Solid Films*, **228** (1993) 210–214.

[6] S. Miyazaki, T. Hashinaga and K. Yumikura, Shape-memory characteristics of sputter-deposited TiNi-base thin film, *Smart Materials SPIE*, **2441** (1995) 156–164.

[7] S. Miyazaki, T. Hashinaga and A. Ishida, Martensitic transformations in sputter-deposited TiNi-Cu shape memory alloy thin films, *Thin Solid Films*, **281–282** (1996) 364–367.

[8] A. Gyobu, Y. Kawamura, H. Horikawa and T. Saburi, Martensitic transformations in sputter-deposited shape memory TiNi films, *Mater. Trans. JIM*, **37** (1996) 697–702.

[9] E. Quandt, C. Halene, H. Holleck *et al.*, Sputter deposition of TiNi, TiNiPd and TiPd films displaying the two-way shape-memory effect, *Sensors and Actuators A*, **53** (1996) 434–439.

[10] T. Matsunaga, S. Kajiwara, K. Ogawa, T. Kikuchi and S. Miyazaki, High strength TiNi-based shape memory thin film, *Mater. Sci. Eng. A,* **273–275** (1999) 745–748.

[11] B. Gabry, C. Lexcellent, V. H. No and S. Miyazaki, Thermodynamic modelling of the recovery strains of sputter-deposited shape memory alloys TiNi and TiNi-Cu thin films, *Thin Solid Films*, **372** (2000) 118–133.

[12] T. Matsunaga, S. Kajiwara, K. Ogawa, T. Kikuchi and S. Miyazaki, Internal structures and shape memory properties of sputter-deposited thin films of TiNi-Cu alloy, *Acta Mater.*, **49** (2001) 1921–1928.

[13] Y. Liu, M. Kohl, K. Okutsu and S. Miyazaki, A TiNiPd thin film microvalve for high temperature applications, *Mater. Sci. Eng. A*, **378** (2004) 205–209.

[14] B. Clemens, Solid-state reaction and structure in compositionally modulated zirconium-nickel and titanium-nickel films, *Phys. Rev. B*, **33** (1986) 7615–7623.

[15] B. Clemens, Effect of sputtering pressure on the structure and solid state reaction of titanium-nickel compositionally modulated film, *J. Appl. Phys.*, **61** (1987) 4525–4529.

[16] J. F. Jongste, M. A. Hollanders, B. J. Thijsse and E. J. Mittemeijer, Solid state amorphization in Ni/Ti multilayers, *Mater. Sci. Eng.*, **97** (1988) 101–104.

[17] A. F. Jankowski and M. A. Wall, Transmission electron microscopy of Ni/Ti neutron mirrors, *Thin Solid Films*, **181** (1989) 305–312.

[18] M. A. Hollanders, B. J. Thijsse and E. J. Mittemeijer, Amorphization along interfaces and grain boundaries in poly crystalline multilayers: an X-ray-diffraction study of Ni/Ti multilayers, *Phys. Rev. B*, **42** (1990) 5481–5494.

[19] A. F. Jankowski, On eliminating deposition-induced amorphization of interfaces in refractory metal multilayer system, *Thin Solid Films*, **220** (1992) 166–171.

[20] C. Sella, M. Massza, M. Kaabouchi, S. El Monkade, M. Miloche and H. Lassri, Annealing effects on the structure and magnetic properties of Ni/Ti multilayers, *J. Magnetism Mag. Mater.*, **121** (1993) 201–204.

[21] A. F. Jankowski and M. A. Wall, Formation of face-centered cubic titanium on a Ni single crystal and in Ni/Ti multilayers, *J. Mater. Res.*, **9** (1994) 31–38.

[22] A. F. Jankowski, P. Sandoval and J. P. Hayes, Superlattice effects on solid state amorphization, *NanoStructured Materials*, **5** (1995) 497–503.

[23] A. F. Jankowski and M. A. Wall, Synthesis and characterization of nanophase face-centered-cubic titanium, *NanoStructured Materials*, **7** (1996) 89–94.

[24] R. Benedictus, K. Han, C. Traeholt, A. Bottger and E. J. Mittemeijer, Solid state amorphization in Ni-Ti systems: the effect of structure on the kinetics of interface and grain-boundary amorphization, *Acta Mater.*, **46** (1998) 5491–5508.

[25] M. Vedpathak, S. Basu, S. Gokhale and S. K. Kulkarni, Studies with Ni/Ti multilayer films using X-ray photoelectron spectroscopy and neutron reflectometry: microscopic characterization of structure and chemical composition, *Thin Solid Films*, **335** (1998) 13–18.

[26] T. Lehnert, S. Tixier, P. Böni and R. Gotthardt, A new fabrication process for Ni-Ti shape memory thin films, *Mater. Sci. Eng. A*, **273–275** (1999) 713–716.

[27] T. Lehnert, H. Grimmer, P. Böni, M. Horisberger and R. Gotthardt, Characterization of shape-memory alloy thin films made up from sputter-deposited Ni/Ti multilayer, *Acta Mater.*, **48** (2000) 4065–4071.

[28] H. Cho, H. Y. Kim and S. Miyazaki, Fabrication and characterization of TiNi shape memory thin film using TiNi multilayer technique, *Sci. Tech. Adv. Mater.*, **6** (2005) 678–683.

[29] H. Cho, H. Y. Kim and S. Miyazaki, Alloying process of sputter-deposited TiNi multilayer thin films, *Mater. Sci. Eng. A*, **438–440** (2006) 699–702.

[30] H. Cho, H. Y. Kim S. Miyazaki, Crystallization of sputter-deposited TiNi amorphous thin films, *Trans. MRS-J*, **28** (2003) 651–654.

[31] J. J. Kim, P. Moine and D. A. Stevenson, Crystallization behaviour of amorphous Ni-Ti alloys prepared by sputter deposition. *Scripta Metallugica*, **20** (1986) 243–248.

[32] X. Wang and J. J. Vlassak, Crystallization of amorphous NiTi shape memory alloy thin films, *Scripta Materialia*, **54** (2006) 925–930.

[33] J. Sakurai, J. I. Kim, H. Hosoda and S. Miyazaki, Shape memory characteristics of sputter-deposited Ti-51.3at%Ni thin film aged at various temperatures, *Trans. MRS-J*, **26** (2001) 315–318.

[34] J. Sakurai, J. I. Kim, H. Hosoda and S. Miyazaki, Microstructure and tensile deformation behaviour of sputter-deposited Ti-51.3at%Ni thin films crystallized at various temperatures, *Trans. MRS-J*, **26** (2001) 319–322.

5 Crystallization and microstructural development

A. G. Ramirez, X. Huang and H.-J. Lee

Abstract

This chapter focuses on the crystallization of TiNi shape memory alloy thin films. These materials are commonly sputter-deposited in an amorphous form and require high-temperature thermal treatments to create their crystalline (actuating) form. The microstructures that emerge during crystallization depend on the nucleation and growth kinetics. This chapter briefly surveys crystallization theory and methods to determine these kinetic parameters such as calorimetry, X-ray diffraction and microscopy. Novel microscopy methods have also been developed to provide a robust description that can give rise to the prediction of microstructures. In addition to presenting these tools, this chapter will also survey various factors that influence crystallization and microstructural development, which include annealing temperature, composition, substrate materials and film thickness.

5.1 Introduction

The crystallization of nickel-titanium thin films is of important scientific concern since the as-deposited form of these films is commonly amorphous and requires a high-temperature annealing step to create their crystalline form. It is well-known that the resulting microstructures dictate the phase transformation behavior of these unique materials [1]. As such, an understanding of the structure–property link can give rise to the development of more reliable devices based on them [1, 2]. Unlike the melting of materials, crystallization does not occur at a unique temperature, but hinges on both temperature and time. For TiNi thin films, the general rule of thumb for crystallization is an annealing step near 500 °C for one hour [3, 4]. However, there is a wide range of processing conditions in the literature from very high temperatures to no annealing at all. Such an assortment in processing conditions produces different films. To comprehend these differences,

Thin Film Shape Memory Alloys: Fundamentals and Device Applications, eds. Shuichi Miyazaki, Yong Qing Fu and Wei Min Huang. Published by Cambridge University Press. © Cambridge University Press 2009.

the impact of crystallization on microstructures and their associated properties must be understood. Hence, this chapter will examine the crystallization behavior of TiNi thin films, the factors that change crystallization, and the microstructures that result from it.

5.2   Crystallization

5.2.1   Crystallization principles

The crystallization of amorphous materials has been thoroughly reviewed in the literature [5]. Consequently, this section will provide a brief summary. In the crystallization process, nucleation and growth initiate once an amorphous material is sufficiently heated. From a thermodynamic vantage, the driving force for an amorphous material to crystallize is the free energy difference between the amorphous and crystalline phases. Figure 5.1 shows a possible free energy curve for a devitrification reaction of a metastable amorphous material [5]. This diagram shows the variation of free energy as a function of composition for an amorphous material and its crystalline counterpart. Concave in shape, since there is no atomic clustering, the higher free energy of the amorphous material is reduced by converting to its lower free energy crystalline phase [6].

It is widely accepted that equiatomic TiNi thin films undergo a polymorphic crystallization, whereby the amorphous material transforms to a crystalline form of the same composition. This simple reaction can be described as polymorphic crystallization:

$$a \rightarrow \alpha(A_x B_y), \tag{5.1}$$

where "a" represents the amorphous form of a material and α represents the crystalline form at a specific composition of $A_x B_y$ [7]. In this model, polymorphic crystallization is fast and assumed to be isotropic, since it is unencumbered by

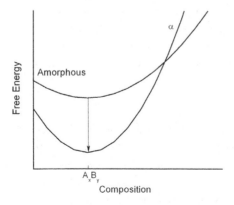

Figure 5.1   Free energy curves for amorphous and crystalline phases showing a potential polymorphic crystallization reaction.

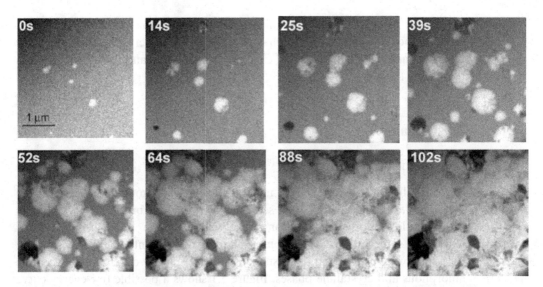

Figure 5.2   Transmission electron micrographs showing the progression of polymorphic crystallization in a TiNi thin film. Grains are spherical and grow isotropically until impingement ([49], with permission).

long-range diffusion. The resulting microstructures at a set temperature can consist of large grains of α, which grow at a constant rate in the amorphous matrix until grains impinge each other. Figure 5.2 displays polymorphic crystallization in a TiNi thin film. Other forms of crystallization include the generation of precipitates from simple intermetallics to more complex Guinier–Preston (GP) zones. All of these microstructures will be discussed in detail in the following sections.

5.2.2   Crystallization theory (Johnson–Mehl–Avrami–Kolmogorov theory)

The crystallization of amorphous materials is a nucleation and growth process that is dependent on temperature and time. Under isothermal conditions, it can be mathematically described by the Johnson–Mehl–Avrami–Kolmogorov (JMAK) theory [8, 9, 10, 11, 12]. The JMAK expression describes the fraction crystallized, Φ, as a function of temperature and time as

$$\Phi = 1 - \exp(-kt^n), \tag{5.2}$$

where k is a temperature-dependent rate constant and n is the Avrami exponent, which is sensitive to the dimensionality of growth as well as the time-dependence of both nucleation and growth. The amount crystallized as a function of time generates a sigmoidal curve as illustrated in Fig. 5.3(a). By simply rearranging Eq. (5.2) and taking the logarithm twice we get the expression

$$\ln[-\ln(1 - \Phi)] = \ln k + n \ln t. \tag{5.3}$$

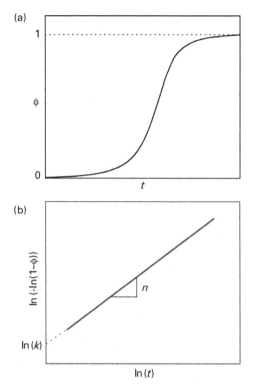

Figure 5.3   (a) A typical Avrami plot of the fraction crystallized, Φ, as a function of time. (b) A graphical construction for obtaining the Avrami exponent, n, and the constant k in Eq. (5.2).

A plot of $\ln[-\ln(1-\Phi)]$ versus $\ln t$ can be used to determine the kinetic parameters k (from the intercept $(\ln(k))$ and n (the slope), as illustrated in Figure 5.3(b). This Avrami analysis can illuminate the type of crystallization mode that is active. However, an unambiguous interpretation of the values of n can only be verified with simultaneous microstructural analysis and an independent determination of the nucleation and growth kinetics [5]. Other limitations of this graphical approach to the JMAK theory are that it doesn't tell us about the individual contributions of nucleation and growth and the theory method does not hold if there is a compositional change in the untransformed volume [13, 14]. Despite its limitations [14, 15], the JMAK method is largely embraced by many researchers.

5.2.2.1   The Avrami exponent, n

The value of the Avrami exponent, n, indicates the type of crystallization that is occurring. Table 5.1 shows a list of values for various nucleation and growth conditions [16]. The exponent n, which generally lies between values of 1.5 and 4, can be written as $n = n_{n} + n_{g}$, where n_{n} describes the nucleation rate, which is either zero or a constant. The growth rate, n_{g}, is proportional to time t or $t^{1/2}$. When thin

Table 5.1   Conditions for transformation and the value of Avrami exponent, n ([14,16])

Transformation condition	n (bulk)	n (thin film)
Nucleation at increasing rate	>4	>3
Constant nucleation and growth (Polymorphic crystallization)	4	3
Nucleation only at start of transformation	3	2
Constant nucleation and diffusion limited growth	2.5	1.5

films meet the condition given by Cahn (where the film thickness is comparable to the growth velocity times time) [14], a thin-film approximation can be taken which removes one dimension of growth and therefore reduces the value of n by one. For example, polymorphic crystallization in bulk materials has a constant nucleation rate, followed by linear growth in all directions. As such, the overall Avrami number for this fast and isotropic process is 4 for bulk, but 3 for thin films.

5.2.2.2   The activation energy

The activation energy of crystallization can be used to describe the thermal stability of a phase, whereby larger numbers indicate higher stability. During a crystallization event, the onset of crystallization is commonly monitored. This onset time is a thermally-activated phenomena that can be describe by an Arrhenius-type expression

$$t = t_0 \exp(-E/RT), \tag{5.4}$$

where t_0 is a constant and E is the activation energy, which is attained from the slope of $\ln(t)$ versus $1/T$ [17].

5.3   Crystallization kinetics of TiNi thin films

TiNi thin films are commonly sputter-deposited, as discussed in Chapter 3, and their amorphous form requires a high-temperature annealing step to crystallize them. Some of the earliest studies in the crystallization behavior examined TiNi ribbons and powders [18, 19, 20, 21, 22, 23]. These materials serve as a good model to aid understanding of the behavior of thin films. However, not all of the lessons from bulk, ribbons, and powders are fully translatable to their thin film counterparts [24].

Researchers have used an array of processing and annealing temperatures in thin films. More recent efforts have systematically studied the kinetic parameters and the effects of various factors on crystallization. Since this work is quite active and still in its early stages, this section will focus on the techniques used for these studies and provide a survey of the current understanding.

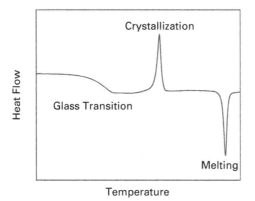

Figure 5.4 A schematic of a DSC curve showing the exothermic peak at a specific temperature that is associated with crystallization.

5.3.1   Experimental techniques

5.3.1.1   Quantitative indirect methods

A change in physical property as a result of crystallization can be used as a way to monitor the progression of crystallization. Electrical resistivity is among the simplest of methods [5]. Physical property changes can also be determined by X-ray diffraction [25], differential scanning calorimetry [5], as well as dilatometric [26], film-stress [27] and elastic modulus [28] measurements. If these changes can be correlated to the transformed (crystallized) fraction, they may be fitted to the model to determine JMAK parameters. From these many techniques, calorimetry and X-ray diffraction have been the key players in TiNi crystallization studies. We will explore these tools in the following sections.

Differential scanning calorimetry

In this method, a small sample (of less then a few milligrams) and a reference sample are placed on their respective thermocouples and subjected to a controlled temperature program. By observing the difference in heat flow between the sample and reference, differential scanning calorimeters (DSCs) are able to measure the amount of heat absorbed or released during transitions, whereby crystallization is observed as an exothermic peak as shown in Fig. 5.4 [29].

Kissinger method. In the Kissinger method, samples are continuously heated at different heating rates, β, as schematically shown in Fig. 5.5 [30]. In the heating experiment, the onset temperature for crystallization, T_x, or the peak temperature, T_p, are noted for each heat rate. The activation energy can be calculated by the expression

$$\ln(\alpha/T_p^2) = C - E/RT_p, \tag{5.5}$$

where R is the gas constant and C is a constant. A plot of $\log T_p^2/\beta$ versus $1/T_p$ should generate a straight line with a slope of E (the activation energy). Enthalpies associated with crystallization can also be determined by DSC [29].

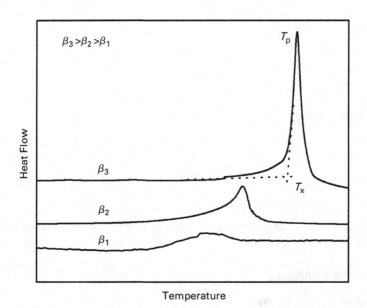

Figure 5.5   A schematic of the Kissinger method showing DSC curves generated at different heat rates, β. The onset and peak temperatures are noted and can be used to solve the activation energy using Eq. (5.5).

Isothermal method. The activation energy and the crystallization mode can be determined with isothermal DSC methods coupled with the JMAK model. The amount crystallized, Φ, corresponds to the fractional area under the crystallization peak, $A(t)/A_{\text{total}}$ [31, 32]. That is, by integrating from the onset time to some later time one can generate a typical Avrami plot, as Fig. 5.6 shows. Using Eq. (5.3), the Avrami exponent, n, can be obtained and the overall crystallization activation energy can be estimated at $t_{50\%}$ by $t_{50\%} = t_0 \exp(-E/kT)$.

X-ray diffraction
X-ray diffraction (XRD) is largely used in conjunction with other techniques to identify the resulting phases during crystallization. It has also been used to determine the kinetics of reaction by measuring the peaks *in situ* during annealing. Sputter-deposited TiNi thin films on silicon are set onto a sample heating stage within the X–ray diffractometer. The $(110)_{\text{B2}}$ austenite peak is monitored as a function of time during isothermal crystallization and the relative amount of austenite formed at a given temperature, W, can be determined from the integrated intensities of the associated diffraction peak by

$$W = \sum_{i=1}^{m} \frac{I_i}{I_{0i}}, \tag{5.6}$$

where I_i is the integrated intensity of the ith diffraction peak, I_{0i} is the same at some reference condition, and m is the number of peaks used in the summation

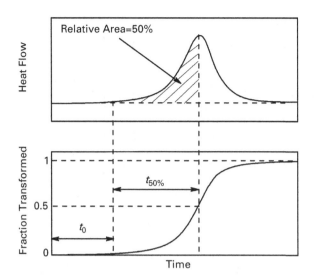

Figure 5.6 In an isothermal DSC experiment, the exothermic peak can be parallel to the Avrami curve. The fractional area under the curve at a certain time, $A(t)/A_{total}$, corresponds to Φ, the fraction transformed. This approach can be used in concert with Eqs. (5.2) and (5.3) to find the activation energy and Avrami exponent, n.

[33, 34]. The reference condition, I_{0i}, is taken for the sample after complete crystallization. To ensure a proper measurement, several peaks should be employed to eliminate the effect of preferred orientation. These data can then be applied to the JMAK expression (in Eq. (5.2)) where the fraction crystallized, Φ, is set equal to W. Using the methods described above, the Avrami exponent, n, and the associated activation energy can be determined.

5.3.1.2   Quantitative direct methods

The JMAK theory describes the overall crystallization process and cannot attain the individual activation energies of growth and nucleation [14]. As such, direct measurements of the crystallization process are a powerful way to describe the crystallization mechanism. With them, one can describe the separate contributions of nucleation and growth. Such work was reported by Lee et al. [35], using in situ TEM, and corroborated by Wang et al. [36], using optical microscopy. Using a collection of images obtained from a video recording, the fraction transformed, Φ, as a function of time can be determined directly and the associated Avrami exponent can be determined with Eq. (5.2) (as depicted in Fig. 5.7). With these same images, the number of crystals, N, and the area fraction crystallized, Φ, can be tracked as a function of time to determine the nucleation rate, J, expressed as

$$J(t_i) = \frac{1}{1 - \Phi(t_i)} \frac{dN(t_i)}{dt} \approx \frac{1}{1 - \Phi(t_i)} \frac{N(t_i) - N(t_{i-1})}{t_i - t_{i-1}}, \qquad (5.7)$$

where t_i is the time of image i. After a short time lag the steady-state value, J_0, is established [37]. The growth rate, V_0, can be attained by analyzing the size of

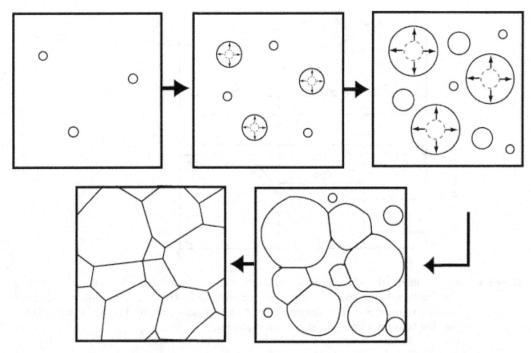

Figure 5.7   With direction observational methods, the nucleation rate, J, can be determined by the number of new crystals per frame; and the growth rate, V, can be determined by the change in size per grain. These values can be inserted into Eq. (5.8) to achieve the final average grain diameter.

individual crystals as a function of time [38]. For a range of temperatures, the expression for the individual nucleation and growth rates as a function of temperature can be obtained by an Arrhenius relationship [35]. In addition, the resulting average grain size, $\bar{d}$, can be expressed as

$$\bar{d} \approx 1.2 \left(\frac{V_0}{J_0} \right)^{1/3}. \tag{5.8}$$

Such a relationship allows researchers to predict the microstructure's final average grain size *a priori* with reasonable agreement with experimental values [35, 36].

5.3.2   A background of TiNi crystallization studies

A glance at the history of TiNi crystallization studies shows that a few key parameters that were established early on are still applied today. Early experiments demonstrated the need for crystallization, with conditions at 500 °C for one hour [39]. Moberly and coworkers later verified with transmission electron microscopy that thin films crystallized to the B2 (CsCl) parent phase and had spheroidal grains, which were

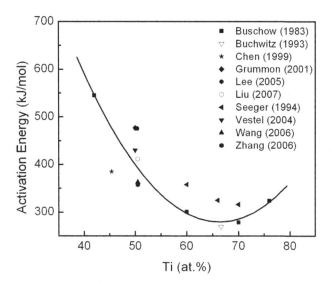

Figure 5.8 A plot of activation energies in TiNi thin film as a function of composition transposed against values for bulk (solid line) as first demonstrated by Buschow [19].

consistent with a constant nucleation and growth mode [40]. This morphology was later called polymorphic crystallization by Chang and Grummon [41]. The activation energy for crystallization is often reported. The first researcher to describe activation energies was Buschow, who investigated the crystallization behavior of bulk Ni_xTi_{1-x} alloys, with a compositional range in x from 0.25 to 0.64 [19]. This work has become the standard with which other work is often compared, as shown in Fig. 5.8. As shown, there is a significant scatter in activation energy values, which have been attributed to the variation in measuring techniques, the difficulty of determining the crystallization start and finish points, and differing final microstructures [42]. From technique-to-technique, activation energies can have a range. Chen and Wu used the Kissinger method, the isothermal method and X-ray diffraction for the same material and found the activation energy from X-ray diffraction was an outlier [42]. They suggested that films constrained by silicon wafers may be experiencing stress, which might alter the crystallization kinetics [42]. They, like others, found less variation between the Kissinger and isothermal methods [43]. At present, the TiNi thin film community readily reports activation energies as an indicator of thermal stability, but it is important to bear in mind its inherit variation and the factors that affect it [20]. One factor that influences the activation energy is the addition of a third element, such as copper. Chen and Wu found the activation energy of $Ni_{50.07}Ti_{49.93}$ and $Ni_{40.09}Ti_{49.96}Cu_{9.95}$ films to be 416 kJ/mol and 388 kJ/mol, respectively, and claimed that amorphous TiNiCu has a lower stability than TiNi [24]. Other ternaries have been reported as well, such as TiNiCu ribbon [44, 45], TiNiHf [46] and NiMnGa [47] films with a range in activation energies of 341, 389 and 234 kJ/mol, respectively.

The underlying crystallization mechanism has received new attention. Two different nucleation mechanisms have been proposed: polymorphic crystallization (where $n = 3$) and site-saturated nucleation (where $n = 2$). Several researchers have demonstrated with various experimental methods that TiNi undergoes polymorphic crystallization. [35, 36, 37, 48, 49, 50, 51]. With *in situ* TEM, Lee *et al.* [49]. determined n as 2.8–3.2 and E as 404.4–567.5 kJ/mol. Using grazing incidence X-ray diffraction (GIXD), Martins and coworkers found their materials behaved polymorphically also [50]. Vestel *et al.* also showed in their comprehensive work with DSC that the Avrami exponent is 2.9–3.1 with an activation energy of $E \sim 406$ kJ/mol [52].

A site-saturated mechanism has been proposed more recently with lower values of n (~ 2.1) and E (~ 352 kJ/mol), reported by Vestel and coworkers [52]. This mechanism is believed to consist of heterogeneous nucleation that occurs at the film surface coupled with lateral interface-controlled growth [52]. Using XRD, Zhang *et al.* obtained the Avrami exponent that corroborates this proposed mechanism with an n between 2.0 and 2.4 and the activation energy of 358.1kJ/mol for a Ti-49.2at%Ni film [53]. It has also been found that amorphous TiNi thin films undergo multiple crystallization events during annealing. With TEM, Vestel *et al.* observed microstructural features to correlate with these n values, i.e. large surface-nucleated columnar grains ($n = 2$) and plate-shaped grains ($n = 3$) [52]. All in all, these numerous observations form a very solid foundation for future work in crystallization and show that this field is still rich with interesting questions.

5.3.3   Factors that influence crystallization

Temperature effects

Microstructures that result from crystallization are dependant on temperature and time. The sensitivity of microstructure, particularly the grain size, to the crystallization temperature has been demonstrated [35]. Lee and coworkers annealed TiNi thin films from 465 °C to 515 °C and found that the average grain size decreased with increasing temperature in a predictable way [35]. They also found that the time to completion change drastically from over 4700 to 10 seconds, as Fig. 5.9 illustrates. These observations have been used to generate a microstructural map that connects processing conditions to their resulting microstructures [35, 37]. Annealing temperatures have also been found to promote (110) texture in the films, under certain elevated temperatures [54, 55].

Compositional effects

The film composition has been found to affect the crystallization behavior. Grummon and Zhang observed that Ni-rich films crystallize faster than their equiatomic counterparts using film stress monitoring methods [56]. It has also been shown by Chang and Grummon that the onset temperature of crystallization, T_x, decreased with increasing Ti-content in their DSC measurements [41].

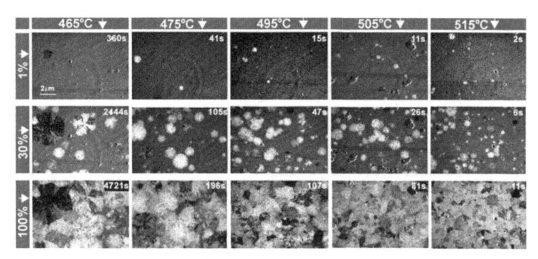

Figure 5.9 A small change in the isothermal heating temperature can give rise to different resulting grain structures that fully crystallize at different times. (Adapted from [35] with permission.)

Using TEM, Ni and coworkers demonstrated that Ti-rich films were very sluggish in comparison to near-equiatomic films and that they did not exhibit martensitic phase transformations at room temperature [57]. It was proposed that diffusion played a role in the delay. The understanding of diffusion in the generation of precipitates and its interplay with crystallization continues to be of scientific interest.

Film thickness effects

There are preliminary reports on the impact of film thickness on crystallization [58, 59, 60]. Wang and coworkers found the growth velocity of thinner films (less than 600 nm) decreased and exhibited an increased activation energy, which is potentially due to compositional shifts and impurities [60]. Han and coworkers analyzed the effects of film thickness with ion-milled thin films (with a thickness range of 0.5 to 4 μm) [58]. They observed two modes of crystallization: inhomogeneous nucleation that produced nanocrystalline grains (5–20 nm) in the ultra-thin region; and polymorphic crystallization with submicron grains in the thicker region. They did not discuss the impact that ion milling might have on these films.

Ishida and coworkers elegantly showed the effects of film thickness using individually deposited Ti-50.0at%Ni films with a thickness range of 0.5 to 7 μm [59]. For all film thicknesses, the average grain size was approximately 5 μm. Films that were 5 μm and more were equiaxed and surrounded by other grains, as Fig. 5.10 illustrates. When films were less than 2 μm thick, the grains encountered the lower film surface. The shape memory behavior is sensitive to film thicknesses below 5 μm, with a maximum transformation strain at 2 μm.

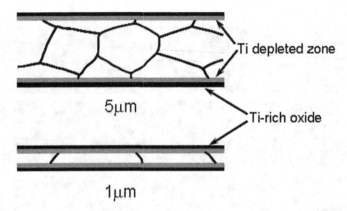

5μm

1μm

Figure 5.10   A schematic of the grain structure as a function of film thickness and the associate layers that exit due to processing.

Ishida *et al.* attributed this effect to two kinds of constraint: a surface oxide layer and neighboring grains. As microelectromechanical systems (MEMS) become more relevant for TiNi thin films, the impact of film thickness will continue to receive attention.

Substrate effects

Surface conditions. Researchers have reported an effect of substrate smoothness on crystallization [50, 52, 61]. Martins and coworkers deposited a rough poly-silicon layer onto silicon and found this layer enhanced crystallization with a shorter crystallization time in comparison to films directly deposited on silicon. This polysilicon layer is proposed to be rougher and generate more nucleation sites. Vestel and coworkers have reported the opposite effect [52]. They found that 22 μm thick films on rough stainless steel lengthen the nucleation rate compared to 13 μm thick films on silicon. This contrasting behavior suggests that other mechanisms may be involved and further investigations are required to clarify the role of substrate roughness.

Interfacial reactions. The adhesion of TiNi onto silicon is important for MEMS devices and has been studied by dynamic mechanical measurements [62]. and peel tests [63]. The reactions between TiNi films and Si substrate were studied intensively by Huang and Mayer, who suggested several Ti-Ni-Si intermetallics might be formed [64]. Using electron diffraction, they identified the ternary silicide as $TiNiSi_2$ and proposed that nickel migrates into Si below 600 °C and that all elements migrate above 600 °C, resulting in various binary or ternary silicides [65]. To prevent this intermixing and to improve adhesion, silicon nitride (SiN_x) can be deposited between TiNi and silicon. This layer can be used sacrificially in the microfabrication process [63] and acts as a good barrier layer [66]. However, the possible reactions between silicon nitride and nickel titanium at annealing conditions still remain unknown.

5.4   Microstructural development

5.4.1   Crystallography of martensite and austenite

Austenite (B2) and martensite (B19) are distinguishable by their spherulitic and twinned features. More details about their phase transformation can be found in Chapter 6 and in the literature [2]. This section will focus on the microstructures that evolve from amorphous materials during crystallization and the impact of film composition and the substrate on these microstructures.

5.4.2   Compositional effects

Ni-rich films

The effect of composition on the final microstructure can be illuminated by the TiNi phase diagram, in Fig. 5.11. This diagram shows a broad range of solubility on the Ni-rich side of the intermetallic region (in gray), and the potential for precipitate formation on cooling. It is interesting to note that although $TiNi_3$ is the stable precipitate, other metastable phases, like Ti_3Ni_4 and Ti_2Ni_3, also exist. In the literature, Ti_3Ni_4 precipitates have attracted much attention since they are useful for changing the shape memory effect and for adjusting the transformation temperature [67]. Using processing conditions used for bulk materials, Ishida and Miyazaki [68] solution-treated and aged Ni-rich thin films (Ti-51.3at%Ni) and found lenticular-shaped Ni_4Ti_3 precipitates similar to those found in bulk [69, 70]. The size of these precipitates increases with higher temperatures and longer aging times. Overall, these precipitates are believed to modify the transformation

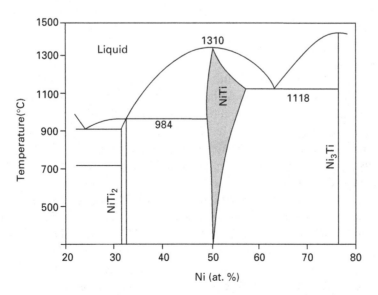

Figure 5.11   A section of the equilibrium TiNi phase diagram.

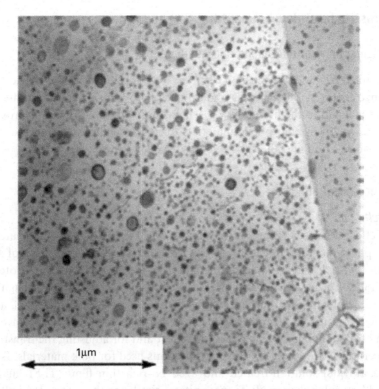

1μm

Figure 5.12   Spherical Ti_2Ni precipitates commonly found in TiNi thin films after a high temperature anneal. (After [72] with permission.)

behavior with their stress fields [71] and effect the crystallization temperature of the amorphous material with the depletion of nickel [68].

Ti-rich films

Unlike Ni-rich films, Ti-rich films exhibit microstructures that differ from bulk [72, 73]. Two types of precipitate have been reported: coherent plate-like [74, 75]. and partially-coherent spherical Ti_2Ni precipitates [76]. The spherical precipitates, as shown in Fig. 5.12, have not been reported in bulk materials and are found to improve the shape memory properties [73, 74, 77, 78, 79, 80, 81]. Various combinations of precipitates can be generated by a host of heat treatments, as Fig. 5.13 displays [76]. With increasing temperatures, the precipitates evolve from plate precipitates (Fig. 5.13(d) and (e)) to plate precipitates with spherical Ti_2Ni precipitates (seen as Moiré fringes in Fig. 13(c)) to spherical Ti_2Ni precipitates (Fig. 5.13(b)) [82]. The structural details of the plate precipitates are still controversial [72]. High-resolution TEM studies have shown a few layers of perfect coherency between the matrix and the precipitate (the precipitates are Guinier–Preston (GP) zone-like) [74, 83], but the structure of the GP zones still remains unknown [76]. The distribution of plate-precipitates is

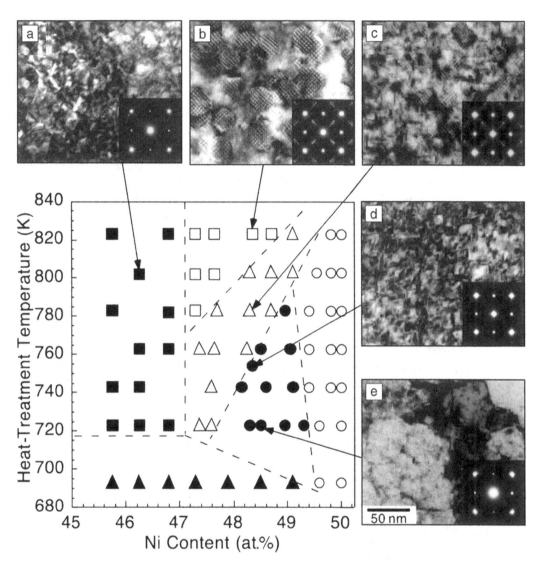

Figure 5.13 A map of precipitate morphology as a function of heat treatments for Ti-rich thin films. (a) randomly oriented Ti_2Ni particles, noted at solid squares (■); (b) Ti_2Ni precipitates oriented with matrix, noted as open squares (□); (c) co-existing plate precipitates and oriented Ti_2Ni precipitates, noted as open triangles (▽); (d) plate precipitates, noted as solid circles (●). Regions for amorphous (O) and no precipitates are also shown. (After [77] with permission.)

found to depend on the heat treatment. There is a uniform precipitate distribution with high temperatures [72]. and a localized distribution with low ones [72]. Overall, these plate precipitates continue to be of interest, since they increase the critical stress for slip with a 6% recoverable strain and 670 MPa recovery strength [74].

5.5   Summary

In sum, understanding the crystallization that takes place in TiNi thin films is tantamount to the development of devices based on them. Several methods of characterizing the crystallization have been used, which include calorimetry and X-ray diffraction. These measurements can be improved when coupled with observational methods, which provide details of the individual activation energies of nucleation and growth and enable the generation of a microstructural map. The microstructures that result from crystallization have been found to be sensitive to annealing temperature, film thickness, substrate and film composition. Our knowledge of bulk TiNi microstructures serves us well in determining the generation of precipitates in Ni-rich thin films. In Ti-rich films, however, the plate-shape precipitates commonly observed in bulk can be accompanied by spherical precipitates, which only seem to occur in thin films. All in all, there have been significant developments in uncovering the link between processing and structure in TiNi thin films. As these materials become more technologically relevant, it is envisioned that studies that link processing, structure and properties will increase.

References

[1] C. M. Wayman and T. W. Duerig. An introduction to martensite and shape memory. In *Engineering Aspects of Shape Memory Alloys*, eds. T. W. Duerig, K. N. Melton, D. Stokel and C. M. Wayman, New York: Butterworth-Heinemann (1990).

[2] K. Otsuka and C. M. Wayman. eds. *Shape Memory Materials*. Cambridge: Cambridge University Press (1998).

[3] J. J. Kim, P. Moine and D. A. Stevenson. Crystallization behavior of amorphous NiTi alloys prepared by sputter deposition. *Scripta Metallurgica* **20**(2) (1986) 243–248.

[4] P. Moine, A. Naudon, J. J. Kim, A. F. Marshall and D. A. Stevenson. Characterization of Ni-Ti alloys synthesized by vapor quenching. *Journal de Physique*. **46**(C-8) (1985) 223–227.

[5] M. G. Scott. Crystallization. In *Amorphous Metallic Alloys*, ed. F. E. Luborsky. London: Butterworths (1983).

[6] D. Gaskell. *Introduction to the Thermodynamics of Materials*, 4th edn., London: Taylor & Francis (1995).

[7] J. W. Martin, R. D. Doherty and B. Cantor. *Stability of Microstructure in Metallic Systems*. New York: Cambridge University Press (1997).

[8] M. Avrami. Kinetics of phase change I–General theory. *J. Chem. Phy.* **7**(12) (1939) 1103–1112.

[9] M. Avrami, Kinetics of phase change II: transformation–time relationships for random distribution of nuclei. *J Chem. Phys* **8** (1940) 212.

[10] M. Avrami. Kinetics of phase change III: granulation, phase change, and microstructure. *J Chem. Phys* **9** (1941) 177.

[11] W. A. Johnson and R. F. Mehl. Reaction kinetics in processes of nucleation and growth. *Am. Inst. Min. Metal Pet. Eng.* **135** (1939) 416.

[12] A. N. Kolmogorov. On the statistical theory of metal crystallization. *Izvestiya Akademii Nauk SSSR*. **3** (1937) 355–359.

[13] R. W. Balluffi, S. M. Allen and W. C. Carter. *Kinetics of Materials*. New York: John Wiley & Sons (2005).

[14] J. W. Cahn. The time cone method for nucleation and growth kinetics on a finite domain. In *Thermodynamics and Kinetics of Phase Transformations Symposium*, eds. J. S. Im, B. Park, A. L. Greer and G. B. Stephenson. Pittsburgh: Mater Res. Soc. (1996). p. 425–437.

[15] N. X. Sun, X. D. Liu and K. Lu. An explanation to the anomalous Avrami exponent. *Scripta Materialia* **34**(8) (1996) 1201–1207.

[16] J. W. Christian. *The Theory of Transformation in Metals and Alloys*. Second edn. New York: Pergamon (1975).

[17] G. Ghosh, M. Chandrasekaran and L. Delaey. Isothermal crystallization kinetics of $Ni_{24}Zr_{76}$ and $Ni_{24}(Zr-X)_{76}$ amorphous alloys. *Acta Metall Mater.* **39**(5) (1991) 925–936.

[18] M. Buchwitz, R. Adlwarth-Dieball P. L. Ryder. Kinetics of the crystallization of amorphous Ti_2Ni. *Acta Metall. Mater.* **41**(6) (1993) 1885–1892.

[19] K. H. J. Buschow. Stability and electrical transport properties of amorphous $Ti_{1-x} Ni_x$ alloys. *J. Phys. F (Metal Physics)* **13**(3) (1983) 563–571.

[20] K. H. J. Buschow. Effect of short-range order on the thermal stability in amorphous Ti-Ni alloys. *J Appl. Phys.* **56**(2) (1984) 304–306.

[21] J. Eckert, L. Schultz, K. Urban Synthesis of Ni-Ti and Fe-Ti alloys by mechanical alloying: formation of amorphous phases and extended solid solutions. *J. Non-Crystalline Solids* **127**(1) (1991) 90–96.

[22] R. B. Schwarz, R. R. Petrich and C. K. Saw. The synthesis of amorphous Ni-Ti alloy powders by mechanical alloying. *J. Non-Crystalline Solids* **76**(2–3) (1985) 281–302.

[23] C. Seeger and P. L. Ryder. Kinetics of the crystallization of amorphous Ti-Ni and Ti-Ni-Si alloys. *Mat. Sci. Eng.* **179** (1994) 641–644.

[24] J. Z. Chen and S. K. Wu. Crystallization temperature and activation energy of rf-sputtered near-equiatomic TiNi and $Ti_{50}Ni_{40}Cu_{10}$ thin films. *J. Non-Crystalline Solids* **288**(1–3). (2001) 159–165.

[25] F. Edelman, T. Raz, Y. Komem, P. Zaumseil, H. J. Osten and M. Capitan. Crystallization of amorphous $Si_{0.5}Ge_{0.5}$ films studied by means of in-situ X-ray diffraction and in-situ transmission electron microscopy. *Phil. Mag. A* **79**(11) (1999) 2617–2628.

[26] D. Turnbull, Kinetics of solidification of supercooled liquid mercury droplets. *J. Chem. Phys.* **20**(3) (1952) 411–424.

[27] D. S. Grummon and J. P. Zhang. Stress in sputtered films of near-equiatomic TiNiX on (100) Si: intrinsic and extrinsic stresses and their modification by thermally activated mechanisms *Physica Status Solidi A–Applied Research* **186**(1) (2001) 17–39.

[28] A. Kursumovic and M. G. Scott. The use of Young's modulus to monitor relaxation in metallic glasses. *App. Phys. Lett.* **37**(7) (1980) 620–622.

[29] W. W. Wendlandt. *Thermal Analysis*. Third edn. New York: Wiley-Interscience (1986).

[30] H. E. Kissinger. Reaction kinetics in differential thermal analysis. *Analytical Chemistry* **29** (1957) 1702–1706.

[31] A. L. Greer. Crystal nucleation and growth in metallic liquids and glasses. In *Acta-Scripta Metallurgica Proceedings Series*, **B**, eds. P. Haasen and R. I. Jaffee, New York: Pergamon Press. pp. 94–107 (1985).

[32] M. G. Scott. The crystallization kinetics of Fe-Ni based metallic glasses. *J. Mater. Sci.* **13**(2) (1978) 291–296.

[33] R. Dahan, J. Pelleg and L. Zevin. Silicide formation by reaction of Ta-Ti thin-films and a Si single-crystal. *J. Appl. Phys.* **67**(6) (1990)2885–2889.

[34] H. P. Klug and L. E. Alexander. X-ray diffraction procedures for polycrystalline and amorphous materials, second edn. New York: Wiley (1974).

[35] H. J. Lee, H. Ni, D. T. Wu and A. G. Ramirez, Grain size estimations from the direct measurement of nucleation and growth. *Appl. Phys. Lett.* **19** (2005) 87.

[36] X. Wang and J. J. Vlassak. Crystallization kinetics of amorphous NiTi shape memory alloy thin films. *Scripta Materialia* **54**(5) (2006) 925–930.

[37] H. J. Lee, H. Ni, D. T. Wu and A. G. Ramirez. A microstructural map of crystallized NiTi thin film derived from in situ TEM methods. *Materials Transactions* **47**(3) (2006) 527–531.

[38] ASTM E112. Standard text methods for determining average grain size (www.astm.org).

[39] J. D. Busch, A. D. Johnson, C. H. Lee and D. A. Stevenson. Shape memory properties in Ni-Ti sputter-deposited film. *J. Appl. Phys.* **68** (12) (1990) 6224–6228.

[40] W. J. Moberly, J. D. Busch, A. D. Johnson and M. H. Berkson. In situ HVEM of crystallization of amorphous TiNi thin films. In *Phase Transformation Kinetics in Thin Films*, eds. M. Chen, M. O. Thompson, R. B.Schwarz and M. Libera. Pittsburgh: Mater Res. Soc. (1992). p 85–90.

[41] L. W. Chang and D. S. Grummon. Structure evolution in sputtered thin films of Ti−x(Ni,Cu)(1−x) 1. Diffusive transformations *Phil. Mag. A* **76**(1) (1997) 163–189.

[42] J. Z. Chen and S. K. Wu. Crystallization behavior of rf-sputtered TiNi thin films. *Thin Solid Films* (1999) 194–199.

[43] K. T. Liu and J. G. Duh. Kinetics of the crystallization in amorphous NiTi thin films. *J. of Non-Crystalline Solids* **353** (2007) 1060–1064.

[44] T-H Nam, S-M Park, T-Y Kim and Y-W Kim. Microstructures and shape memory characteristics of Ti-25Ni-25Cu(at.%) alloy ribbons. *Smart Materials and Structures* **14**(5) (2005) 239–240.

[45] S. K. Wu, S. H. Chang and H. Kimura. Annealing effects on the crystallization and shape memory effect of $Ti_{50}Ni_{25}Cu_{25}$ melt-spun ribbons. *Intermetallics* **15** (2007) 233–240.

[46] Y. X. Tong, Y. Liu, J. M. Miao and L. C. Zhao. Characterization of a nanocrystalline NiTiHf high temperature shape memory alloy thin film. *Scripta Materialia* **52**(10) (2005) 983–987.

[47] S. K. Wu, K. H. Tseng and J. Y. Wang. Crystallization behavior of RF-sputtered near stoichiometric Ni_2MnGa thin films. *Thin Solid Films*. **408**(1–2) (2002) 316–320.

[48] H. J. Lee, H. Ni, D. T. Wu and A. G. Ramirez. Experimental determination of kinetic parameters for crystallizing amorphous NiTi thin films. *Appl. Phys. Lett.* **87**(11) (2005).

[49] H-J Lee and A. G. Ramirez. Crystallization and phase transformations in amorphous NiTi thin films for microelectromechanical systems. *Appl. Phys. Lett.* **85**(7) (2004) 1146–1148.

[50] R. M. S. Martins, F. M. Braz Fernandes, R. J. C. Silva, *et al*. The influence of a poly-Si intermediate layer on the crystallization behaviour of Ni-Ti SMA magnetron sputtered thin films. *Appl. Phys. A (Materials Science Processing)* **A83**(1) (2006) 139–145.

[51] A. G. Ramirez, H. Ni and H. J. Lee. Crystallization of amorphous sputtered NiTi thin films. *Mat. Sci. Eng. A* **438** (2006) 703–709.

[52] M. J. Vestel, D. S. Grummon, R. Gronsky and A. P. Pisano. Effect of temperature on the devitrification kinetics of NiTi films. *Acta Materialia* **51**(18) (2003) 5309–5318.

[53] L. Zhang, C. Y. Xie and J. S. Wu. In situ X-ray diffraction analysis on the crystallization of amorphous Ti-Ni thin films. *Scripta Materialia* **55**(7) (2006) 609–612.

[54] K. R. C. Gisser, J. D. Busch, A. D. Johnson and A. B. Ellis. Oriented nickel-titanium shape memory alloy films prepared by annealing during deposition. *Appl. Phys. Lett.* **61**(14) (1992) 1632–1634.

[55] L. Hou and D. S. Grummon. Transformational superelasticity in sputtered titanium-nickel thin films *Scripta Metallurgica et Materialia* **33**(6) (1995) 989–995.

[56] D. S. Grummon and J. Zhang. Stress in sputtered films of near-equiatomic TiNiX on (100) Si: intrinsic and extrinsic stresses and their modification by thermally activated mechanisms. *Physica Status Solidi A* **186**(1) (2001) 17–39.

[57] H. Ni, H-J Lee and A. G. Ramirez. Compositional effects on the crystallization kinetics of nickel titanium thin films. *J. Mat. Res.* **20**(7) (2005) 1728–1734.

[58] X. D. Han, S. C. Mao, Q. Wei, Y. F. Zhang and Z. Zhang. In-situ TEM study of the thickness impact on the crystallization features of a near equal-atomic TiNi thin film prepared by planar magnetron sputtering. *Materials Transactions* **47**(3) (2006) 536–539.

[59] A. Ishida and M. Sato. Thickness effect on shape memory behavior of Ti-50.0at.%Ni thin film. *Acta Materialia* **51**(18) (2003) 5571–5578.

[60] X. Wang, M. Rein and J. J. Vlassak. Crystallization kinetics of amorphous equiatomic NiTi thin films: effect of film thickness. *J. Appl. Phys.* **103**(8) 023501-1 to 023501-6 (2002).

[61] Q. M. Su, S. Z. Hua and M. Wuttig. Nondestructive dynamic evaluation of thin NiTi film adhesion. *J. Adhesion Science and Technology* **8**(6) (1994) 625–633.

[62] R. H. Wolf and A. H. Heuer. TiNi (shape memory) films on silicon for MEMS applications. *J. Microelectromech. Syst.* **4**(4) (1995) 206–212.

[63] L. S. Hung and J. W. Mayer. Interactions of 4 metallic compounds with Si substrates. *J. Appl. Phys* **60**(3) (1986) 1002–1008.

[64] S. K. Wu, Z. Chen, Y. J. Wu *et al.* Interfacial microstructures of rf-sputtered TiNi shape memory alloy thin films on (100) silicon. *Phil Mag A.* **81**(8) (2001) 1939–1949.

[65] X. Wang, A. Lai, J. J. Vlassak and Y. Bellouard. Microstructure evolution of on-substrate NiTi shape memory alloy thin films. Boston, MA, USA: Mater Res. Soc: (2004). p. 275–280.

[66] K. Otsuka and T. Kakeshita. Science and technology of shape-memory alloys: new developments. *MRS Bulletin* **27**(2) (2002) 91–100.

[67] A. Ishida and S. Miyazaki. Microstructure and mechanical properties of sputter-deposited Ti-Ni alloy thin films. *J. Eng. Mat. Techn. Transactions of the ASME* **121**(1) (1999) 2–8.

[68] A. Ishida, A. Takei, M. Sato and S. Miyazaki. Stress–strain curves of sputtered thin film of Ti-Ni. *Thin Solid Films* **281–282** (1996) 337–339.

[69] S. Miyazaki and A. Ishida. Martensitic transformation and shape memory behavior in sputter-deposited TiNi-base thin films. *Mat. Sci. and Eng. A.* **15** (1999). 106–133.

[70] L. Bataillard and R. Gotthardt. Influence of thermal treatment on the appearance of a three step martensitic transformation in NiTi. *Journal de Physique IV.* **5**(C8) (1995) 647–652.

[71] A. Ishida and V. Martynov Sputter-deposited shape-memory they thin films: properties and applications. *MRS Bulletin* (2002) 111–114.

[72] S. Kajiwara, K. Ogawa, T. Kikuchi, T. Matsunaga and S. Miyazaki. Formation of nanocrystals with an identical orientation in sputter-deposited Ti-Ni thin films. *Phil. Mag. Lett.* **74**(6) (1996) 395–404.

[73] S. Kajiwara, T. Kikuchi, K. Ogawa, T. Matsunaga and S. Miyazaki Strengthening of Ti-Ni shape-memory films by coherent subnanometric plate precipitates. *Phil. Mag. Lett.* **74**(3) (1996) 137–144.

[74] K. Ogawa, T. Kikuchi, S. Kajiwara, T. Matsunaga and S. Miyazaki. Coherent sub-nanometric plate precipitates formed during crystallization of as-sputtered Ti-Ni films *Journal de Physique IV* **7**(C5) (1997) 221–226.

[75] Y. Kawamura, A. Gyobu, T. Saburi and M. Asai. Structure of sputter-deposited Ti-rich Ti-Ni alloy films. *Shape Memory Materials* **27** (2000) 303–306.

[76] T. Kikuchi, K. Ogawa, S. Kajiwara, T. Matsunaga, S. Miyazaki and Y. Tomota. High-resolution electron microscopy studies on coherent plate precipitates and na-nocrystals formed by low-temperature heat treatments of amorphous Ti-rich Ti-Ni thin films. *Phil. Mag. A.* **78**(2) (1998) 467–489.

[77] T. Lehnert, S. Crevoiserat and R. Gotthardt. Transformation properties and micro-structure of sputter-deposited Ni-Ti shape memory alloy film *J. Mater. Sci.* **37**(8) (2002) 1523–1533.

[78] T. Matsunaga, S. Kajiwara, K. Ogawa, T. Kikuchi and S. Miyazaki. High strength Ti-Ni-based shape memory thin films. *Mater. Sci. Eng. A* **275** (1999) 745–748.

[79] T. Matsunaga, S. Kajiwara, K. Ogawa, T. Kikuchi and S. Miyazaki. Internal struc-tures and shape memory properties of sputter-deposited thin films. of a Ti-Ni-Cu alloy *Acta Materialia* **49**(11) (2001) 1921–1928.

[80] K. Nomura, S. Miyazaki and A. Ishida Effect of plastic strain on shape memory characteristics in sputter-deposited Ti-Ni thin films. *Journal De Physique IV* **5**(C8) (1995) 695–700.

[81] Y. Nakata, T. Tadaki, H. Sakamoto, A. Tanaka and K. Shimizu. Effect of heat treatments on morphology and transformation temperatures of sputtered Ti-Ni thin films. *Journal de Physique IV* **5**(C8) (1995) 671–676.

6 Mechanical properties of TiNi thin films

A. Ishida

Abstract

This chapter is devoted to the mechanical properties of TiNi thin films. In this chapter, the shape memory behavior, two-way shape memory effect, super-elasticity and stress–strain curves of sputter-deposited TiNi films are discussed with special attention to unique microstructures in Ti-rich TiNi films.

6.1 Shape memory behavior of TiNi thin films

TiNi bulk alloys are the most widely used shape memory alloys (SMA) in commercial applications [1] and their shape memory behavior has been well investigated (see Chapter 2). The development of SMA thin films, therefore, started with TiNi binary alloys [2, 3] and there are already many papers on the shape memory effect of TiNi thin films. Most work has been carried out on films attached to Si substrates from the viewpoint of practical application (see Chapter 8), but their shape memory behavior is severely affected by stress from the substrates. To understand the shape memory behavior of thin films themselves, study on freestanding films is essential [4, 5]. In this chapter, the shape memory behavior of freestanding TiNi films is discussed.

All the films in this chapter were sputter deposited on glass substrates and then peeled off the substrates before heat treatment. Unless the substrates are intentionally heated, as-sputter-deposited films are amorphous. Usually heat treatment above 773 K is required to produce crystallization. This heat treatment provides us with an opportunity to control the structure of crystallized films and thus their shape memory behavior. The resulting shape memory behavior was evaluated by measuring the strain during cooling and heating under a constant stress. The specimens used for these measurements were typically 5 mm long (gauge portion), 0.4 mm wide and 7 μm thick. Figure 6.1 presents an example of such a strain

Thin Film Shape Memory Alloys: Fundamentals and Device Applications, eds. Shuichi Miyazaki, Yong Qing Fu and Wei Min Huang. Published by Cambridge University Press. © Cambridge University Press 2009.

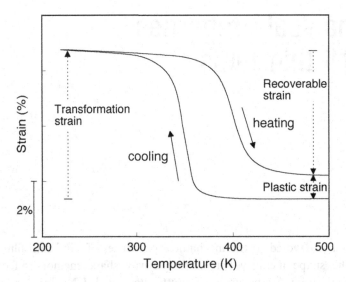

Figure 6.1   Strain–temperature curve at 360 MPa for Ti-50at%Ni thin film annealed
at 773 K for 3.6 ks.

temperature measurement. The measurement was performed for a Ti-50 at %Ni
thin film annealed at 773 K for 3.6 ks [6]. The martensitic transformation from
austenite (B2 phase) to martensite (B19' phase) takes place on cooling, and simultaneously the transformation strain appears. This strain should be completely
recovered by the reverse martensitic transformation on heating. However, it is seen
that a significant residual strain remains in this film. This strain is ascribed to a
plastic deformation introduced during the transformations. Obviously, Ti-50at%Ni
film is susceptible to plastic deformation because of its single phase. Needless to say,
desirable features such as an actuator are a small plastic strain and a large transformation strain, and a high martensitic transformation temperature and a narrow
temperature hysteresis. In order to obtain a perfect shape memory effect, that is, no
detectable plastic strain, a TiNi phase (B2 phase) should be strengthened by a
second phase, as discussed in the following sections.

6.2   Shape memory behavior of Ni-rich Ti-Ni thin films

Figure 6.2 presents the phase diagram near the equiatomic composition of the
TiNi alloy system [7] (This figure is schematic only. The detailed phase diagram
experimentally obtained is seen in [1].) On the basis of this diagram, a solution
aging treatment of Ni-rich TiNi films is applicable to suppress plastic deformation, which is the most common heat-treatment in bulk alloys. Figure 6.3 provides a typical microstructure of aged Ni-rich TiNi films [8]. The film composition
is Ti-51.3at%Ni. From the diffraction pattern, the precipitates are identified to be
Ti_3Ni_4. These precipitates are known to form on {111} planes of the B2 matrix,

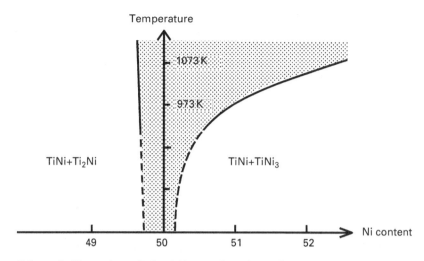

Figure 6.2   Schematic illustration of Ti–Ni binary alloy phase diagram.

Figure 6.3   Microstructure of Ti-51.3at%Ni thin film aged at 673 K for 36 ks after solution treatment at 973 K for 3.6 ks. The inset is the corresponding electron diffraction pattern of the $[111]_{B2}$ zone showing extra spots due to Ti_3Ni_4. (After Ishida *et al.* [8] with kind permission of Springer Science and Business Media.)

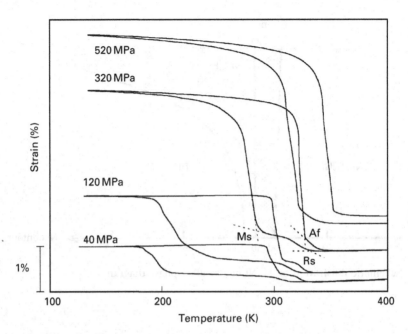

Figure 6.4   Strain–temperature curves under various stresses of Ti-51.3at%Ni thin film aged at 673 K
for 36 ks after solution treatment at 973 K for 3.6 ks. (After Ishida *et al.* [8] with kind
permission of Springer Science and Business Media.)

having a lenticular shape, instead of $TiNi_3$, and play an important role in
strengthening Ni-rich bulk alloys (see Chapter 2).

Figure 6.4 shows the strain–temperature curves of the film of Fig. 6.3. A series of
strain–temperature measurements under various constant stresses was carried out
with an identical specimen varying the stress from 40 to 120 MPa in steps of
20 MPa, and then 120 to 600 MPa in steps of 40 MPa, though only four stresses are
shown in the figure. This test involved loading the sample at a high temperature,
cooling it down to 143 K at the rate of 10 K min^{-1} and heating it back to the
original temperature at the same rate. During the thermal cycle the displacement of
the sample was measured. Under low stresses, the film exhibits a two-stage
elongation on cooling. The first small elongation arises from the R-phase transformation (B2 phase to R phase), whereas the second large one is from the martensitic transformation (R phase to B19' phase). The R-phase transformation is
commonly observed in age-treated Ni-rich TiNi alloys, although it is not detected in
solution-treated ones [9]. Figure 6.4 demonstrates that the shape memory behavior
is significantly affected by an applied stress. This is one of the important features of
SMA, suggesting that the performance of SMA actuators can or should be controlled by an applied stress called "bias stress".

In order to discuss the dependence of shape memory characteristics on stress,
the temperatures for the R-phase transformation start (R_s), the martensitic
transformation start (M_s), and the reverse martensitic transformation finish (A_f)

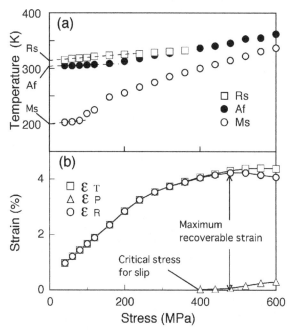

Figure 6.5   (a) Transformation temperatures and (b) transformation strain (ε_T), plastic strain (ε_P) and recoverable strain (ε_R) as a function of stress for a Ti-51.3at%Ni thin film aged at 673 K for 36 ks. (After Ishida *et al.* [8] with kind permission of Springer Science and Business Media.)

were measured by the tangential extrapolation method, as shown in Fig. 6.4, and plotted as a function of stress in Fig. 6.5(a). All the transformation temperatures are found to increase with the increasing stress. This stress dependence is expected from the Clausius–Clapeyron equation [9]. The M_s temperature shows stronger dependence on stress than the A_f temperature and, as a result, the temperature hysteresis between the forward and reverse transformations becomes small with increasing stress, as shown in Fig. 6.4. The R_s temperature is also less sensitive to stress than the M_s temperature. This is the reason that the deformation mode changes from a two-stage deformation (B2→R→B19′) to a single-stage deformation (B2→B19′) when the stress is high in Fig. 6.4. Basically, the R-phase transformation is less sensitive to stress than the martensitic transformation, since the lattice deformation associated with the former transformation is only one tenth of that of the latter transformation.

Figure 6.5(b) shows the stress dependence of the transformation strain (ε_T), plastic strain (ε_P), and recoverable strain (ε_R). The recoverable strain is defined as ($\varepsilon_\mathrm{T} - \varepsilon_\mathrm{P}$). The plastic strain is not detected under low stresses, but it appears at a certain stress. This stress is defined as "a critical stress for slip". This critical stress is considered to correspond to the maximum generative stress of an SMA thin-film actuator in that it can recover its original shape up to this stress. As the stress increases, the transformation strain increases gradually and then levels off. Since

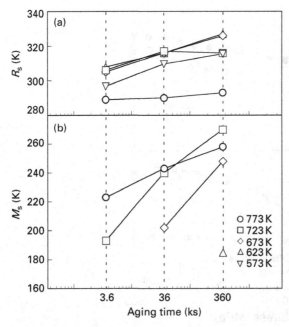

Figure 6.6   (a) R-phase transformation start (R_s) and (b) martensitic transformation start (M_s) temperatures of Ti-51.3at%Ni thin films aged at various temperatures as a function of aging time. (After Ishida *et al.* [8] with kind permission of Springer Science and Business Media.)

the plastic strain increases above the critical stress, the recoverable strain decreases after reaching a maximum.

Similar measurements were carried out for Ti-51.3at%Ni thin films aged at various temperatures (573–773 K) for three different times (3.6, 36 and 360 ks) after solution treatment at 973 K for 3.6 ks [8]. The summary is presented in Figs. 6.6–6.8.

Figure 6.6 shows the effects of aging on the transformation temperatures. The R_s and M_s temperatures were obtained by extrapolating the temperature–stress plot to zero stress, as shown in Fig. 6.5(a). All the transformation temperatures increase with the increasing aging time, but their dependence on aging tempera-ture is a little complicated. They increase with the increasing aging temperature from 573 to 723 K, but the further increase from 723 to 773 K causes a decrease in the transformation temperatures. These effects are explained in terms of two factors, that is, the composition of the B2 matrix and the precipitation morph-ology of the second phase. As the precipitation process proceeds, the Ni content of the matrix decreases owing to the formation of Ti_3Ni_4. The transformation temperature of a TiNi phase is known to decrease with increasing Ni content [10]. Therefore the composition change of the matrix causes the transformation tem-peratures to increase. However, when the composition of the matrix approaches the equilibrium one after a long-time aging, the transformation temperatures of

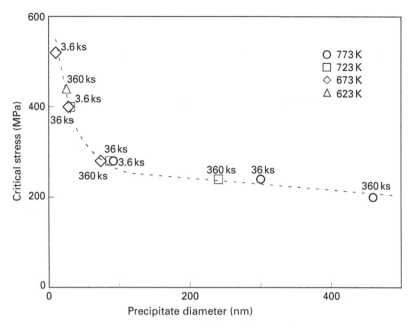

Figure 6.7 Relationship between critical stress for slip and precipitate diameter in Ti-51.3at%Ni thin films aged at various temperatures for various times. (After Ishida *et al.* [8] with kind permission of Springer Science and Business Media.)

films aged at 773 K can be lower than those of films aged at 723 K. This occurs because the equilibrium Ni content in the matrix phase increases with increasing temperature, as shown in Fig. 6.2. The transformation temperatures are affected by the microstructure as well as the matrix composition. The fine precipitates strengthen the matrix and thus suppress the shape changes associated with the transformations. This results in lowering transformation temperatures. As seen in Fig. 6.6, this effect is strong for the martensitic transformation compared with the R-phase transformation because of its large lattice deformation.

As either aging temperature or time increases, the critical stress for slip decreases. In Fig. 6.7, the critical stresses are plotted against the precipitate size. This figure suggests that the precipitation hardening due to the Ti_3Ni_4 phase plays an important role in increasing the critical stress for slip while the precipitate diameter is less than 100 nm.

Figure 6.8 shows the maximum recoverable strain of the aged films. The maximum recoverable strain decreases with increasing aging temperature and time. The film aged at 773 K for 360 ks, however, shows a larger strain than the film aged at 723 K for the same time probably due to the smaller amount of precipitates (refer to Fig. 6.2).

Figures 6.6–6.8 show that the shape memory characteristics of a Ti-51.3at%Ni film can be varied by aging treatment. The aging effects of Ni-rich films are almost the same as those reported in bulk alloys [11].

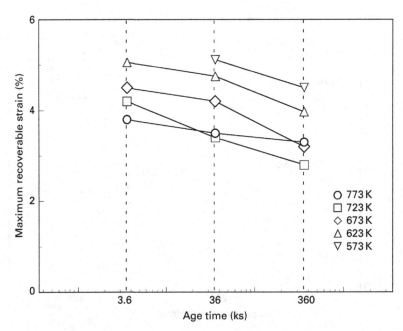

Figure 6.8   Maximum recoverable strain of Ti-51.3 at %Ni thin films aged at various temperatures as a
function of aging time. (After Ishida *et al.* [8] with kind permission of Springer Science
and Business Media.)

6.3   Shape memory behavior of Ti-rich TiNi thin films

In contrast to Ni-rich films, Ti-rich TiNi films exhibit different microstructures
and shape memory behaviors from those of bulk alloys. This section discusses
their microstructures and shape memory behaviors, which are peculiar to thin
films.

It has been known that Ti-rich TiNi alloys exhibit a high martensitic transformation temperature compared with Ni-rich TiNi alloys [10]. This should be an
advantage, but Ti-rich bulk alloys have not been used for practical applications
and even the shape memory behavior has not been measured except for near
equiatomic compositions. The reason is their coarse structure formed by casting.
In bulk alloys, coarse particles of Ti_2Ni form only along grain boundaries, leading
to their brittleness [12]. As seen in Fig. 6.2, the steep solubility line on the Ti-rich
side of the TiNi single-phase region prevents any heat treatment from changing
this cast structure. However, the crystallization process of Ti-rich amorphous
films produces various kinds of interesting microstructures effective in improving
their shape memory behavior [13, 14].

Figure 6.9 presents the representative structures observed in Ti-rich films.
Figure 6.9(a)–(c) show the precipitation process at 773 K for a Ti-48.2at%Ni
film; the annealing times of (a)–(c) are 0.3, 3.6 and 36 ks, respectively. The
microstructure in the B2 matrix evolves in the following sequence [15]: (a) plate

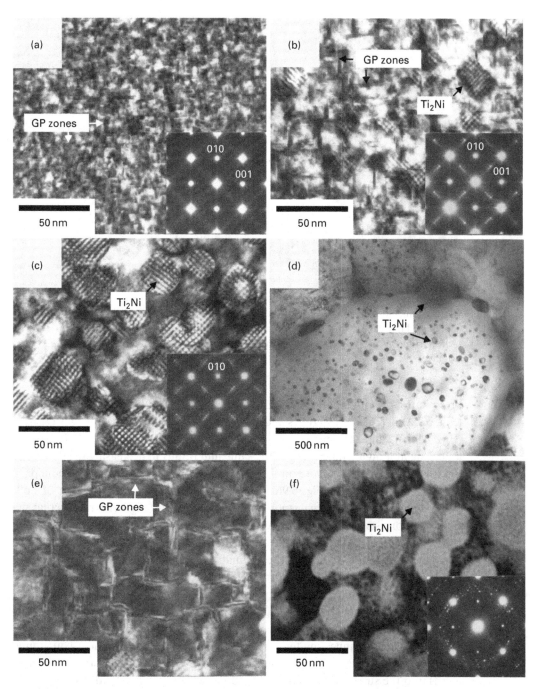

Figure 6.9   Various types of microstructures observed in Ti-rich TiNi (Ti-48.2 at %Ni) thin films:
(a) 773 K, 0.3 ks; (b) 773 K, 3.6 ks; (c) 773 K, 36 ks; (d) 973 K, 360 ks; (e) 703 K, 10.8 ks;
and (f) 773 K, 3.6 ks (this picture was taken from Ti-45.2 at%Ni). (After Ishida *et al.* [7]
with kind permission of the Japan Institute of Metals.)

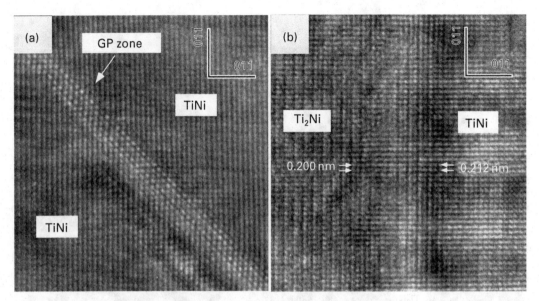

Figure 6.10   High-resolution electron micrographs showing (a) coherent interface between GP zone and TiNi phase, and (b)semicoherent interface between Ti_2Ni and TiNi phases. (After Ishida *et al.* [5] with kind permission of American Society of Mechanical Engineers.)

precipitates of Guinier–Preston (GP) zones along {100} planes of the matrix (Fig. 6.9(a)) – [16], (b) GP zones and spherical Ti_2Ni precipitates with Moiréfringes (Fig. 6.9(b)), and (c) spherical Ti_2Ni precipitates alone (Fig. 6.9(c)). Prolonged annealing at a high temperature of the same film results in decreasing the density of the spherical Ti_2Ni precipitates in the grain interiors and enhancing grain boundary precipitation, as seen in Fig. 6.9(d) [17]. This structure is similar to that of bulk alloys. High-resolution transmission electron microscopy (Fig. 6.10) reveals that the GP zone consists of several atomic layers and lies on {100} planes with perfect coherency to the matrix [18]. On the other hand, the spherical Ti_2Ni precipitates have the same orientation as that of the B2 matrix, but the interface between them is semicoherent [15].The microstructure of the Ti-rich films is affected by the crystallization process as well as the precipitation process. The homogeneous distribution of GP zones seen in Fig. 6.9(a) suggests that they were generated from a supersaturated B2 matrix with the same composition as that of the amorphous phase. However, when an annealing temperature is low, B2 phase with a near-equiatomic composition nucleates from the amorphous matrix and grows, ejecting surplus Ti atoms into the surrounding matrix. This process results in localized distribution of GP zones, as shown in Fig. 6.9(e) [19–21]. Furthermore, Ti_2Ni particles without any specific orientation relationship are found in films with higher Ti content, as seen in Fig. 6.9(f) [22]. In this Ti-45.2at%Ni film, Ti_2Ni particles crystallize first, and then the B2 phase surrounding them crystallizes.

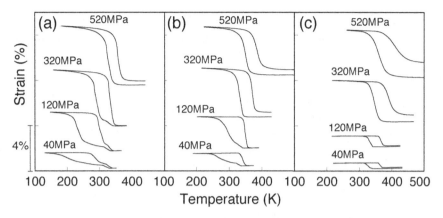

Figure 6.11 Strain–temperature curves under constant stresses of Ti-48.2at%Ni thin films annealed (a) at 773 K for 0.3 ks, (b) at 773 K for 36 ks, and (c) at 978 K for 360 ks. (After Ishida *et al.* [5] with kind permission of American Society of Mechanical Engineers.)

The diversity in microstructure, shown in Fig. 6.9, affects the shape memory behavior of Ti-rich TiNi films. Figures 6.11(a)–(c) presents the shape memory behaviors under various constant stresses of the annealed Ti-48.2at%Ni films; Figs. 6.11(a)–(c) corresponds to the films in Figs. 6.9(a), (c) and (d), respectively [23]. Figure 6.11(a) shows a two-stage elongation in the cooling curve and a two-stage contraction in the heating curve under constant stresses of 40 and 120 MPa. These two-stage shape changes are explained by the R-phase and martensitic transformations. In Fig. 6.11(b), a two-stage elongation is still detected in the cooling curve, but a two-stage contraction does not occur on heating. This change in the transformation behavior results from the rise in the martensitic transformation temperature. A noticeable change in shape memory behavior is seen in Fig. 6.11(c); the R-phase transformation disappears, resulting in a single-stage transformation attributable to the martensitic transformation. In this film Ti_2Ni precipitates are observed along the grain boundaries, like bulk alloys. The absence of the R-phase transformation in this film agrees with the transformation behavior reported in bulk Ti-rich TiNi alloys [12]. The R-phase transformations of Ti-rich TiNi thin films in Figs. 6.11(a) and (b) are, therefore, characteristic of thin films [17, 24]. In addition to the difference in transformation behavior, Fig. 6.11 shows another significant difference in plastic strain. The film in Fig. 6.11(a) exhibits a perfect shape memory effect under a constant stress of 320 MPa, whereas the film in Fig. 6.11(b) shows a plastic strain under the same constant stress. The GP zones produce coherency strain fields around them and, as a result, suppress plastic deformation [18]. On the other hand, Ti_2Ni precipitates have semicoherent interfaces and the strengthening effect is not so high as that of GP zones. Compared with these two films, the film in Fig. 6.11(c) shows a considerably

Table 6.1 Microstructures of Ti-rich TiNi thin films annealed at various temperatures for 3.6 ks

Annealing temperature	Ni content (at%)				
	45.2	46.1	47.0	47.9	48.5
873 K	(f)	(c)	(c)	(c)	(c)
823 K	(f)	(c)	(c)	(c)	(c)
773 K	(f)	(c)	(c)	(b)	(b)

(b), (c) and (f) represents the structure type (b), (c) and (f) shown in Fig. 6.9. (After Ishida *et al.* [28] with kind permission of The Japan Institute of Metals.)

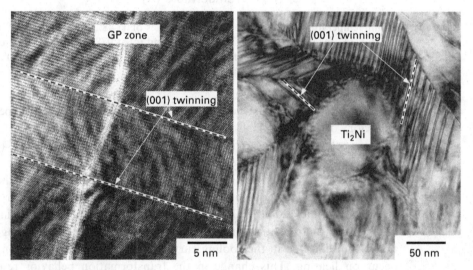

Figure 6.12 (a) GP zone (high-resolution electron micrograph) and (b) Ti_2Ni precipitate in martensite phase (electron beam $[1\bar{1}0]_{B19'}$). (After Zhang *et al.* [27] with kind permission of Elsevier.)

large plastic strain because of the preferential precipitation of Ti_2Ni along the grain boundaries.

It should also be noted in Fig. 6.11 that the film in Fig. 6.11(a) exhibits a large transformation strain when compared with the other films in Figs. 6.11(b) and (c) [18]. This is an advantage of GP zones. The large transformation strain of films containing GP zones can be understood by Fig. 6.12. This figure shows the effects of GP zones and Ti_2Ni precipitates on (001) twinning in the martensite, which is a dominant twinning mode in films with either a large amount of GP zones or Ti_2Ni precipitates [25, 26, 27], In Fig. 6.12(a), the (001) twin plate intersects the existing GP zone, suggesting that GP zones can be elastically deformed by the shear deformation [28]. In contrast, it is clear from Fig. 6.12(b) that the Ti_2Ni precipitates can not deform by the shear deformation associated with the (001) twinning, implying a small transformation strain. Figure 6.13 summarizes the

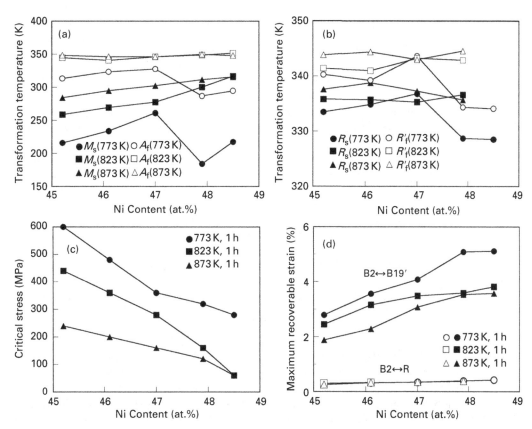

Figure 6.13 (a) Martensitic transformation start (M_s) and reverse martensitic transformation finish (A_f) temperatures, (b) R-phase transformation start (R_s) and reverse R-phase transformation finish (R'_f) temperatures, (c) critical stress for slip and (d) maximum recoverable strain of annealed Ti-rich TiNi thin films as a function of Ni content. (After Ishida *et al.* [28] with kind permission of The Japan Institute of Metals.)

shape memory behavior of Ti-rich TiNi thin films [29]. Films of Ti-45.2, 46.1, 47.0, 47.9 and 48.5at%Ni were annealed either at 773, 873 or 973 K for 1 h and their shape memory characteristics were measured in the same manner as described in Section 6.2. Table 6.1 shows the microstructures observed in the annealed films. (A similar structure map [13] is seen in Chapter 6, but there is a discrepancy in the boundary composition of the randomly oriented Ti_2Ni region between them, probably owing to different composition analysis methods.) According to Fig. 6.13, the dependence of the shape memory characteristics on annealing temperature and Ni content for films with Ti_2Ni precipitates alone is summarized as follows: (1) M_s temperature decreases with decreasing Ni content and annealing temperature; (2) R_s temperature is almost constant irrespective of Ni content, but slightly increases with increasing annealing temperature; (3) critical stress for slip increases with decreasing Ni content and annealing temperature; (4) maximum recoverable strain decreases with decreasing Ni content and increasing annealing temperature. On the other hand, films containing GP zones show characteristic behaviors such as low transformation temperatures, a high critical stress for slip and a large maximum recoverable strain.

The small temperature hysteresis of the R-phase transformation is of practical importance to increase the response speed of an actuator. Specifically the R-phase transformation of Ti-rich TiNi thin films is attractive, since their transformation temperatures are usually higher than those of Ni-rich Ti-Ni thin films.

6.4   Stability of shape memory behavior

High stability of shape memory behavior against cyclic deformation is important to apply SMA thin films to actuator applications. The stability depends on the nature of transformation [30], the composition and structure of a film [31], the level of an applied stress [32] and so on.

Thermal cycle tests were carried out for a Ti-43.9at%Ni film annealed at 973 K for 3.6 ks [30]. The shape memory characteristics of the R-phase transformation did not change after 100 thermal cycles, but those of the martensitic transformation significantly changed: under a constant stress of 250 MPa, the transformation temperature increased by 23 K, temperature hysteresis decreased from 51 to 33 K and recovery strain increased from 1.3 to 1.8%. As stress increased, the degree of these changes increased.

These changes in the martensitic transformation are caused by the internal stress field formed by the introduction of dislocations during the cyclic deformation. The high stability of the R-phase transformation is due to its small transformation strain. During the R-phase transformation, plastic deformation hardly occurs and, besides, the R-phase transformation involving a small strain is not so sensitive to stress. The changes of the shape memory characteristics were large in the initial stage, but became unchanged after a few tens of cycles owing to work hardening. This stabilization is called "a training effect".

The shape memory effect is also stabilized by precipitation hardening. A Ti-52 at%Ni thin film annealed at 773 K for 3.6 ks showed no plastic strain after 100 thermal cycles under 120 MPa and its shape memory characteristics were almost constant irrespective of the number of cycles [5].

6.5   Two-way shape memory effect

Ni-rich Ti-Ni thin films exhibits an aging effect due to the Ti_3Ni_4 precipitates, as described in Section 6.2. Therefore, the films are expected to show a two-way shape memory effect if they are aged under constraint. Figure 6.14 shows a

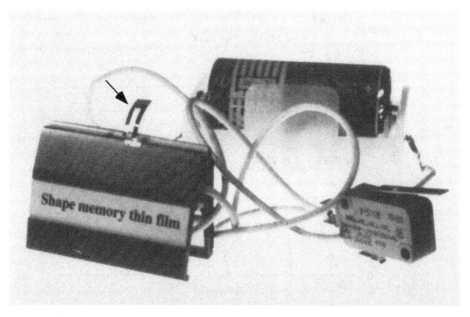

Figure 6.14   Microactuator made of Ti-51.3at%Ni thin film aged at 673 K for 360 ks under constraint. The arrow indicates the thin film. (After Sato *et al.* [32] with kind permission of Elsevier.)

simple actuator made of a Ti-51.3at%Ni thin film[33]. This film of 8.5 μm thickness was aged at 673 K for 360 ks after being wound on a stainless steel pipe (7.5 mm in outside diameter). Both the legs of the film indicated by the arrow are connected to the 1.5 V battery through leading wires. When the switch is on, electric current flows through the film and heats it. Then the film returns to the original shape. Since the original shape of the film has an opposite curvature to that of the film at room temperature, the film bends forward and backward reversibly by turning on and off the battery repeatedly, as first demonstrated by Kuribayashi [34]. The two-way shape memory effect has been reported for Ti-rich Ti–Ni films [35] as well as Ni-rich TiNi films [36].

6.6   Superelasticity

Superelasticity is another useful property of SMAs and has been also reported for thin films [37, 38, 39]. Figure 6.15 presents the stress–strain curves at various temperatures of a Ti-51.4at%Ni film aged at 773 K for 3.6 ks after solution treatment at 973 K for 3.6 ks [5]. All the films show yielding followed by a plateau region in the stress–strain curves. The strain in the plateau region completely recovers on unloading at deformation temperatures higher than the reverse martensitic transformation finish temperature, A_f ("superelasticity"), whereas, at

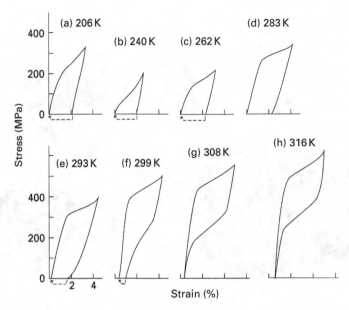

Figure 6.15   Stress–strain curves at various temperatures of Ti-51.4at%Ni thin film aged at 773 K for 3.6 ks after solution treatment at 973 K for 3.6 ks (Af: 306 K). (After Ishida *et al.* [5] with kind permission of American Society of Mechanical Engineers.)

deformation temperatures lower than A_f, the strain remains after unloading, but completely recovers after subsequent heating ("shape memory effect"). A perfect superelasticity is generally more difficult to achieve than a perfect shape memory effect[39], since the transformation is induced at a high stress level, as seen in the figure. Figure 6.15 demonstrates that this film possesses a perfect superelasticity as well as a perfect shape memory effect.

6.7   Stress–strain curves of TiNi thin films

Figure 6.16 presents the stress–strain curves of TiNi thin films. The Ti-50.0at%Ni film shows an elongation of more than 50%, whereas the Ti-51.5 at %Ni film containing a large amount of Ti_3Ni_4 shows a tensile fracture stress as high as 1.6 GPa [40]. These values are comparable to those in bulk alloys. In particular, the Ti-48.3at%Ni film annealed at a low temperature (773 K) shows an elongation of almost 20%. It is known that a bulk alloy with the same composition does not show any plastic deformation. This improvement comes from the suppression of coarse grain-boundary precipitates, which cause premature fracture in bulk alloys. Indeed, the film with a large amount of grain boundary precipitates, annealed at 873 K for 3.6 ks, shows no plastic deformation [41]. The stress–strain curve of shape memory alloys is very sensitive to a deformation temperature, as shown in Fig. 6.15. The detailed stress–strain

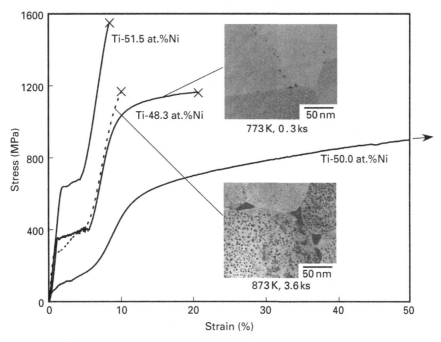

Figure 6.16   Stress–strain curves of Ti-48.3, 50.0 and 51.5at%Ni thin films (deformation temperature: 315 K for Ti-48.3 and 51.5 at%Ni; 345 K for Ti-50.0 at%Ni). The Ti-50.0 at%Ni film was annealed at 773 K for 3.6 ks, and the Ti-51.5at%Ni film aged at 673 K for 3.6 ks after solution treatment at 973 K for 3.6 ks. The Ti-48.3 at%Ni film was annealed either at 773 K for 0.3 ks or at 873 K for 3.6 ks. × means fracture. (After Ishida *et al.* [39, 40].)

curves have been reported in Ref. [40] and the related deformation mechanism was also discussed in Refs. [42, 43]. According to these, microscopic plastic deformation already starts in the linear stage after the plateau-region, in addition to the macroscopic plastic deformation starting at around 10%. This result should be taken account of in designing an actuator [44].

6.8   Thickness effect of shape memory behavior

The thicknesses of the films used for the measurements of the shape memory characteristics described in the previous sections were 7 μm or higher, but a thickness effect can not be ignored, if the thickness is less than the mean grain diameter. Figure 6.17 shows such a thickness effect on the shape-memory behavior of thin films [45]. As the thickness decreases from 5 μm, the transformation strain and plastic strain increase gradually and then decrease rapidly after reaching a maximum around 1–2 μm. These variations are explained in terms of two kinds of constraint from neighboring grains and surface oxide layers: with decreasing thickness, the former effect decreases, but the latter effect

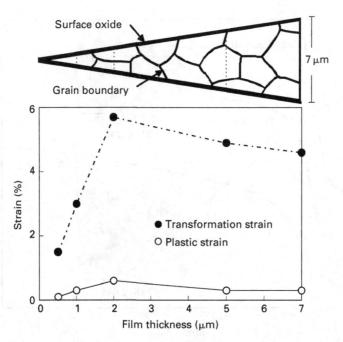

Figure 6.17   Effect of film thickness on transformation and plastic strains (at 200 MPa) of Ti-50.0at%Ni thin films. (After Ishida *et al.* [44] with kind permission of Elsevier.)

increases. In addition, surface oxidation lowers the transformation temperature owing to the consumption of Ti in the film when the thickness is less than 1 μm [45]. Though the mean grain size of the Ti-50 at %Ni film in Fig. 6.17 is about 5 μm, those of Ti-rich and Ni-rich films are generally less than 1 μm, being much smaller than that of bulk alloys, typically several tens of micrometers. Therefore, it is considered that the thickness effect can be ignored when the film thickness is more than 2 μm.

6.9   Summary

As shown in this chapter, TiNi thin films formed by sputtering possess a stable shape memory effect and reliable mechanical properties comparable or even superior to those of bulk alloys. Sputtering and successive heat-treatment provide unique microstructures which cannot be realized in cast alloys: (1) small grain size, (2) formation of unstable phases, (3) homogeneous distribution of fine precipitates and (4) suppression of grain boundary precipitates. These microstructures characteristic of thin films all contribute to the improvement of their shape memory behavior and mechanical properties. Since this advantage is applicable not only to TiNi binary alloy films but also to films containing more than two constituents, new SMAs with different structures and compositions from those of conventional

SMAs are expected and this diversity of sputter-deposited thin films seems to make fundamental studies and practical applications interesting [46].

References

[1] K. Otsuka and T. Kakeshita, Science and technology of shape-memory alloys: new developments, *MRS Bull.*, **27** (2002) 91–98.

[2] A. D. Johnson, Vacuum-deposited TiNi shape memory film: characterization and applications in microdevices, *J. Micromech. Microeng.*, **1** (1991) 34–41.

[3] A. Ishida, A. Takei and S. Miyazaki, Shape memory thin film of Ti-Ni formed by sputtering, *Thin Solid Films*, **228** (1993) 210–214.

[4] S. Miyazaki and A. Ishida, Martensitic transformation and shape memory behavior in sputter-deposited TiNi-base thin films, *Mater. Sci. Eng.*, **A273–275** (1999) 106–133.

[5] A. Ishida and S. Miyazaki, Microstructure and mechanical properties of sputter-deposited Ti-Ni alloy thin films, *J. Eng. Mater. Technol.-Trans. ASME*, **121** (1999) 2–8.

[6] A. Ishida, A. Takei, M. Sato and S. Miyazaki, Shape memory behavior of Ti-Ni thin films annealed at various temperatures, *Mater. Res. Soc. Proc.*, **360** (1995) 381–386.

[7] A. Ishida, Progress on sputter-deposited shape memory thin films, *Materia Japan*, **40** (2001) 44–51 [in Japanese].

[8] A. Ishida, M. Sato, A. Takei, K. Nomura and S. Miyazaki, Effect of aging on shape memory behavior of Ti-51.3at. pct Ni thin films, *Metall. Mater. Trans. A*, **27A** (1996) 3753–3759.

[9] S. Miyazaki and A. Ishida, Shape memory characteristics of sputter-deposited Ti-Ni thin films, *Mater. Trans. JIM*, **35** (1994) 14–19.

[10] T. Honma, TiNi-based shape memory alloys, in *Shape memory alloys*, ed. H. Funakubo, New York: Gordon and Breach Science Publishers (1987) p. 92.

[11] T. Saburi, Ti-Ni shape memory alloys, in *Shape Memory Materials*, ed. K. Otsuka and C. M. Wayman, Cambridge: Cambridge University Press (1998) p. 49.

[12] H. C. Lin, Shyi-Kaan Wu and J. C. Lin, A study of the martensitic transformation in Ti-rich TiNi alloys, *Proc. Int. Conf. on Martensitic Transformations (ICOMAT92)*, Monterey, 1992, 875–880.

[13] A. Ishida and V. Martynov, Sputter-deposited shape-memory alloy thin films: properties and applications, *MRS Bulletin*, **27** (2002), 111–114.

[14] Y. Kawamura, A. Gyobu, T. Saburi and M. Asai, Stucture of sputter-deposited Ti–rich Ti–Ni alloy films, *Mater. Sci. Forum* **327–328** (2000) 303–306.

[15] A. Ishida, K. Ogawa, M. Sato and S. Miyazaki, Microstructure of Ti-48.2at. pct Ni shape memory thin films, *Metall. Mater Trans. A*, **28A** (1997) 1985–1991.

[16] Y. Nakata, T. Tadaki, H. Sakamoto, A. Tanaka and K. Shimizu, Effect of heat treatments on morphology and transformation temperatures of sputtered Ti-Ni thin films, *J. Phys. IV*, 5 (1995) C8–671–6.

[17] A. Ishida, M. Sato, A. Takei and S. Miyazaki, Effect of heat treatment on shape memory behavior of Ti-rich Ti-Ni thin films, *Mater. Trans. JIM*, **36** (1995) 1349–1355.

[18] S. Kajiwara, T. Kikuchi, K. Ogawa, T. Matsunaga and S. Miyazaki, Strengthening of Ti-Ni shape-memory films by coherent subnanometric plate precipitates, *Philos. Mag. Lett.*, **74** (1996) 137–144.

[19] T. Kikuchi, K. Ogawa, S. Kajiwara, T. Matsunaga and S. Miyazaki, High-resolution electron microscopy studies on coherent plate precipitates and nanocrystals formed by low-temperature heat treatments of amorphous Ti-rich TiNi thin films, *Philos. Mag. A*, **78** (1998) 467–489.

[20] S. Kajiwara, K. Ogawa, T. Kikuchi, T. Matsunaga and S. Miyazaki, Formation of nanocrystals with an identical orientation in sputter-deposited Ti-Ni thin films, *Philos. Mag. Lett.* **74** (1996) 395–404.

[21] T. Matsunaga, S. Kajiwara, K. Ogawa, T. Kikuchi and S. Miyazaki, Effect of Ti content on nanometric substructure and shape memory property in sputter-deposited Ti-rich Ti-Ni thin films, *Mater. Sci. Forum*, **327–328** (2000) 175–178.

[22] Y. Kawamura, A. Gyobu, H. Horikawa and T. Saburi, Mortensitic transformations and shape memory effect in TiNi sputter-deposited thin films, *J. Phys. IV*, **5** (1995) C8–683–8.

[23] A. Ishida, M. Sato and S. Miyazaki, Microstructure of Ti-rich Ti-Ni thin films, *Proc. Int. Conf. on Shape Memory and Superelastic Technologies (SMST-97)*, Pacific Grove, 1997, 161–166.

[24] A. Gyobu, Y. Kawamura, H. Horikawa and T. Saburi, Martensitic transformations in sputter-deposited shape memory Ti-Ni films, *Mater. Trans., JIM*, **37** (1996)697–702.

[25] J. X. Zhang, M. Sato and A. Ishida, Influence of Guinier–Preston zones on deformation in Ti-rich Ti-Ni thin films, *Philos. Mag. Lett.*, **82** (2002) 257–264.

[26] J. X. Zhang, M. Sato and A. Ishida, Structure of martensite in sputter-deposited Ti–Ni thin films containing Guinier–Preston zones, *Acta Mater.* **49** (2001) 3001–3010.

[27] J. X. Zhang, M. Sato and A. Ishida, Structure of martensite in sputter-deposited Ti-Ni thin films containing homogeneously distributed Ti$_2$Ni precipitates, *Philos. Mag. A*, **82** (2002) 1433–1449.

[28] J. X. Zhang, M. Sato and A. Ishida, On the Ti$_2$Ni precipitates and Guinier–Preston zones in Ti-rich Ti-Ni thin films, *Acta Mater.*, **51** (2003) 3121–3130.

[29] A. Ishida, M. Sato, T. Kimura and T. Sawaguchi, Effects of composition and annealing on shape memory behavior of Ti-rich Ti-Ni thin films formed by sputtering, *Mater. Trans.*, **42** (2001), 1060–1067.

[30] K. Nomura and S. Miyazaki, The stability of shape memory characteristics against cyclic deformation in Ti-Ni sputter-deposited thin films, *Smart Materials, SPIE Proc. Ser.*, **2441** (1995) 149–155.

[31] K. Nomura, S. Miyazaki and A. Ishida, Cycling effect on the shape memory characteristics in sputter-deposited Ti-Ni alloy thin films, *Proc. 5th Int. Conf. on New Actuators(ACTUATOR-96)*, Bremen, 1996, 417–420.

[32] K. Nomura, S. Miyazaki and A. Ishida, Effect of plastic strain on shape memory characteristics in sputter-deposited Ti-Ni thin films, *J. Phys. IV*, **5** (1995) C8–695–700.

[33] M. Sato, A. Ishida and S. Miyazaki, Two-way shape memory effect of sputter-deposited thin films of Ti-51.3 at.%Ni, *Thin Solid Films*, **315** (1998) 305–309.

[34] K. Kuribayashi, Reversible SMA actuator for micron sized robot, *Proc. IEEE Micro Electro Mechanical Systems Workshop*, Napa Valley, 1990, 217–221.

[35] A. Gyobu, Y. Kawamura, T. Saburi and M. Asai, Two-way shape memory effect of sputter-deposited Ti-rich Ti-Ni alloy films, *Mater. Sci. Eng.*, **A312** (2001) 227–231.

[36] A. Gyobu, Y. Kawamura, H. Horikawa T. Saburi, Martensitic transformation and two-way shape memory effect of sputter-deposited Ni-rich Ti-Ni alloy films, *Mater. Sci. Eng.*, **A273–275** (1999)749–753.

[37] S. Miyazaki, K. Nomura and H. Zhirong, Shape memory effect and superelasticity developed in sputter-deposited Ti-Ni thin films, *Proc. Int. Conf. on Shape Memory and Superelastic Technologies(SMST-94)*, Pacific Grove, CA, 1994, p. 19.

[38] L. Hou and D. S. Grummon, *Progress on sputter-deposited thermotractive titanium-nickel films. Scripta Metall. Mater.*, **33** (1995) 989–995.

[39] A. Ishida, A. Takei, M. Sato and S. Miyazaki, Stress–strain curves of sputtered thin films of Ti-Ni, *Thin Solid Films*, **281–282** (1996) 337–339.

[40] A. Ishida, M. Sato, T. Kimura and S. Miyazaki, Stress–strain curves of sputter-deposited Ti-Ni thin films, *Philos. Mag. A*, **80** (2000) 967–980.

[41] A. Ishida, M. Sato and S. Miyazaki, Mechanical properties of Ti-Ni shape memory thin films formed by sputtering, *Mater. Sci. Eng.*, **A273–275** (1999) 754–757.

[42] J. X. Zhang, M. Sato and A. Ishida, The effect of two types of precipitates upon the structure of martensite in sputter-deposited Ti-Ni thin films, *Smart Mater. Struct.* **13** (2004) N37–42.

[43] J. X. Zhang, M. Sato and A. Ishida, Deformation mechanism of martensite in Ti-rich Ti-Ni shape memory alloy thin films, *Acta Mater.*, **54** (2006) 1185–1198.

[44] A. Ishida, M. Sato, W. Yoshikawa and O. Tabata, Graphical design for thin-film SMA microactuators, *Smart Mater. Struct.*, **16** (2007) 1672–1677.

[45] A. Ishida and M. Sato, Thickness effect of shape memory behavior of Ti-50.0 at. %Ni thin film, *Acta Mater.*, **51** (2003) 5571–5578.

[46] A. Ishida, Shape-memory alloy thin films, in *Comprehensive Microsystems*, eds. Y. B. Gianchandani, O. Tabata and H. Zappe, Oxford: Elsevier (2007) p. 61.

7 Stress and surface morphology evolution

Y. Q. Fu, W. M. Huang, M. Cai and S. Zhang

Abstract

The residual stress in a film is an important topic of extensive studies as it may cause film cracking and peeling off, or result in deformation of the MEMS structure, deterioration of the shape memory and superelasticity effects. In this chapter, the stress in thin film shape memory alloys is characterized. The recovery stress, stress rate and stress–strain relationship during phase transformation are introduced. The dominant factors affecting stress evolution during deposition, post-annealing and phase transformation are discussed. The mechanisms for the stress-induced surface relief, wrinkling and reversible trench have also been studied. New methods for characterization of stress induced surface morphology changes have been introduced, including atomic force microscopy (AFM) and photoemission electron microscopy (PEEM).

7.1 Introduction

One of the major obstacles to the wide application of TiNi based shape memory alloy (SMA) thin films is stress as a result of processing. The stress generated and left in the film is the so-called residual stress. Residual stresses may cause creep, cracking and even peeling-off of films (with one example shown in Fig. 7.1), and also result in deformation of the MEMS structure and deterioration of the mechanics and thermodynamics of displacive transformation [1, 2, 3].

In general, residual stresses are composed of [4] intrinsic stresses and extrinsic stresses. The intrinsic stresses, called growth stresses, are strongly dependent on the materials involved and growth conditions (such as temperature, pressure, growth flux, etc.). The intrinsic stresses in a thin film come from [5]:

- surface/interface stress;
- cluster coalescence and grain growth;

Thin Film Shape Memory Alloys: Fundamentals and Device Applications, eds. Shuichi Miyazaki, Yong Qing Fu and Wei Min Huang. Published by Cambridge University Press. © Cambridge University Press 2009.

Figure 7.1   Fracture and cracking of TiNi film. Detailed observation reveals that the cracks initiate from the film surface, propagate into the Si substrate near to the TiNi film, instead of at the interface between the TiNi film and Si substrate. The fracture morphology shows wavy features, and cracks propagate through the (111) crystalline plane, along the [110] directions of the Si (100) wafer.

- defects of generation and annihilation such as impurities, voids, vacancies;
- precipitation formation;
- structural damage due to sputtering or other energetic particle impingement.

Extrinsic stresses include stress changes due to variation in the physical or chemical environment of film materials after deposition. In shape memory alloy films, the extrinsic stresses mainly include the following origins:

- thermal stress due to post-annealing (or temperature changes);
- martensitic and reverse phase transformation;
- plastic or creep formation at high temperature;
- electrostatic or magnetic field;
- chemical reactions.

The growth of a thin film can be simply described in the following sequence [4]: (1) initial nucleation of atomic islands on the substrate; (2) island growth; (3) contact and coalescence of the islands; (4) continuous growth and formation of grain boundaries; (6) grain coarsening and formation of defects. The role of stress in these stages of microstructure evolution is not yet fully understood. In the initial nucleation of islands, compressive stress generates and rises because the newly formed crystallites try to firmly attach to the substrate and the total surface area increases [6]. Upon further growth, the internal elastic strain in the crystallites tends to relax due to increase in size/volume. When the crystals impinge on each other or the islands coalesce, tensile stress is generated [7]. In further growth, the tensile stress component decreases, due to the excess number of atoms comprising the film or incorporation of excess atoms into grain boundaries [8]. Grain growth and grain boundary reduction during further growth cause the formation of tensile stress, and the stress decreases as the film thickness increases [9].

The sputtered films normally have large compressive stress, due to the high kinetic energy of atoms arriving at the growth surface, bombardment of the growth surface by the energetic atoms and the so-called atomic peening effect [10]. The generation of point defects or interstitials is another contribution to compressive stress in the film. Both the atomic weight of the deposited species and the chamber pressure have significant effects on the film stress because they will influence the kinetic energy of the atoms arriving on the film surface. Low pressure or a low deposition rate leads to the generation of large compressive stress. Increase of the gas pressure could result in a higher chance of target atoms colliding with gas molecules, thus lowering the kinetic energy of atoms arriving on the film surface; the film compressive stress could be lower, even changing into tensile stress [10]. In the next section, the methodology for measuring thin film stresses will be discussed.

7.2   Film stress: measurement and characterization

There are many different techniques for measuring the film stress, for example curvature measurement, X-ray diffraction and Raman spectroscopy.

7.2.1   Curvature method

One popular method for stress measurement is the wafer curvature technique. This is a simple, inexpensive and non-destructive method that obtains stress from the changes in curvature of the wafer (used as the substrate). According to Stoney's study [11], stress in the film is related to the amount of bending in the substrate, which is typically described using the following equation:

$$\kappa = \frac{1}{R} = \frac{6\sigma_f h}{\hat{E} H^2},$$

(7.1)

in which κ is curvature and R is the radius of curvature. H is the beam thickness, and h is the thin film thickness ($h \ll H$). σ_f is the film stress in the as-deposited state. $\hat{E} = E_b/(1-v)$ and E_b is the beam modulus. v is the Poisson ratio.

The curvature method is applicable to both crystalline and amorphous films. Methods to measure the changes in substrate (normally silicon wafer) curvature during stress evolution can be classified into the following groups: mechanical method, capacitance method and optical method [4]. All these methods provide information on the out-of-plane deflection of the curved film/substrate system. The most common mechanical method of estimating curvature is using a stylus (from a profilometer) to scan the film surface and record the position. The residual stress σ_f is then calculated from the radius of curvature using:

$$\sigma_f = \frac{E_{Si} H^2}{(1 - v)6h} \left(\frac{1}{R_1} - \frac{1}{R_2} \right),$$

(7.2)

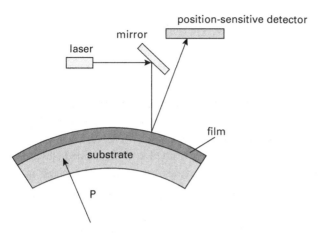

Figure 7.2   Experimental set-up for the curvature measurement based on laser scanning.

where $E_{Si}/(1-\nu)$ is the biaxial elastic modulus of the Si substrate. Since the original substrate is not generally flat, it is essential to measure the substrate radii of curvature R_1 and R_2 before and after film deposition to use in the calculation of film stress. The problem with this method is that the force applied on the stylus could damage the film surface or slightly change the wafer position during measurement.

The capacitance method involves a non-contact probe that records the changes in the capacitance between a reference point and the surface it scans [4]. The variation of capacitance changes is then converted to out-of-plane displacement.

Both the above two methods are not suitable for measuring curvature change at an elevated temperature or during temperature fluctuation. Also, they are not suitable for use in a closed chamber, for example, a film deposition chamber or furnace. This can be overcome by optical methods. The most common optical method is using a laser beam to scan on a substrate surface, as schematically shown in Fig. 7.2 [4]. The angular deflection of the reflected beam from the incident is measured as a function of distance from the starting point using a positive-sensitive detector. However, there are still some limitations for the optical measurement method [4]. For example, the measured stress is an average value of local stress because it assumes that stress is constant across the whole wafer. Since it employs monochromatic beams, it has a problem in measuring transparent films with thickness of one-fourth of the wavelength [4]. This method also requires good reflective surfaces, as for curvature measurement, and it is inaccurate for a very rough surface.

7.2.2   X-ray diffraction

X-ray diffraction (XRD) is a viable method to measure the stress in thin crystalline films on substrates [12]. The normal spacing d_{hkl} between the adjacent crystallographic planes within a family of planes of indices (*hkl*) is determined by Bragg's law:

$$d_{hkl} = \frac{n\lambda}{2\sin\theta},$$ (7.3)

where λ is the wavelength of the X-ray incident on the film, n is an integer determined by the order and θ is the angle of incidence. The strain in the film alters the interplanar spacing from its unstressed state, d_0. The difference in the d-spacing of crystallographic planes between the strained and unstrained states of a thin film, as determined by X-ray diffraction, provides a measure of the stress in the film. The stress in the film is calculated using the equation

$$\sigma_f = -\frac{E_f}{2\upsilon_f}\left(\frac{d_{hkl} - d_0}{d_0}\right).$$ (7.4)

However, the existence of mismatch strain distribution in the film causes X-ray peak broadening and a more diffuse intensity of peaks, which causes a problem in the precise calculation of the film stress. An improved method is to use grazing incidence X-ray scattering [4]. The incident X-ray beams impinge on the film surface at a very low angle of incidence from 0.2 to 10 degrees. The low angle of incidence results in reflection of the X-rays out of the film rather than penetrating through the film and into the substrate. By varying the grazing angle, strains over different depths can be obtained by the XRD method [4]. The use of the X-ray diffraction method is restricted to crystalline films. The accuracy of the standard X-ray diffraction units may not be adequate for the level of precision required in many applications, for example, in patterned films. Cost and safety issues are the other concerns.

7.2.3   Micro-Raman spectroscopy

Micro-Raman spectroscopy has evolved as a possible experimental method for estimation of local stresses in non-planar, submicron scale MEMS structures [13, 14]. A laser beam, typically an argon–ion laser, is focused on a local region. The signal of the excited Raman fluorescence is gathered by the objective lens and then passed through a monochromator for frequency analysis. The shift in the Raman spectroscopy peak can be used to calculate the film stress [13]. A big problem of this method lies in its inability to be applied to metallic materials due to the limitation of the Raman spectroscopy. Therefore, it is inapplicable for crystalline shape memory TiNi thin films.

7.3   Stress and strain evolution in TiNi based films

7.3.1   Stress evolution in shape memory events

Using the curvature method, stress change as a function of temperature can be measured *in situ* with change in temperature. The martensitic transformation

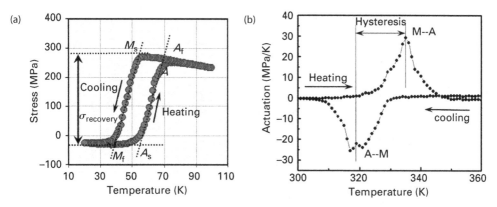

Figure 7.3 The stress vs. temperature curve (a), and stress increase rate vs. temperature (b) curve for a TiNiCu$_4$ film showing the sequence of martensitic transformation.

temperatures and hysteresis, multi-stage transformation, and magnitude of shape recovery can be easily obtained from the stress–temperature curves [15, 16, 17]. Figure 7.3(a) shows a typical curve of the measured stress of a Ti50Ni46Cu4 film as a function of temperature up to 100 °C. The stress vs. temperature plot forms a closed hysteresis loop shape. At room temperature, the stress is compressive with a low value. During heating, the tensile stress increases significantly due to the phase transformation from martensite to austenite. Above the austenite transition start temperature (A_s), the stress increases linearly until the temperature reaches the austenite transition finish temperature (A_f). With a further increase of temperature, the transformation completes and thermal stress is generated with the stress values decreasing linearly due to the difference in coefficient of thermal expansion (CTE) between the TiNiCu film ($a_{TiNiCu}=15.4\times10^{-6}/°C$) and Si substrate ($a_{Si}=3\times10^{-6}/°C$). The theoretical slope of stress vs. temperature due to pure thermal effect can be calculated using the following equation:

$$d\sigma/dT = [E_{TiNiCu}/(1 - \upsilon_{TiNiCu})](a_{Si} - a_{TiNiCu}) \qquad (7.5)$$

in which E_{TiNiCu} is the Young's modulus of TiNiCu film (about 78 GPa according to nano-indentation results); υ_{TiNiCu} is the Poisson ratio of TiNiCu film (about 0.33). The calculated data of $d\sigma/dT$ is about –1.44 MPa/°C, which matches well with the experimental data of –1.38 MPa/°C.

During cooling, tensile thermal stress develops in the TiNiCu films at a rate of –1.38 MPa/K. When the temperature is just above the martensitic transition start temperature (M_s), the residual stress reaches its maximum value. Cooling below M_s, the martensitic transformation occurs and the tensile stress decreases significantly due to the formation and alignment of twins (shear-induced phase transformation) [18]. The recovery stress, σ_{rec}, is given by the difference between the stress at A_f and that at M_f (see Fig. 7.3(a)). For example, from Fig. 7.3(a), the recovery stress value is about 330.7 MPa. Recovery stress reflects the significance

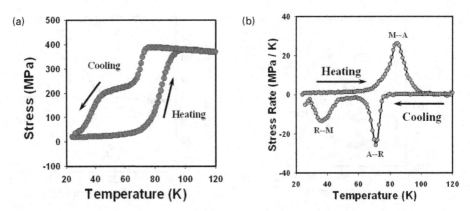

Figure 7.4   The stress vs. temperature curve (a) and stress increase rate vs. temperature curve (b) for a Ti50Ni50 film showing the sequence of martensitic transformation.

of the shape memory effects. Transformation temperatures can also be obtained, as illustrated in Fig. 7.3(a).

Figure 7.3(b) shows the stress rate (i.e., the differentiated stress vs. temperature curve derived from Fig. 7.3(a)) during thermal cycling. This is quite similar to those curves from a differential scanning calorimeter (DSC). The stress rate corresponds to the generation rate for actuation (or actuation speed). A higher stress rate during phase transformation is promising for actuation application due to its fast response. From the stress rate vs. temperature curve, the transformation hysteresis can also be obtained, as shown in Fig. 7.3(b).

A typical stress evolution vs. temperature curve of a Ti50Ni50 film on Si is shown in Fig. 7.4(a). It shows a one-stage transformation corresponding to martensite (B19′) to austenite (B2) transformation during heating, and a two-stage transformation during cooling corresponding to transformations among martensite, R and austenite phases. The occurrence of R-phase transformation during cooling can be clearly observed in the stress increase rate vs. temperature curve in Fig. 7.4(b). From this curve, the transformation between R phase and austenite has a much larger stress rate (or higher actuation speed) than transformation between R phase and martensite.

7.3.2   Factors affecting stress evolution

Stress in sputter-deposited TiNi based films is governed by the competition among [3, 17]:

- compressive stress induced by atomic peening;
- tensile stress caused by the coalescence process during film formation;
- relaxation of stress caused by the microstructure evolution and defect formation during film growth and post-annealing.

The stress in the as-sputtered TiNi films is normally compressive stress. The high kinetic energies of atoms or ions bombarding the growth surface lead to the

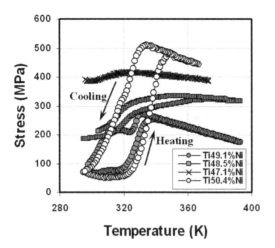

Figure 7.5 Stress evolution for TiNi films with different Ti/Ni ratios deposited at a temperature of 450 °C [17]. (Reproduced with permission from Elsevier, UK.)

generation of large compressive stress, which is dependent on (1) the energy of the arriving atoms (plasma power); (2) the background pressure of the inert gas. The stress generation and relaxation behaviors are significantly affected by film composition, deposition and/or annealing temperatures, which strongly control the intrinsic stress, thermal stress and phase transformation behaviors. The following sections summarize the influence of the process parameters on the film stress evolution.

7.3.2.1 Ti/Ni ratio

Figure 7.5 shows the stress evolution curves for the post-annealed (at 600 °C) films with different Ti/Ni ratios [17]. For the film 47.8% Ti, there is only a small hump in the stress evolution curve, indicating that the martensite transformation is not significant. One possible reason for the high residual stress in the film and weak phase transformation could be related to high Ni content in films and/or R-phase transformation. With the increase of Ti content to 48.5%, both the transformation temperature and recovery stress increase, and the stress evolution curve shows a partially closed loop. Heating and cooling curves are not close at room temperature, indicating that phase transformation does not finish when cooling down to room temperature. Large stress is only partially released, and the residual film stress is quite large. For the film with a Ti content of 49.1%, an apparent two-step transformation during both cooling and heating can be observed, as shown in Fig. 7.5, corresponding to martensite and R-phase transformations. For the film with a Ti content of 50.2%, the stress vs. temperature plot shows a closed hysteresis loop, i.e., perfect shape memory effect. At room temperature, the residual stress is quite low due to the significant relaxation of stress by the martensitic transformation.

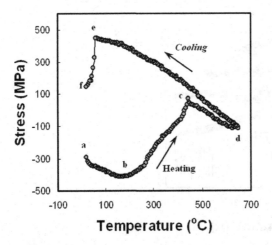

Figure 7.6 Stress evolution of TiNiCu film annealed up to 650 °C using the curvature measurement method [19]. (Reproduced with permission from Elsevier, UK.)

7.3.2.2   Post annealing

During annealing, stress evolution is governed by the competition among [3, 17]:

- residual stress in as-deposited films, which is normally large compressive;
- thermal stress changes (compressive stress generation during heating, while tensile during cooling);
- tensile stress due to densification of structure and film crystallization;
- relaxation of stress due to martensitic transformation;
- internal stress changes due to reduction of defects, or formation of precipitates, which is rather difficult to evaluate.

The stress evolution of an amorphous TiNiCu film during annealing up to 650 °C is shown in Fig. 7.6 [19]. In the beginning, below 150 °C (from a to b), a net compressive stress increases linearly, indicating that thermal stress is at play: compressive stress results because the film expands more than four times that of the substrate (the coefficient of thermal expansion (CTE) of TiNiCu film, a_f, is about 15.4×10^{-6}/K and that of the Si substrate, a_{Si}, is 3×10^{-6}/K). However, above 120 °C (i.e., after b), the tensile component prevails such that the net stress becomes less compressive as a result of densification with increasing temperature. With further heating to about 435 °C (point c), crystallization of TiNiCu film occurs and densification ends [19]. After c, heating generated stress (compressive) prevails, resulting in an almost linear increase in net stress. From d to e, cooling-induced thermal stress (now tensile) increases linearly with decreasing temperature. At e, martensitic transformation starts (M_s). With further decrease in temperature, significant decrease in stress occurs (from e to f). Figure 7.7 clearly shows that an appropriate annealing temperature is needed to promote film crystallization, thus the phase transformation can occur above

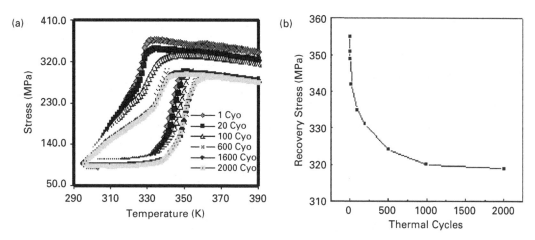

Figure 7.7   (a) Hysteresis evolution of Ti52.5Ni film on an Si substrate after thermal cycling in different cycles and becoming stable after 2000 cycles; (b) effect of thermal cycles on the changes of recovery stress [3]. (Reproduced with permission from Elsevier, UK.)

room temperature and the large thermal stress generated during cooling can be released significantly.

7.3.2.3   Thermal cycling

Stability and fatigue have always been concerns in TiNi thin film applications. Fatigue of TiNi films is related to the non-durability and deterioration of the shape memory effect after millions of cycles. The repeated phase changes will alter the microstructure and hysteresis of the transformation and in turn will lead to changes in transformation temperatures, recovery stresses and strains. The performance degradation and fatigue of thin films are influenced by a complex combination of internal (alloy composition, lattice structure, precipitation, defects, film/substrate interface) and external parameters (thermomechanical treatment, applied maximum stress, stress and strain rate, the amplitude of temperature change, cycling frequency) after long term thermal–mechanical cycles [3]. For freestanding films, there are some studies using tensile tests to characterize the fatigue problems, and results indicated that there need to be tens of cycles before the stability of shape memory is effected [20]. Fu *et al.*, [3] studied the fatigue of constrained TiNi films using the curvature method by investigating changes of recovery stress during thermal cycling. Results show that the recovery stress of TiNi films from curvature measurement decreases dramatically in the first tens of cycles, and becomes stable after thousands of cycles (with one example shown in Fig. 7.7) [3]. This reduction of the recovery stress is believed to result from the dislocation movement, grain boundary sliding, void formation, or partial de-bonding at the film/substrate interfaces, non-recoverable plastic deformation, changes in stress, etc. Transformation temperatures also change dramatically during cycling. The repeated phase changes will alter the microstructure and hysteresis of the transformation and in turn lead to changes in transformation temperatures, recovery stresses and strains.

7.3.2.4   Film thickness

Since the as-deposited TiNi films need to be annealed for crystallization to show a shape memory effect, large stress evolution in the annealing process could significantly affect phase transformation behavior and shape memory effects. Ti 50.2Ni 49.8 films with different film thickness were prepared by co-sputtering of a Ti55Ni45 (at.%) target (RF, 400 W) and a pure Ti target (70 W, DC), and then post annealed at 650 °C. Figures 7.8(a) and (b) show the measured stress vs. temperature curves, residual stress and recovery stress of the as-annealed TiNi films [21]. Residual stress decreases and recovery stress increases drastically as film thickness increases to the range of a few hundred nanometers. As we know, residual stress in SMA films is related to the film intrinsic stress, thermal stress and phase transformation stress. At the same deposition and annealing temperatures, thermal stresses of all the films would be similar. The estimated thermal stress is very large (about 900 MPa) as a result of cooling from the annealing temperature of 650 °C to room temperature of 20 °C. For a film with a thickness of tens of nanometers, due to the formation of surface oxide and an oxygen diffusion layer as well as a thick interfacial diffusion layer, the Ti/Ni stoichiometry would be significantly altered, thus the shape memory effect of the films deteriorates (i.e., there is a lower recovery stress or even loss of the shape memory effect) [21]. Therefore, large intrinsic and thermal stress generated after high temperature post-annealing could not be significantly released due to insignificant phase transformation, resulting in the large residual stress in the film.

From results shown in Fig. 7.8, a minimum thickness (about 100 nm) is necessary to guarantee an apparent shape memory effect in the TiNi films. Surface oxide and oxygen diffusion layers as well as an interfacial diffusion layer are dominant in films with thickness of tens of nanometers [21]. For films with a thickness of 50 nm, the combined constraining effects from both surface oxide and interfacial diffusion layers will be detrimental to the phase transformations among austenite, R phase and martensite, giving rise to the degraded phase transformation and shape memory performance. As the film thickness increases above a few hundred nanometers, the effects of the surface oxide, oxygen diffusion layer and inter-diffusion layer become relatively insignificant. Therefore, phase transformation becomes significant and the recovery stress increases as thickness increases. Due to the significant phase transformation effect, thermal and intrinsic stresses in the films are drastically relieved, resulting in significant decreases in residual stress. The recovery stress peaks at a certain film thickness (for example, 820 nm for recovery stress in Fig. 7.8(a)), and then decreases slightly with further increase in film thickness. The decrease in recovery stress results from the constraining effects from the neighboring grains, as illustrated in Fig. 7.8(a). With the increase in film thickness, more and more grain boundaries form in the films. The grain boundaries are the weak points for generation of large distortion and twinning processes. Therefore, as the

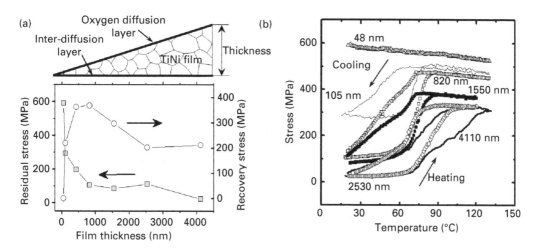

Figure 7.8 (a) Residual stress and recovery stress for films with different thickness; (b) stress–temperature evolution curves for the TiNi films with different thickness [21]. (Reproduced with permission from Elsevier, UK.)

film thickness increases, the constraining effect from the neighboring grains becomes more and more significant, causing a decrease in recovery stress [22].

In brief, to minimize the residual stress, while maintaining a large recovery stress during martensitic transformation for actuation purposes, it is necessary to:

- deposit film with thickness above a few hundred nanometers;
- precisely control the Ti/Ni ratio; deposit films at a possible lower pressure;
- select a suitable deposition temperature or annealing temperature, with a compromise between thermal stress and intrinsic stress;
- use some interlayers (with possible compressive stress) to reduce large tensile stress in some TiNi films.

7.4 Stress induced surface morphology changes

7.4.1 Transition between surface relief and wrinkling

Films of a nominal composition of Ti50Ni47Cu3 with a thickness of 3.5 microns were sputter-deposited on a 4-inch (100) silicon wafer 500 microns thick (without intentional substrate heating) by magnetron sputtering of a Ti(55at%)Ni(45at%) target (3-inch diameter, RF, 400 W) and a pure Cu target (3-inch diameter, DC, 2 W). The films were then annealed in vacuum (2×10^{-7} torr) at a temperature of 450 °C for one hour for crystallization. As-deposited amorphous films are smooth and reflective to the naked eye. Annealing at 450 °C for one hour significantly changes the surface morphology. At the edge of the annealed 4-inch wafer, the film surface remains reflective and shiny. However, analysis using optical microscopy and scanning electron microscopy (SEM) showed that some circular crystals with size ranging from a few microns to 30 microns diameter are embedded inside a featureless matrix (see Figs. 7.9 (a) and (b)) [23]. Further XRD analysis confirmed that these circular patterns are single crystals and the surrounding matrix is amorphous.

The film surface becomes opaque and cloudy to the naked eye in regions closer to the wafer center. The number and average size of the martensite crystals increase dramatically, with some crystals of relatively small size also observed, as shown in Figs. 7.9(c) to (e) [23]. When the two adjacent clusters contact each other, some martensite plates with similar orientation appear to coalesce into one large plate, so that there is no clear boundary between these adjacent clusters (see Figs. 7.9(c) to (e)). The morphology of the edge of a martensite crystal is quantified by AFM, with an image shown in Fig. 7.9(c). The undulating and rough edge is due to the formation of martensitic striations. In the center area of the film, the entire film surface is covered by significant relief morphology (interweaving martensite plates), as shown in Fig. 7.9(e). XRD analysis in these areas confirms the fully crystalline and martensite-dominant structure.

Surface relief morphology changes in different regions are due to the variation in film composition. Since the deposition method is based on co-sputtering from two targets, slight misalignment of the two-targets with respect to the center of the wafers could affect the film composition uniformity. This is the intrinsic problem for the co-sputtering deposition method. EDX results confirmed that along the radial direction of the substrate holder, Ti contents can vary by about 2 at% over the whole Si wafer. The crystallization temperature varies according to the Ti/Ni ratio in the film, for example, crystal nucleation and growth rate is much slower and crystal size is much smaller in Ti-rich film than in a stoichoimetric Ti50Ni50 film [23]. The annealing temperature used in this study was 450 °C, which is about the crystallization temperature of the TiNiCu films. Due to the composition effect, annealing at this temperature, the edge of the film (Ti rich) was only partially crystallized, whereas the center of the film became fully crystallized. Depending on the residual stress and the nuclei distribution, some grains may nucleate and then grow significantly before the other grains nucleate. During subsequent cooling after annealing, martensite only forms inside these nucleated grains, resulting in the structures shown in Fig. 7.10.

In situ optical microscopy observation with a substrate heated up to 100 °C revealed that the interweaving martensite plate structure (significant ridge and valley patterns) gradually disappeared (see Figs. 7.10(a) to (d)) [23]. The film surface became relatively smooth. With further increasing temperature, only some large martensite plates can be observed, and radial wrinkling patterns appear within large circular patterns (see Figs. 7.10(c) and (d)). Profilometer results showed that above 65 °C, the surface becomes wavy again and surface roughness also increases, which is mainly due to the formation of wrinkling patterns. Further heating up to 300 °C did not lead to much change in these wrinkling patterns. Upon subsequent cooling to room temperature, the twinned martensite plates or bands reformed as before thermal cycling. Many small round islands of less than 10 to 20 microns did not

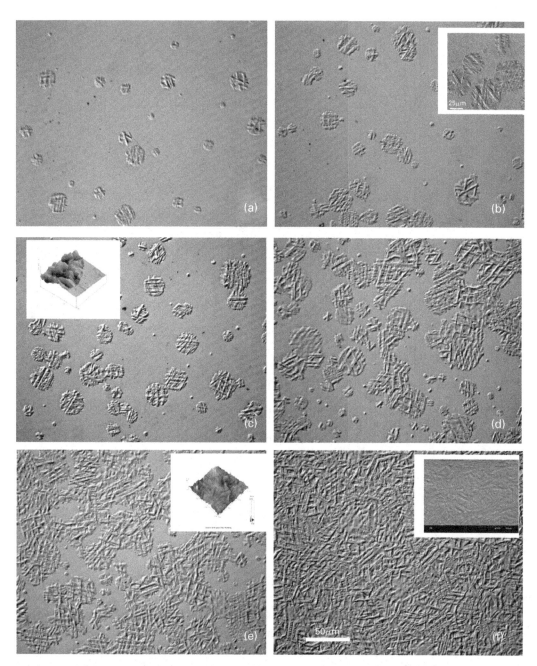

Figure 7.9   Optical microscopy images showing the changes of surface morphology; (a) and (b) crystals with surface relief morphology; (c) and (d) martensite crystal clusters and their coalescence; (e) and (f) near to the center of 4-inch Si wafer, dense martensitic relief morphology [23]. (Reproduced with permission from *Scientific American*, USA.)

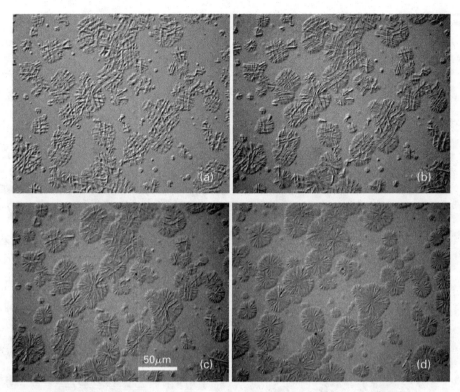

Figure 7.10   Surface morphology evolution with temperature for TiNiCu film (a) surface relief
morphology (twinning structure) at 25 °C; (b) 55 °C and (c) 75 °C; and (d) surface
wrinkles at 100 °C [23]. (Reproduced with permission from *Scientific American*, USA.)

show any wrinkling patterns at a high temperature, which may indicate that the
formation of wrinkling patterns is size dependent, and smaller islands can relax by
in-plane expansion without wrinkling.

7.4.2   Reversible trenches

$Ti_{50}Ni_{47}Cu_3$ shape-memory thin films of about 4 μm thick were co-sputter
deposited on standard 4-inch (100) 450 μm thick silicon wafers at a nominal
temperature of 450 °C. The deposited film shows a nanocrystalline structure with a
certain amount of amorphous structure, confirmed by the X-ray diffraction test.
Upon thermal cycling, no apparent change in surface morphology, but only a
slight variation in surface roughness, can be found in the film. The film was left in
air for a few days before annealing in a vacuum chamber at 650 °C for one hour.
Observation using a temperature-controlled AFM reveals that at room temperature, there are many trenches on the film surface, with a typical one shown in
Fig. 7.11(a). However, upon heating to a high temperature, e.g., 100 °C in this
particular thin film, all trenches disappear (Fig. 7.11(b)) [24]. When the sample is
cooled back to room temperature, almost identical trenches re-occupy the surface

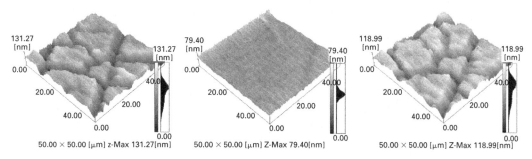

Figure 7.11 Reversible surface trench morphology upon thermal cycling [24]. (Reproduced with permission from Institute Of Physics, UK.)

as shown in Fig. 7.11(c). This phenomenon is repeatable after many thermal cycles. These trenches are neither grain boundaries nor the surface relief induced by the martensitic transformation, as evidenced by the surface electron microscopy (SEM) observation. A cross-sectional view of a typical trench upon heating suggests the occurrence of a kind of elastic buckling at room temperature, which is seemingly similar to the thermally induced ratcheting that requires: (1) an elastic thin layer, which is under compression; (2) a relatively soft elastic–plastic metallic layer underneath the elastic thin layer [25].

Similar to the distribution of the morphology of surface wrinkling, surface trenches also change significantly in different regions on a four-inch Si wafer. This is also attributed to the variation in film composition. The trench distribution over a four-inch Si wafer is similar to those of cracking morphology on a wafer (Fig. 7.1(a)). The position where the large-area cracking initiates is the exact position where the largest networking trenches appear. This clearly indicates that stress is the dominant factor for the formation of these surface trenches.

7.4.3 Theoretical analysis of stress-induced wrinkling and trenches

As the shape-memory thin film is much thinner than the substrate, the stress and strain in the thin film (σ_f and ε_f, respectively) are almost uniform and can be obtained from the measured curvature (ρ) versus temperature variation (ΔT) relationship. The annealing temperature will be taken as a reference point as this is the equilibrium state. In the analysis of the formation of surface wrinkling and trenches, the contribution of the oxide layer will be considered. The total strain in the oxide layer (ε_0) can be expressed as (hereinafter, subscript o stands for oxide layer)

$$\varepsilon_o = \varepsilon_f + \varepsilon_{th} + \varepsilon_e, \tag{7.6}$$

where ε_{th}, thermal mismatch strain, $\varepsilon_{th} = (a_0 - a_f)\Delta T$, is due to the difference in the coefficient of thermal expansion between the oxide layer and SMA thin film, and

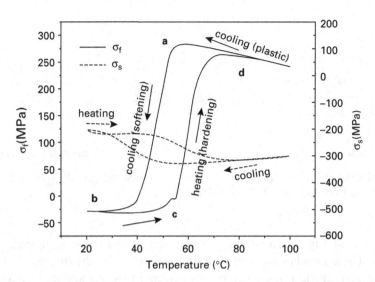

Figure 7.12   Evolution of stresses in SMA thin film and TiO$_2$ upon thermal cycling.

ε_e ($= \sigma_0/\bar{E}_0$) is the elastic strain in the oxide layer. The TiNi film strain can be estimated using Eq. (7.7) [26]:

$$\varepsilon_f = -\frac{1}{6}K\left[4t_s + t_f\left(5 + \frac{t_f}{t_f + t_s}\right)\right] + a_s\Delta T. \tag{7.7}$$

Taking the annealing temperature as the initial equilibrium state, using the measured radius of curvature vs. T data to calculate the strain in TiO$_2$ and the DSC result of SMA thin film for the transformation progress, we can estimate the stress evolution in the SMA thin film and TiO$_2$, respectively, upon thermal cycling. The results are plotted in Figure 7.12. From Fig. 7.12, it can be confirmed that the top TiO$_2$ layer is always under compression (between -200 and -300 MPa within the temperature range in thermal cycling), although the Ti-NiCu film is under tensile stress. In brief, the elastic buckling of the oxide layer (TiO$_2$) on the top surface of shape-memory thin films is the origin of the reversible wrinkles and trenches in TiNi based thin films. The existence of a TiO$_2$ oxide layer (naturally or purposely involved), which is under compression, is proved to be another condition for these phenomena to appear.

The mechanisms behind wrinkling formation in thin films have been extensively studied in recent years [25, 27]. Both experimental evidence and numerical simulation confirmed that large in-plane compressive stress in a thin elastic film on a soft, plastic or viscous substrate can lead to spontaneous circular wrinkled patterns in order to release the large compressive stress [25,27]. The film stress evolution as a function of annealing temperature up to 650 °C has been shown in Fig. 7.6. During heating, the net thermal compressive stress increases significantly up to a maximum value of 410 MPa. During cooling, tensile thermal stress increases linearly with decreasing temperature until the martensitic

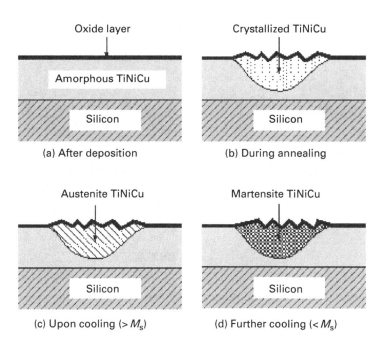

Figure 7.13 Schematic drawing showing the formation of wrinkling patterns and transition to surface relief morphology [28] (Reproduced with permission from American Institute of Physics, USA.)

transformation occurs, at which point the tensile stress decreases sharply due to the formation of twinned martensite. Wrinkling may occur during nucleation of crystals within the amorphous matrix in high-temperature annealing. Once the nucleated crystalline TiNiCu forms, the compressive stress will apply on the thin surface oxide layer, thus surface wrinkling would occur by the elastic buckling of the elastic oxidation layer on top of the crystallized TiNiCu film (see Fig. 7.13) [28]. The wrinkles can grow radially along the growth direction of crystals. During cooling, the martensitic transformation occurs, and significant surface relief appears inside these wrinkles.

7.5 Novel methods in surface morphology characterization

Currently, differential scanning calorimeter (DSC), X-ray diffraction, electrical resistivity, acoustic emission, stress, strain and dilatation measurement are standard techniques used to determine temperature-induced phase changes of thin film SMAs. In this section, we describe new experimental techniques for surface morphology characterization of structural transitions in thin film SMAs – atomic force microscopy (AFM) and photoemission electron microscopy (PEEM).

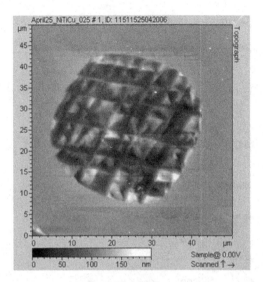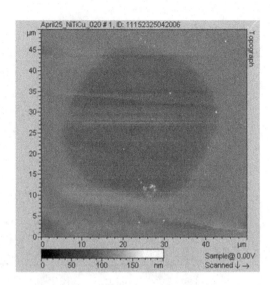

Figure 7.14   Morphological change of one martensite crystal measured with AFM: (a) martensite; (b) austenite [34]. (Reproduced with permission from Elsevier, UK.)

7.5.1   Atomic force microscopy

Atomic force microscopy provides 3-D topographical information of a solid surface at atomic resolution by scanning a sharp tip across the surface [29]. Depending on the materials, surface structure and purpose, both contact mode and non-contact mode have been practiced for measuring thin film surfaces. AFM data provides quantitative measurements of surface relief, surface roughness and morphological evolution during martensitic transformation. In addition, AFM is sensitive to lattice structure and can distinguish the twinned structure of martensite plates, the habit plane structure between martensite and austenite, and even the lattice correspondence variants of martensite.

AFM was applied to study the surface morphology changes in shape memory materials undergoing various processes, for example to examine the surface oxide layer on TiNi [30], focused ion beam processed TiNi thin film [31], ion implantation and annealing of TiNi [32]. The surface microstructure changes in shape memory thin film due to the thermally induced martensitic transformation can also be probed by *in situ* AFM. Figure 7.14 shows that the twinned structure in the TiNiCu martensite phase disappears during heating up to the austenite finish temperature. There is also a noticeable decrease in surface roughness (RMS) from 35 nm to 9.8 nm in the transition from martensite to austenite [33].

Based on the dramatic surface relief changes associated with martensitic transformation in bulk and thin film shape memory materials, temperature controllable AFM can be used as an alternative for determining transformation temperature [34]. Further, an environment-equipped AFM can be employed to characterize mechanical and chemical behaviors of SMAs at small length scales.

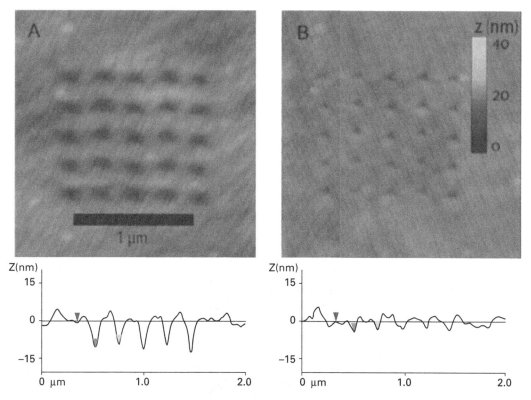

Figure 7.15 An indent data array (A) before and (B) after heating past the martensitic transformation temperature. Inserts below the images show profiles drawn through a row of indentions [36]. (Reproduced with permission from Wiley-VCH GmbH & Co.)

More recently, nano-indentation has been widely applied to study the localized mechanical properties of shape memory materials, or thin films. The shape memory effect of those materials at the indented area has been quantitatively studied by AFM [35, 36, 37]. The nature of the two-way shape recovery in indents and indent arrays on shape memory alloys and their thin films was explored as an alternative approach for high-density data storage (Fig. 7.15) and controllable reversible surface protrusions (Fig. 7.16) [38].

It can be concluded that AFM is a robust tool in the quantitative investigation of structural evolution during surface engineering and surface relief during martensitic transformation. The *in situ* AFM data provides a quantitative description of the shape memory effect at micrometer and nanometer scales.

7.5.2   Photoemission electron microscopy

PEEM adopts direct-imaging technique by projecting the surface-excited photoelectrons to an image screen. A photoemission electron microscope is composed of a light source (e.g. using an incoherent mercury lamp, laser or synchrotron

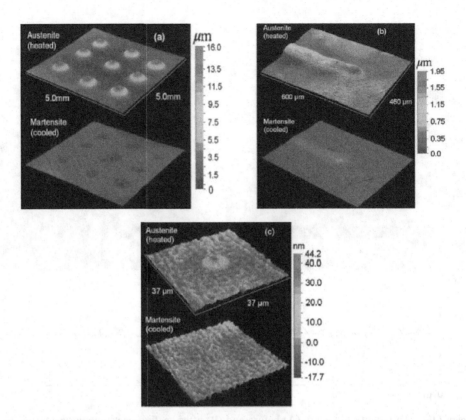

Figure 7.16   Three-dimensional profiles of reversible surface protrusions. (a) A 3×3 matrix of circular protrusions on the surface of the austenite phase of TiNi at high temperature which disappears when the sample is cooled to the martensite phase. (b) A protruding line on the surface of the austenite phase that nearly disappears in the martensite phase. (c) A nanoscale reversible protrusion on the surface a TiNi film [38]. (Reproduced with permission from Wiley-VCH GmbH & Co.)

radiation sources), a sample chamber and manipulation system, a series of projection lenses, a fast-speed CCD camera attached to the collinear column, a controlling system and high voltage power supply, etc. The improvement of PEEM instrumentation makes it most suitable for surface science analysis and currently gives a lateral resolution of about 10 nm. Figure 7.17 shows the schematic of a PEEM and an ELMITEC GmbH – PEEM III microscope. The emission of low-energy photoelectrons is extremely sensitive to a variety of surface dynamic phenomena, such as surface adsorption, desorption and surface chemical reactions [39]. PEEM is particularly useful for studying surface dynamics in metallic materials when the surface work function varies with chemical composition, crystal orientation, surface dipole, and phase [40, 41]. PEEM has been readily applied to studying the thermally-induced martensitic transformations in CuZnAl- and TiNi-based SMAs [42]. This technique takes advantages of the direct imaging capability of PEEM and its sensitivity to the work function difference

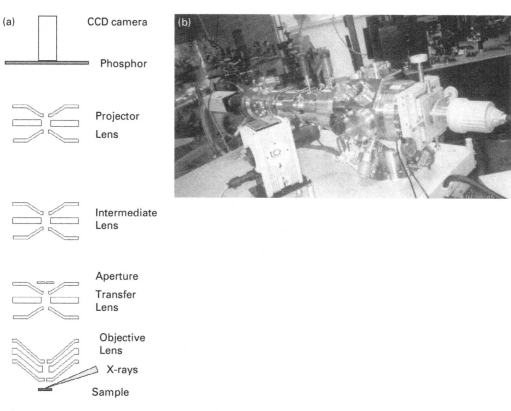

Figure 7.17 Schematic of PEEM (a); and (b) an ELMITEC GmbH – PEEM III microscope located at EMSL of the Pacific Northwest National Laboratory (PNNL), USA.

between the martensite and austenite to characterize these transformations *in situ* and in real-time.

Polished CuZnAl shape memory alloy before and after the phase transformation during heating has been probed by PEEM, as shown in Fig. 7.18 [39]. The spear-like martensite plates within a single martensite grain can be clearly resolved. The contrast mechanism in resolving the martensite variants is: (a) surface relief contrast, and (b) surface work function due to the misorientation between martensite lattice correspondence variants.

The 3.5 μm thick TiNiCu thin film on a Si(100) wafer was prepared by magnetron sputtering at 450 °C. Samples were then post-annealed in vacuum at 650 °C for 1 hour. Figure 7.19 shows the TiNiCu shape memory thin film surface microstructure at 25 °C (martensite phase) and at 100 °C (austenite phase) probed by PEEM [42]. Except for the image contrast change (due to the surface work function difference), the most prominent difference is the wrinkle-like microstructure at low temperature (Fig. 7.19(a)) and its absence at high temperature (Fig. 7.19(b)). The transition between the wrinkled and smooth states is remarkably sharp. Sequences of PEEM images indicate that the wrinkles disappear at 73.1 °C during heating and reappear at 52.4 °C during cooling. The appearance and

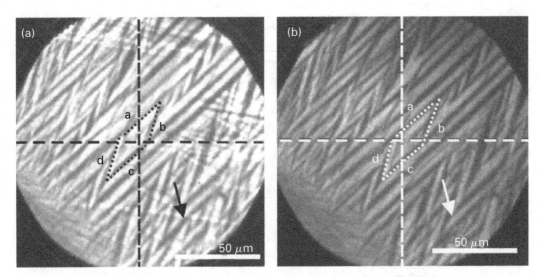

Figure 7.18   PEEM images of a CuZnAl surface showing (a) the low temperature martensite phase and (b) the high temperature austenite phase in the same area. The position of a polishing scratch, visible on the martensite surface but not on the austenite surface, is marked by the arrow. The diamond marks a set of four self-accommodated variants that are displaced by the transformation [39]. (Reproduced with permission from the American Institute of Physics, USA.)

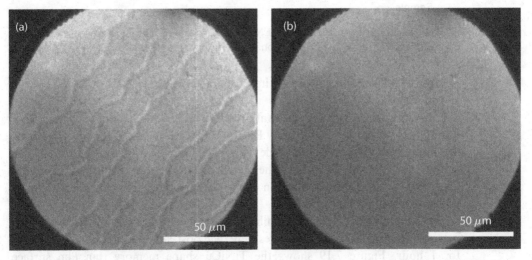

Figure 7.19   *In situ* PEEM images of a TiNiCu thin film at (a) 25 °C (martensite) and (b) 100 °C (austenite) [42]. (Reproduced with permission from Wiley-VCH GmbH & Co.)

disappearance of these features are closely related to the martensitic transformation, as explained in Section 7.4.

Corresponding quantitative measurements of the integrated image intensities during a complete thermal cycle are plotted in Fig. 7.20 [42]. The PEEM intensity

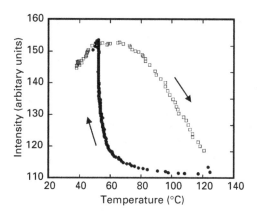

Figure 7.20 Integrated PEEM intensities as a function of temperature for TiNiCu during complete thermal cycles. The arrows indicate the temporal order of data collection [42]. (Reproduced with permission from Wiley-VCH GmbH & Co.)

increases initially (30 °C – 45 °C) during heating, remains nearly constant until 70 °C, and then decreases monotonically until 120 °C. The onset of the monotonic decrease near 70 °C corresponds roughly to the onset of the phase transformation in film stress measurements. During cooling, the photoelectron intensity increases slowly from 120 °C to 55 °C, then rises rapidly to the level prior to heating between 55 and 50 °C. The rapid rise in photoemission intensity near 55 °C corresponds to the martensitic transformation during cooling. This sudden change accompanying the re-appearance of wrinkles indicates the martensitic transformation is a first-order process with only temperature-dependence. Since the photoelectrons originate within nanometers of the surface, PEEM intensities are especially sensitive to conditions along the surface.

7.6   Summary

1. A wide range of residual stresses exist in the sputter-deposited TiNi films, either compressive or tensile. The stress generation and relaxation behaviors were significantly affected by film composition, deposition and/or annealing temperatures, which strongly control the intrinsic stress, thermal stress and phase transformation behaviors.

2. The TiNi and TiNiCu shape-memory thin films sputter-deposited atop a silicon wafer may have different types of thermo-induced reversible surface morphologies. Irregular surface trenches (with a width of tens of microns and depth of hundreds of nanometers) may appear atop the surface of post-annealed and fully crystallized thin films, but they disappear when heated above the temperatures for phase transformation. On the other hand, in partially crystallized thin films, the crystalline parts (islands) appear in chrysanthemum-shape at high temperature; while at room temperature, the surface morphology within the

islands changes to standard martensite striations. Similar to that of surface relief in shape memory thin films, both phenomena are fully repeatable upon thermal cycling as well. Under the large compressive stress during annealing, elastic buckling of the oxide layer (TiO_2) on the top surface of shape memory thin films is the origin for these phenomena.

3. Both AFM and PEEM provide quantitative, spatial resolved information on the evolution of the surface microstructure as the transformation in shape memory thin film proceeds. Real-time PEEM shows great promise for the quantitative measurement of the kinetics of phase transformations in a wide range of materials and thermal-mechanical processes. The transition temperatures inferred by real-time PEEM observation are consistent with DSC measurements and wafer curvature (stress) measurements for thin-film TiNiCu, when account is taken of the surface sensitivity of the PEEM technique. The novel characterization technique by combining AFM and PEEM provides in-depth information about martensitic transformation in thin films, for example the physical, chemical and electronic properties.

Acknowledgement

The authors would like to thank S. Sanjabi and Z. H. Barber from the University of Cambridge, UK; Dr. Xu Huang, Dr. Mingjie Wu and Prof. H. J. Du from Nanyang Technological University, Singapore; Dr. Wayne P. Hess from Pacific Northwest National Laboratory and Prof. J. Thomas Dickinson from Washington State University, US.

References

[1] D. S. Grummon, J. P. Zhang and T. J. Pence, Relaxation and recovery of extrinsic stress in sputtered titanium-nickel thin films on (100)-Si, *Mater. Sci. Engng.*, **A273–275** (1999) 722–726.

[2] C. L. Shih, B. K. Lai, H. Khan, S. M. Philips and A. H. Heuer, A robust co-sputtering fabrication procedure for TiNi shape memory alloys for MEMS, *J. MEMS*, **10** (2001) 69–79.

[3] Y. Q. Fu, H. J. Du, W. M. Huang, S. Zhang and M. Hu, TiNi-based thin films in MEMS applications: a review, *Sensors & Actuators: A.* **112** (2004) 395–408.

[4] L. B. Freund and S. Suresh, *Thin Film Materials: Stress, Defects Formation and Surface Evolution*, Cambridge: Cambridge University Press (2003).

[5] M. F. Doerner and W. D. Nix, Stresses and deformation processes in thin films on substrates, *CRC Critical Review in Solid State and Materials Sciences*, **14** (1988) 225–268.

[6] R. C. Cammarata, T. M. Trimble and D. J. Srolovitz, Surface stress model for intrinsic stresses in thin films, *J. Mater. Res.*, **15** (2000) 2468–2474.

[7] W. D. Nix and B. M. Clemens, Crystallite coalescence: a mechanism for intrinsic tensile stresses in thin films, *J. Mater. Res.*, **14** (1999) 3467–3473.

[8] R. Abermann, Measurements of the intrinsic stress in thin metal films, *Vacuum*, **41** (1990) 1279–1282.

[9] P. Chaudhari, Grain growth and stress relief in thin films, *J. Vacuum Science and Technology*, **9** (1972) 520–522.

[10] J. A. Thornton, J. Tabock and D. W. Hoffman, Internal stresses in metallic films deposited by cylindrical magnetron sputtering, *Thin Solid Films*, **64** (1979) 111–119.

[11] G. Stoney, The tension of metallic films deposited by electrolysis, *Proceedings of the Royal Society (London)*, **A82** (1909) 172–175.

[12] R. Spolenak, W. L. Brown, N. Tamura, *et al.* Local plasticity of Al thin films as revealed by X-ray microdiffraction, *Phys. Rev. Lett.* **90** (2003) 096102.

[13] Y. Q. Fu, H. J. Du and C. Q. Sun, Interfacial structure, residual stress and adhesion of diamond coating on titanium, *Thin Solid Films*, **424** (2003) 107–114.

[14] K. Kobayashi, Y. Inoue, T. Nishimura, *et al.* Local oxidation induced stresses measured by Raman microscopy, *J. Electrochemical Society*, **137** (1990) 1987–1988.

[15] Y. Q. Fu and H. J. Du, Relaxation and recovery of stress during martensite transformation for sputtered shape memory TiNi film, *Surf. Coat. Technol.*, **153** (2002) 100–105.

[16] Y. Q. Fu and H. J. Du, Effects of film composition and annealing on residual stress evolution for shape memory TiNi film, *Mater. Sci. Engng.*, **A 342** (2003) 236–244.

[17] Y. Q. Fu, H. J. Du and S. Zhang, Sputtering deposited TiNi films: relationship among processing, stress evolution and phase transformation behaviors, *Surf. Coat. Technol.*, **167** (2003) 120–128.

[18] T. W. Duerig, and C. M. Wayman, in *Engineering Aspects of Shape-Memory Alloys*, eds. T. W. Durerig, K. N. Melton, D. Stockel, London: Butterworth-Heinemann, (1990) p. 3.

[19] Y. Q. Fu, H. J. Du, S. Zhang and Y. W. Gu, Stress and surface morphology of TiNiCu thin films: effect of annealing temperature, *Surf. Coat. Technol.*, **198** (2005) 389–394.

[20] S. Miyazaki and A. Ishida, Martensitic transformation and shape memory behavior in sputter-deposited TiNi-base thin films, *Mater. Sci. Engng.*, **A 273–275** (1999) 106–133.

[21] Y. Q. Fu, S. Zhang, M. J. Wu, *et al.* On the lower thickness boundary of sputtered TiNi films for shape memory application, *Thin Solid Films*, **515** (2006) 80–86.

[22] A. Ishida and M. Sato, Thickness effect on shape memory behavior of Ti-50.0at.%Ni thin film, *Acta Mater.*, **51** (2003) 5571–5578.

[23] Y. Q. Fu, S. Sanjabi, W. M. Huang, *et al.*, Surface relief and surface wrinkling in TiNiCu shape memory thin films, *J. Nanosci. Nanotechnol.*, **8** (2008) 2588–2596.

[24] M. J. Wu, W. M. Huang and F. Chollet, In situ characterization of TiNi based shape memory thin films by optical measurement, *Smart. Mater. Struct.*, **15** (2006) N29–N35.

[25] S. H. Im and R. Huang, Ratcheting-induced wrinkling of an elastic film on a metal layer under cyclic temperatures, *Acta Mater*, **52** (2004) 3707.

[26] W. M. Huang, Y. Y. Hu and L. An, Determination of stress versus strain relationship and other thermomechanical properties of thin films, *Appl. Phys. Lett.*, **87** (2005) 201904.

[27] J. Liang, R. Huang, H. Yin, J. C. Strum, K. D. Hobart and Z. Suo, Relaxation of compressed elastic islands on a viscous layer, *Acta Materialia*, **50** (2002) 2933.

[28] Y. Q. Fu, S. Sanjabi, Z. H. Barber, *et al.*, Evolution of surface morphology in TiNiCu shape memory thin films, *Appl. Phys. Lett.*, **89** (2006) 171922.

[29] L. A. Bottomley, J. E. Coury and P. N. First, Scanning probe microscopy, *Anal. Chem.* **68** (1996) 185R–230R.

[30] D. A. Armitage and D. M. Grant, Characterisation of surface-modified NiTi alloys, *Mater. Sci. Eng.*, **A 349** (2003) 89–97.

[31] D. Z. Xie, B. K. A. Ngoi, Y. Q. Fu, A. S. Ong and B. H. Lim, Etching characteristics of TiNi thin film by focused ion beam, *Appl. Surf. Sci.* **225** (2004) 54–58.

[32] R. W. Y. Poon, J. P. Y. Ho, C. M. Y. Luk, *et al.* Improvement on corrosion resistance of NiTi orthopedic materials by carbon plasma immersion ion implantation, *Nucl. Instrum. Methods Phys. Res.*, **B 242** (2006) 270–274.

[33] M. Cai, Y. Q. Fu, S. Sanjabi, Z. H. Barber and J. T. Dickinson Effect of composition on surface relief morphology in TiNiCu thin films, *Surf. Coat. Technol.* **201** (2007) 5843–5849.

[34] Q. He, W. M. Huang, M. H. Hong, *et al.* Characterization of sputtering deposited NiTi shape memory thin film using temperature controllable atomic force microscope, *Smart Mater. Struct.* **13** (2004) 977–982.

[35] Y. Zhang, Y.-T. Cheng and D. S. Grummon, Two-way indent depth recovery in a NiTi shape memory alloy, *Appl. Phys. Lett.* **88** (2006) 131904.

[36] Y. Zhang, Y.-T. Cheng and D. S. Grummon, Shape memory surfaces, *Appl. Phys. Lett.* **89** (2006) 041912.

[37] M. Cinchetti, A. Gloskovskii, S. Nepjiko, G. Schonhense, H. Rochholz and M. Kreiter, Photoemission electron microscopy as a tool for the investigation of optical near fields, *Phys. Rev. Lett.* **95** (2005) 047601.

[38] G. A. Shaw, J. S. Trethewey, A. D. Johnson, W. J. Drugan and W. C. Crone, Thermomechanical high-density data storage in a metallic material via the shape-memory effect, *Advanced Materials*, **17** (2005) 1123–1127.

[39] G. Xiong, A. G. Joly, K. M. Beck, *et al.* In-situ photoelectron emission microscopy of thermally-induced martensitic transformation in a CuZnAl shape memory alloy, *Appl. Phys. Lett.* **88** (2006) 091910.

[40] T. C. Leung, C. L. Kao, W. S. Su, Y. J. Feng and C. T. Chan, Relationship between surface dipole, work function and charge transfer: some exceptions to an established rule, *Phys. Rev. B*, **68** (2003) 195408.

[41] W.-C. Yang, B. J. Rodriguez, A. Gruverman and R. J. Nemanich, Photo electron emission microscopy of polarity-patterned materials, *J. Phys. Condens. Matter*, **17** (2005) S1415–S1426.

[42] M. Cai, S. C. Langford, M. Wu, *et al.* Study of martensitic phase transformation in a NiTiCu thin film shape memory alloy using photoelectron emission microscopy, *Adv. Funct. Mater.* **17** (2007) 161–167.

8 Ion implantation processing and associated irradiation effects

T. Lagrange and R. Gotthardt

Abstract

In this chapter, we describe the influence of ion implantation on the microstructural modifications in TiNi SMA thin films. We focus on investigations involving 5 MeV Ni ion irradiation since it can be used as a means to selectively alter the transformation characteristics and to develop NiTi based thin film actuator material for MEMS devices. The primary effects of ion implantation on microstructure are summarized.

8.1 Introduction

This chapter discusses the influence of high energy particle irradiation on the shape memory properties of shape memory alloy (SMA) thin films. In general, SMAs, like TiNi, are very sensitive to high energy particle irradiation and undergo structural changes that can suppress martensitic transformations and thus their shape memory properties. The simplified explanation for this sensitivity to particle irradiation is that the martensitic transformation relies on the local atomic order of the crystal. The introduction of defects by high energy particle irradiation, such as in ion implantation, can destroy the local structural and chemical order that suppresses the SMA's ability to undergo a martensitic transformation and the related shape memory properties. It is these detrimental effects that limit the use of SMA in harsh environments, like space, where the defect production from cosmic radiation can render SMA actuators inactive.

The negative effects of particle irradiation on shape memory properties can, in fact, be useful. If the high energy particle material interactions are well understood, they can be used to tailor the shape memory response and engineer monolithic SMA actuators. That is, the use of particle irradiation, such as ion implantation, to locally suppress the martensitic transformation and shape

Thin Film Shape Memory Alloys: Fundamentals and Device Applications, eds. Shuichi Miyazaki, Yong Qing Fu and Wei Min Huang. Published by Cambridge University Press. © Cambridge University Press 2009.

memory effect (SME) can be used as a processing technique to make SMA based actuators, where motion is biased by ion beam damage.

The first part of the chapter discusses the microstructural modifications and irradiation induced phase transformations that result from high energy (5 MeV Ni) ion implantation, and the influence of the implantation parameters and film deformation prior to irradiation on the microstructure. The observed phase transformation is described through damage resulting by a combination of ion nuclear stopping and electronic stopping effects. The latter part of the chapter discusses an ion implantation processing technique that can be used to make an out-of-plane bending, SMA actuator and how the implantation process parameters and temperature cycling affect the film's two-way bending motion.

8.2   Ion irradiation of SMA TiNi films

8.2.1   Physics of ion irradiation

Radiation damage arises from the interactions of energetic particles with a target material as they are "slowed down" and transfer their energy through a series of nuclear collisions and energy losses through electron excitations with target atoms. In the beginning of the slowing-down process when the ion possesses a high kinetic energy, the ion is slowed down mainly by an electronic stopping processes, and its trajectory is relatively unchanged. When the ion has slowed down sufficiently, the collisions with nuclei (the nuclear stopping) become more and more probable, finally dominating the slowing down. When atoms of the solid receive significant recoil energies when struck by the ion, they will be removed from their lattice positions, and produce a cascade of further collisions in the material. This series of nuclear collisions results in atomic displacements and generates defects that alter microstructure and mechanical properties, such as Frenkel pairs (self-interstitial-vacancy pairs), anti-site defects (wrong atoms on an alloy constituent sublattice), crowdions, diluted zones, defects clusters, etc. [1]. However, in the latter part of this section, we show how the electronic stopping process of MeV heavy ion can also alter the microstructure and affect the shape memory properties of TiNi alloys.

8.2.1.1   Nuclear stopping and cascade formation

Nuclear stopping results from the elastic collisions between the ion and atoms in the sample. The stopping power is a function of the repulsive potential $V(r)$ between two atoms. Nuclear stopping increases when the mass of the ion increases and dominates the stopping at low energy. For very light ions slowing down in heavy materials, the nuclear stopping is weaker than the electronic stopping at all energies. The nuclear collisions create a sequence of atomic displacements (cascades) that produce a localized high concentration of defects. The atomic displacement sequence starts with the creation of a primary knock-on atom

(PKA), which is the atom struck by a bombarding irradiation particle. The PKAs recoil with energies that can be orders of magnitude higher than that of their lattice binding energy, and this residual recoil energy can, in turn, be imparted to neighboring target atoms, producing secondary recoil events. This leads to a collisional sequence of excited target atoms that has a "branching tree"-like structure, which is termed a displacement cascade. Since the cross-section for atomic collisions increases steeply with decreasing energy, the distance between successive collisions becomes progressively shorter. For recoils with energies below a few hundred eV, the mean free path between collisions is of the order of a couple of atomic distances. Thus, a large amount of energy is deposited in a small volume, generating a high concentration of defects in a localized region.

8.2.1.2 Electronic stopping

Electronic stopping refers to a process by which the ion is slowed and its energy is transferred to the target material through inelastic collisions between bound electrons in the medium and the ion moving through it. The term inelastic is used to signify that the collisions may result both in excitations of bound electrons of the medium and in excitations of the electron cloud of the ion. In most materials, these electronic excitations have little effect on damage production, and energy dissipation effects are neglected in most cases, such as in ion implantation of metallic systems. However, electronic stopping may play an important role in the displacement process.

Since the number of collisions an ion experiences with electrons is large, and since the charge state of the ion while traversing the medium may change frequently, it is very difficult to describe all possible interactions for all possible ion charge states, and thus, unlike the nuclear stopping power, there is no universal function describing variation in the stopping power with different ion–target combinations, even though it does not vary greatly. Electronic stopping increases linearly with the square root of the ion energy, and at lower ion energies, the nuclear stopping becomes more relevant. Therefore, near the projected range (average implantation depth of given ion and energy), where the ion has lost most of its initial energy, nuclear stopping energy transfer dominates over electronic stopping. In contrast, swift ions, which possess sufficiently high energy, such that they are not stopped and pass through the target, are slowed principally by the electronic stopping processes.

8.2.1.3 Calculation of the stopping powers and radiation dose

Conventional methods used to calculate ion ranges and stopping powers are based on the binary collision approximation (BCA) [2, 3]. In these methods the atomic displacements and the ion trajectory are treated as a succession of individual collisions between the recoil ion and atoms in the sample. For each individual collision the classical scattering integral is solved by numerical integration. The impact parameter p in the scattering integral is determined either from a stochastic distribution or in a way that takes into account the crystal structure of the sample.

There are a few commercial, BCA-based computer simulations available for ion implantation, such as the transport and range of ions in matter (TRIM) code developed by Ziegler *et al.* [4].

TRIM simulations can give fairly accurate values for the stopping powers and ion penetration range. TRIM can also give a crude estimate of the crystal damage from the amount of displaced atoms from nuclear stopping events (upper limit). However, it does not take into account point defect recombination, effects of overlapping damage zones, or addition effects that occur during cascade cooling, such as defect annealing and migration associated with structural relaxation during and after the thermal spike. More importantly, TRIM neglects electronic stopping entirely, treating it as frictional term in the ion slowing process. In this case, molecular dynamic simulations offer a much more realistic picture of irradiation damage, but are more computationally intensive.

8.2.2   Heavy ion irradiation (5 MeV Ni)

The following paragraphs discuss in detail the irradiation damage observed in TiNi SMA films occurring from 5 MeV Ni ion implantation. These studies were performed to assess the effect of ion implantation on both the microstructure and shape memory properties. The reason that 5 MeV Ni ions were chosen for these investigations was that the energy and ion mass created sufficient damage at high doses to completely suppress the martensitic transformation in an implanted layer. The localized frustration of the martensitic transformation can be used to tailor the shape recovery and develop a preferred two-way shape memory effect. An actuator concept based on the 5 MeV Ni ion implantation is discussed in Section 8.3. Essentially, a two-way shape memory effect (TWSME) can be created by suppressing the transformation in a partial thickness of a deformed martensitic TiNi film. The TWSME, of course, is highly dependent on the irradiated microstructure, and the optimization of actuators based on this ion implantation process requires detailed knowledge of how irradiation parameters affect both the microstructure and shape memory properties. The three main irradiation and sample parameters which have a strong effect on the irradiation induced microstructure are implantation dose, temperature during irradiation and the amount of martensite deformation prior to irradiation. The subsequent paragraphs summarize the effect of these parameters that are discussed in more detail in the following references [5, 6, 7, 8, 9, 10, 11, 12, 13, 14, 15].

8.2.2.1   Implantation dose and temperature effects on the irradiated microstructure

Binary TiNi thin films with 6 μm thickness (fabricated by d.c. magnetron sputtering, as described in Ref. [16]) were used for these studies. The nominal transformation temperatures after substrate removal and heatment were as follows: $M_s \sim 45\,°C$, $M_f \sim 9\,°C$, $A_s \sim 57\,°C$, and $A_f \sim 87\,°C$. Prior to implantation, the freestanding martensitic films were deformed in a micro-tensile machine to strains between 4 and 5%. The deformed films were carefully mounted (so no additional deformation was introduced) to a copper heat sink with conductive carbon cement and irradiated at two different temperatures, 25 °C and −145 °C, to total fluencies of 1×10^{15}, 1×10^{14}, 5×10^{13} and 1×10^{13} ions/cm^2 with 5 MeV Ni ions. It should be noted that the material was quenched in liquid N_2 before irradiation, and thus the material was fully martensitic prior to irradiation. The temperature rise associated with the ion implantation (+20 °C measured on the heat sink surface) did not thermally induce reverse transformation to austenite. To examine the effects of the irradiation, the material was sectioned and analyzed using grazing incidence X-ray diffraction and in a transmission electron microscope (TEM).

8.2.2.2 X-ray studies

The most obvious trend in the X-ray spectra (Fig. 8.1) was an increasing volume fraction of amorphous material with increasing fluence. Even in the lowest applied dose specimen, considerable microstructural changes resulted from irradiation as observed from XRD patterns, where martensite has been partially transformed to austenite (indicated by (110) austenite reflection at 42.4 °). In high dose specimens (above 10^{14} ions cm^{-2}), the faint diffraction peaks from the martensite, austenite and Ti_2Ni phase are superimposed on the broad amorphous peak. These peaks are fairly broad and have widths corresponding to nanocrystalline sized grains. They could have originated from the diffraction of the nanocrystalline phases embedded in amorphous material or the underlying crystalline material at depths below 2 μm.[1]

At lower irradiation temperatures (−145 °C), the material was amorphized significantly at lower fluencies than the high temperature irradiation (see Figure 8.1(b)). Comparing 5×10^{13} ions/cm^2 fluence between the two irradiation temperatures, the transformations and amorphization induced by the low temperature irradiation had a diffraction that resembles the higher 10^{14} ions/cm^2 at 27 °C, which is about twice that of 25 °C irradiation. The enhancement in the amorphization mechanisms at low temperatures can be attributed to the reduced defect annealing as a result of the limited diffusion at cryogenic temperatures.

The spectra of the high dose −145 °C irradiated specimens, however, have similar features as those at 25 °C irradiation, showing superimposed martensite and austenite peaks on the broad amorphous peak. In contrast, the austenite peak observed in the lowest dose specimen was the convolution of two peaks. Judging by the peak positions in the inset of Fig. 8.1(b), the peaks are, in fact, related to the R phase, and the austenite peak is overlapped by and lies below these principal R-phase peaks (as shown in the inset). The mixed-phase microstructure was confirmed, in part, by the TEM observations, showing that all three phases were present in the implanted layer at room temperature for this dose.

---

[1] Given the grazing incidence angle of 4 °, some fraction of the underlying material can contribute to the diffraction signal. Thus, the diffraction peaks observed in the X-ray spectra can be from material below the implanted layer.

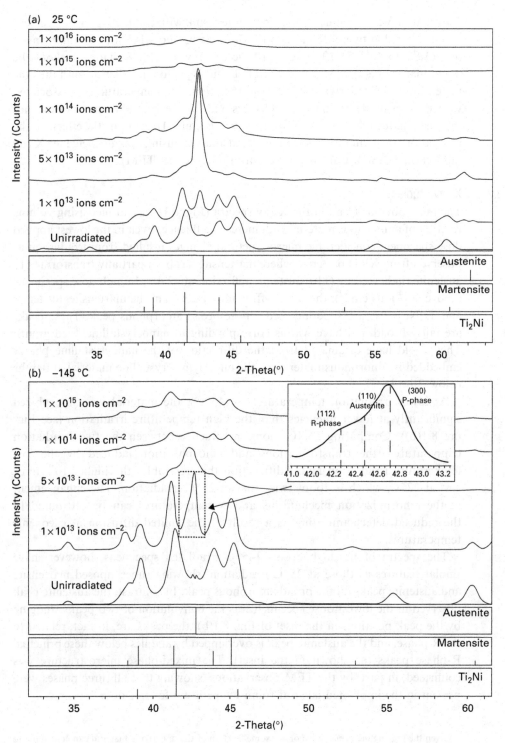

Figure 8.1   X-ray diffraction spectra showing the effect of ion fluence on TiNi SMA film microstructure implanted at (a) 25 °C (b) −145 °C [8].

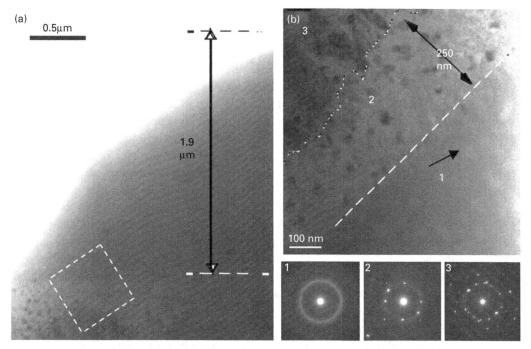

Figure 8.2 TEM cross-sectional image of a specimen implanted to 10^{16} ions/cm^2 at 25 °C: (a) low magnification bright-field image of the region, (b) enlargement of the region marked by the white box of the microstructure near the ion penetration range. The numbered diffraction patterns correspond to the numbered regions in the image.

8.2.2.3 TEM observations of as-irradiated microstructure
Dose $> 10^{15}$ ions cm^{-2}, irradiated at 25°C

The XRD studies of the high dose (10^{15} and 10^{16} ions cm^{-2}) specimens concluded that the pre-irradiated martensite structure was significantly amorphized. TEM observations of the implanted layer showed that a continuous amorphous layer extending from the irradiation surface down to the ion projected range of 1.9 μm was produced, which corresponded well with the TRIM predicted ion projected range of ~2 μm, as shown in Fig. 8.2 (a). A closer look at the amorphous material in the implanted layer showed two distinctly different areas, a light gray, circular area (one such region is marked by an arrow in Fig. 8.2 (b)) on a darker gray background. This would indicate that the amorphous matrix is composed of two different types of amorphous material with different densities.[2] Using EDS (energy dispersive spectroscopy), these light gray, amorphous regions were determined to be Ti-rich TiNi zones in the stoichiometric TiNi amorphous matrix (the darker gray background). Since these Ti-rich regions have a size of 30–50 nm diameter and

---

[2] This is assuming that contrast results from mass-thickness variations in TEM foil, and for the regions with roughly the same thickness.

circular shape, they are most likely residual Ti-rich amorphous regions left by prior Ti$_2$Ni precipitates that existed in the pre-irradiated microstructure and have also been amorphized at these fluencies.

Figure 8.2(b) shows the microstructure near the ion projected range, which has a steep gradation in morphology from a continuous amorphous matrix (1) that extends down to a depth of roughly 1.8 µm, followed by a ~250 nm thick austenite layer (2) that is between the damaged layer and the underlying pristine martensite (3). This austenitic interlayer is severely defected, containing numerous amorphous zones, as indicated by the diffuse ring in the diffraction pattern number 2. The amorphous matrix (region 1) terminates abruptly with a well-defined interface between it and the austenite layer. The interface between austenite and martensite (regions 2 and 3) was noticeably different, showing a wavy, less distinct transition, which is more indicative of ion straggling effects, but inhomogeneities in the local martensite structure can also cause this jagged interface.

Although the low magnification bright-field image suggests that the implanted layer is uniform and completely amorphous, there still exists a small variation in the irradiated microstructure. High resolution transmission electron microscopy (HRTEM) investigations of the implanted layer have shown 2–3 nm sized austenitic crystallites [5, 6, 8, 10]. At depths further from the surface in the middle of the implanted layer, the structure was completely amorphous and devoid of any nanocrystalline phases. The presence of nanocrystals in the amorphous layer may have implications for its mechanical strength (this will be discussed further in Section 8.3.3).

Dose: 10^{14} ions cm^{-2}, irradiated at 25 °C

Typically, microstructural changes from energetic heavy ions vary with depth due to the heterogeneous damage profiles, and the microstructure of the material irradiated to 1×10^{14} ions cm^{-2} at 25 °C exhibited varied damage distributions, as shown in the dark-field TEM image montage of Figure 8.3. There is a strong variation in irradiated microstructure with depth, which is clearly exhibited in diffraction patterns that were taken approximately at the position indicated by the numbered, white encircled areas. Near the surface (<875 nm), the martensite was transformed into a partially amorphized austenite. At greater depths, the material was almost completely amorphous, containing a few remnant austenitic nanocrystalline zones. Beyond this region and at depths >1500 nm, the irradiation damage and transformations were not as severe, and the material was austenitic and completely crystalline. The microstructure at depths beyond 1900 nm showed a jagged interface with nominal undamaged martensitic structure, similar to that observed at higher doses.

This depth dependence in microstructure and change in amorphous volume fraction can be illustrated further by calculating the intensity ratio between the amorphous ring and (110) austenite fundamental reflection from the diffraction patterns in Fig. 8.3, as a function of depth, which is shown in Fig. 8.4.

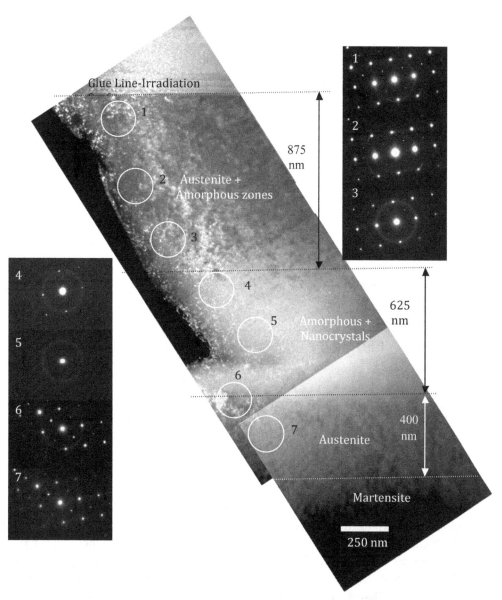

Figure 8.3   Dark-field TEM cross-sectional view of 1×10^{14} ions/cm^2 fluence irradiated at 25 °C. The (100) Z.A. austenite diffraction patterns show the gradation in the irradiation damage. The numbered, encircled areas correspond to the rough position where diffraction patterns were taken [14].

The extent of microstructural modification (1.9 µm) corresponds well with the projected ion range calculated from the TRIM code [4]. The peak damage (atomic displacements from nuclear stopping) calculated in the simulation occurs at depths between 1.5 and 2 µm. If it is inferred that amorphization occurs in regions with the highest incurred damage, then there is a large

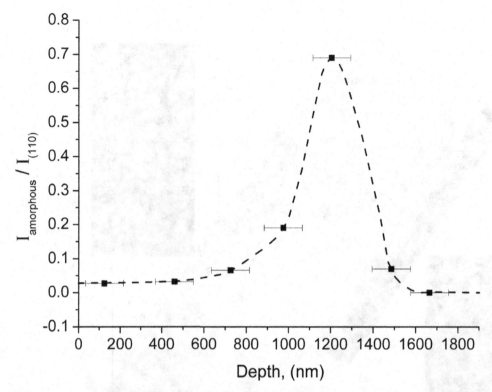

Figure 8.4   Plot of the intensity ration between the amorphous ring and (110) austenite reflection as a function of depth from the irradiation surface in 1×10^{14} ions cm^{-2} fluence irradiated at 25 °C [14].

discrepancy between those depths where there was significant amorphization (1.1–1.4 µm) as was observed in the cross-sectional TEM studies, and the damage peak in the simulation (1.9 µm). In fact, beyond 1.5 µm, the irradiation induced amorphization was minimal. This trend appears to be consistent with the high dose specimens (1×10^{16} ions cm^{-2}); the surface regions should not have been amorphized, given the damage calculations inferred from TRIM and the fact that there are a limited number of nuclear stopping events. This may also suggest that there are additional mechanisms involved which are not accounted for in the TRIM simulation, such as electronic stopping effects. A more detail discussion occurs later in the chapter where electronic stopping effects are discussed and linked to the observed microstructure resulting from 5 MeV Ni ions.

Dose: 5×10^{13} ions cm^{-2}, irradiated at 25 °C
At relatively low fluencies (5×10^{13} ions cm^{-2}), the implantation induces a phase transformation of the martensitic structure into the high temperature austenite phase. Such microstructures are normally observed at temperature above 100 °C, and austenite is not stable at room temperature. The austenite grains extend

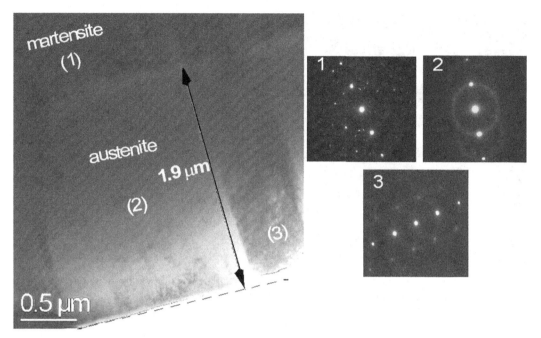

Figure 8.5   TEM cross-sectional bright-field image of the specimen irradiated to 5×10^{13} ions cm^{-2} at 25 °C, showing the implantation induced changes to the martensitic structure. The numbered diffraction patterns next to the image were taken at the approximated numbered position indicated in the bright-field image.

down from the surface to a depth of 1.9 μm, which again is consistent with the ion projected range, as shown in Fig. 8.5.

Interestingly, the austenite grains observed in this image are on the order of 1 to 2 microns wide and terminate abruptly at the ion range, beyond which is underlying, undamaged martensite (as seen from diffraction pattern 1). The columnar shape and size of these grain are similar to the austenite grains observed at high temperatures (125 °C) in the pre-irradiated microstructure. The appearance of such grains suggests that, in the implanted layer, the transformation temperatures are suppressed at these doses, stabilizing the austenite at room temperature (25 °C). Thus, the observed austenite is most likely a "coherent transformation" product of the pre-existing martensite. That is, the observed austenite results from the accumulation of subsequent cascade damage that suppresses the austenite finish temperature below room temperature, causing the transformation. This mechanism differs from phase transformations that occur directly from a quenched atomic displacement cascade (nucleation and growth within the cascade volume).

These austenite grains are not entirely crystalline, which is evident from the diffuse rings observed in diffraction patterns 2 and 3, and contained amorphous zones. High resolution TEM observations showed that the austenite contained large irregular-shaped zones at depths from 1 μm (region 2 in Fig. 8.5) down to 1.9 μm

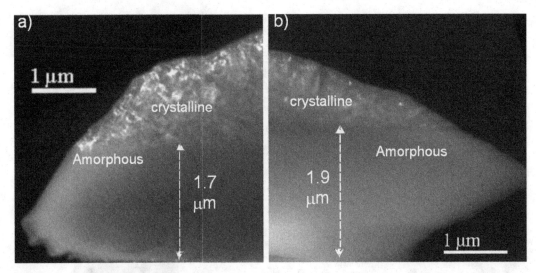

Figure 8.6   Low magnification dark-field TEM images of the implanted layer irradiated at temperatures below $-145\,°C$ with 5 MeV Ni ions, (a) 1×10^{14} ions cm^{-2}, (b) 1×10^{15} ions cm^{-2} [8].

(region 3). The large amorphous regions are most likely the agglomeration of smaller zones that overlapped to form a larger, more continuous amorphous region. It can be conjectured that with increasing dose, these zones can grow, consuming the austenite crystal and forming a continuous amorphous matrix.

8.2.2.4   Effect of irradiation temperature

TEM observations of the material irradiated at temperatures $<-145\,°C$ to fluencies of 1×10^{14} and 1×10^{15} ions cm^{-2} are shown in Fig. 8.6. Similar to high temperature irradiation, the implantation at these doses has a profound effect on the amorphization of the initial martensite structure. A sharp interface between amorphous (damaged layer) and crystalline material (undamaged) was observed. The amorphous, beam-damaged layer had a thickness of $1.7\,\mu m$ in the case of 1×10^{14} ions cm^{-2} and $1.9\,\mu m$ for 1×10^{15} ions cm^{-2} fluence.

The implanted layer in the 1×10^{14} ions cm^{-2} specimen irradiated at $-145\,°C$ was markedly more amorphous than that of the specimen irradiated at $25\,°C$, in accordance with X-ray diffraction studies. Near $1.7\,\mu m$, there was a 150 nm thick transition region observed between the highly amorphous matrix and crystalline, undamaged material, which was not completely amorphous, containing a high density of austenitic nanocrystals, and resembled the transition regions observed in the 1×10^{16} ions cm^{-2} specimen irradiated at $25\,°C$.

8.2.2.5   Effect of pre-straining the martensite on amorphization processes

Since pre-straining the martensite is a necessary step in the actuator production, it is useful to discuss the effect that martensite deformation has on the irradiation induced phase transformations. The shape memory effect depends on specialized plastic deformation mechanisms, where strains induced must be recovered upon heating to the austenite. Likewise, irradiation mechanisms highly depend on specific microstructural aspects, such as grain size and dislocation density. Changes in defect concentrations lower densities, and residual stress can affect the irradiation mechanisms, producing inherently different microstructure after irradiation. Since the actuator motion depends on microstructure in the implanted layer, knowledge of these effects is essential for tailoring the properties of the actuator.

To study the effect of pre-straining, $Ti_{52}Ni_{48}$ thin films with thicknesses of 2 μm have been chosen such that the whole specimen was volume ion implanted and modified by the ion irradiation. These films were cut into 3×10 mm strips, and for pre-straining, films were deformed in uniaxial tension to six different strains: 0, 1, 2, 3, 4 and 5% (residual strain) using a microtensile machine at a constant strain of 2×10^{-4}. The deformed films were then irradiated with 5 MeV Ni ions at room temperature (25 °C) at two fluencies of 1×10^{14} and 5×10^{13} ions cm^{-2}. The irradiated induced microstructure was analyzed by both X-ray diffraction and TEM.

Results from the X-ray diffraction

After irradiation, the samples no longer exhibited an observable martensitic phase transformation in the DSC test temperature range from −150 to 100 °C, which means the irradiation has either completely suppressed the martensitic transformation at these doses or the transformation occurs below the measurement capabilities of the DSC. X-ray diffraction also showed the stabilization of the high temperature austenite phase at room temperature. Figure 8.7 shows an X-ray diffraction montage of the effect of martensite straining on the irradiation-induced phase transformations.

At 0% strain, the high intensity austenite [110] peak at a 2-theta of 42.5° is surrounded by a less intense, overlapping martensite and Ti_2Ni peaks that are superimposed on a broad amorphous peak. It should be noted that it is difficult to separate the individual phase contributions of the martensitic and Ti_2Ni peaks, and it is assumed that the pre-strain only affects the martensitic phase. Thus, the observed reduction in these less intense peaks surrounding the austenite with this amount of pre-strain is related to an increase in the irradiation-induced phase transformations of the martensite into austenite and the amorphous phase.

Figure 8.8 shows the amorphization volume fraction as a function of pre-strain for the two doses used in this study. The amorphous volume fraction was calculated from the X-ray diffraction integrated intensity[3] of a broad amorphous peak, which is situated under the major martensitic and austenitic diffraction peaks at 2-theta angles between 40 and 45 degrees. Initially, the percentage of amorphous phase increased rapidly with pre-strain, then saturated asymptotically

---

[3] The integrated intensity was calculated from peak fitting using a pseudo-Voight algorithm included in the commercial software package JADE® 6.0.

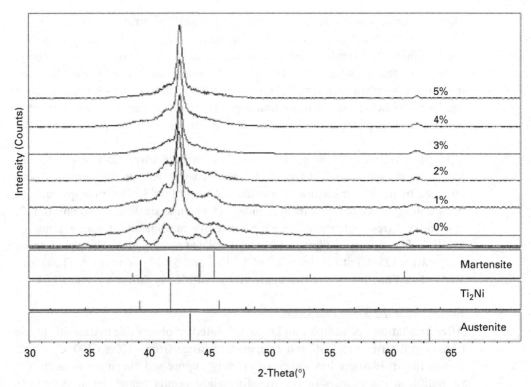

Figure 8.7 X-ray spectra of samples with various levels of pre-strain, irradiated to 1×10^{14} ions cm^{-2} at 25 °C. The bottom-most spectrum is non-irradiated [9].

to a constant value (~66%) at higher strains for the high dose case, while in the low dose, the amorphous phase fraction increased linearly with the pre-strain amount.

TEM observations

TEM observations confirmed the X-ray results that higher pre-strains enhance irradiation induced amorphization (for details see Ref. [9]). Here, it is useful to mention the unique microstructure of the 10^{14} ion cm^{-1} dose with 5% pre-strain shown in Fig. 8.9. The dark field image in Fig. 8.9(a) was generated by using the diffuse crystalline spot ((110) austenite) and a portion of the amorphous ring, as indicated in the inset diffraction pattern. The bright speckles on the light gray background are related to nanocrystalline regions (5–10 nm in size) that are embedded in an amorphous matrix. Only half of the micrograph is illuminated with bright speckles, which originate from a set of nanocrystals that have the same zone axis and are related to a pre-existing austenite grain. The area with no illuminated nanocrystals contains nanocrystals with a different zone axis related to a different pre-existing austenite grain (the proposed high grain boundary is indicated by a dotted line), and therefore do not show any bright speckle

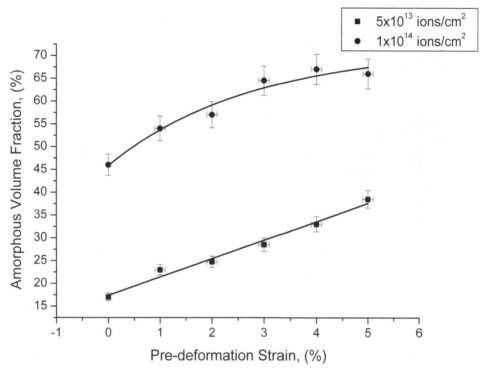

Figure 8.8   Comparison of amorphous phase volume fraction with pre-strain for two different fluencies [9].

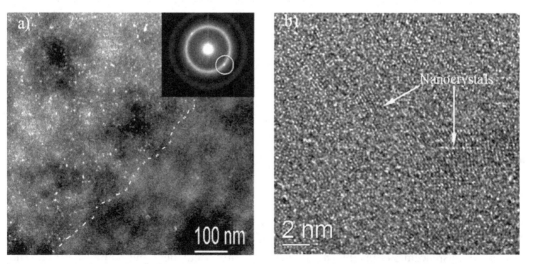

Figure 8.9   (a) Dark-field images of 5% specimens irradiated at a fluence of 1×10^{14} ions cm^{-2} at 25 °C. The dotted line is the proposed position of the pre-existing high angle grain boundary. (b) HRTEM image of nanocrystals in an amorphous matrix [9].

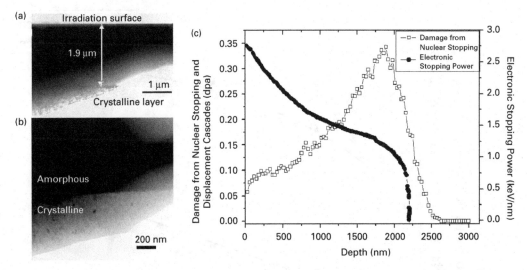

Figure 8.10 (a) Low magnification bright-field TEM images of the implanted layer (5×10^{15} ions cm^{-2}), (b) bright-field image showing the sharp interface between implanted and un-damaged zones, (c) TRIM (2003 stopping powers) simulation of damage at a fluence of 1×10^{14} ions cm^{-2} as a function of the depth in the TiNi thin film [14].

contrast in the dark-field image. The amorphous material appears in the form of isolated zones and, as the dose increases, the number and size of these zones increases until impingement. The amorphous matrix observed at 10^{14} ions cm^{-2} and pre-strains can be imagined as the result of the interconnection of a large number of amorphous zones that join to form a continuous matrix, where the nanocrystals are from remnant crystalline regions, most likely being the more irradiation resilient Ti$_2$Ni phase precipitates [16, 14].

8.2.3 High energy ion irradiation – electronic stopping effects

As was shown in Section 2.2, irradiation of martensitic TiNi films to high doses ($>1 \times 10^{14}$ ions cm^{-2}) with 5 MeV Ni^{2+} produces a homogeneous amorphous matrix extending from the free surface inward to the ion penetration range of 1.9 μm (as shown in Fig. 8.10(a)), where a sharp transition to the underlying crystalline layer was observed (Fig. 8.10(b)). This was surprising since the damage produced by ion irradiation should be more heterogeneously distributed. Nuclear stopping events alone do not explain the observed homogeneous appearance of the damage layer, especially near the surface where the cross-section for nuclear stopping damage is low. The TRIM simulation, shown in Fig. 8.10(c), further illustrates this concept, where the majority of the atomic displacements produced by nuclear events occur near the stopping range. The TEM observations, however, showed that the highest amorphous fraction (which is assumed to be the principal damage) occurs at a depth of 1.2 μm, which is about 0.7 μm shallower than the principle damage predicted by TRIM. The excitations from electronic stopping can produce damage and may offer a possible

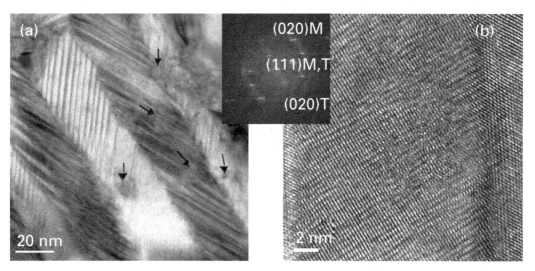

Figure 8.11 (a) Bright-field image of an undeformed martensite irradiated with 350 Au ions to a fluence of 5×10^{11} ions cm^{-2}, showing several circular ion tracks (marked by arrows), (b) HRTEM image of one observable track in (111) type 1 twin, as indexed from the inset FFT.

explanation for this shift in the principal amount of amorphization to surface regions, where electronic stopping is high.

In an extensive study on the effects of electronic stopping in bulk TiNi alloys, Barbu *et al.* [17] found that amorphous tracks only occurred in the martensite phase at electronic stopping powers $[dE/dx]_e$ higher than 40 keV nm^{-1}. The authors conjectured that the microstructure of these tracks consisted of an amorphous core surrounded by austenite. At lower stopping powers of 32 keV nm^{-1}, latent tracks[4] composed of austenite were observed within the martensite variants, indicating that the martensite to austenite transformation temperature had been depressed, stabilizing austenite within the track. Below 17 keV/nm, no ion-induced modification of the microstructure was observed.

In more recent studies [17, 14], slightly different results and conclusions were obtained. Firstly, in films irradiated with 350 MeV Au ions[5] (an electronic stopping power of 43 keV nm^{-1}) at low doses of $\sim 5 \times 10^{11}$ ions cm^{-2}, amorphous tracks were formed in the martensite and Ti$_2$Ni precipitates (Fig. 8.11(a)). However, HRTEM investigations (Fig. 8.11(b)) of these tracks clearly showed that austenite does not surround the tracks. Another interesting fact is that the track core was not totally amorphous, whereas some exterior portions appeared distorted (indicated by the wavy appearance of the lattice planes) and influenced

[4] Latent tracks are cylinders of material that contain different microstructure than the matrix, which are created by the passage of a single high energy ion and typically have diameters on a nanometric scale.

[5] The films were 8 μm thick and given 350 MeV Au ions, the ions were not nuclear stopped in the film thickness. Thus, nuclear stopping effects can be neglected.

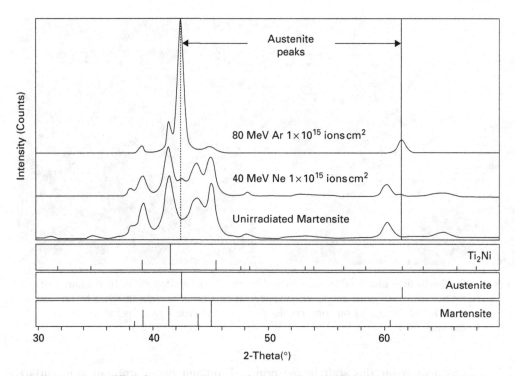

Figure 8.12 X-ray diffraction spectra for a specimen irradiated with 80 MeV Ar and 40 MeV Ne up to a fluence of 1×10^{15} ions cm^{-2} compared with the as-received films [14].

by the ion passage, but were not completely amorphous. At higher doses (10^{12} ions cm^{-2}), the material surrounding the tracks transformed to austenite, and in material irradiated at fluence of 10^{13} ions cm^{-2}, large amorphous zones were observed which originated by the growth of initial amorphous ion tracks. The trend in the observed microstructure suggested that the induced phase transformations result from the accumulation of damage of overlapping ion tracks. Microstructural differences between the bulk (Barbu *et al.* [17]) and thin film material (LaGrange *et al.* [14]), e.g., film can have a supersaturation of Ti in the matrix and finer microstructure containing non-equilibrium Ti$_2$Ni precipitates, may explain the discrepancies between the studies. These differences can, of course, strongly influence the irradiation damage mechanisms. The most notable influence on the irradiation damage was the amount of martensite deformation prior to irradiation, which, as in the 5 MeV Ni ion experiments, enhanced the amount of irradiation induced phase transformations.

As stated above, Barbu *et al.* [17] did not observe any microstructural modification in the samples at stopping powers below 17 keV/nm. They also suggested that conduction electrons dissipate the electronic excitation before it is transferred and coupled to the phonon system, thus limiting the amount of atomic displacements. However, LaGrange *et al.* [14] conducted experiments using 80 MeV Ar (9 keV/nm) and 40 MeV Ne (4 keV/nm) ions, and clearly showed that the Ti-rich TiNi thin films were sensitive to the irradiation effects at these electronic stopping powers and that significant damage was produced. Figure 8.12 shows the grazing incidence X-ray diffraction spectra for 8 μm thick TiNi films, which were deformed to a 4% tensile strain prior to irradiation and were irradiated with 80 MeV Ar and 40 MeV Ne ions to fluencies of 1×10^{15} ions/cm^2. Both irradiations induced a martensite (B19′) to austenite (B2) transformation in the samples, and considerable amorphization (37 vol. %) was observed in the case of the 80 MeV Ar ion irradiation ((110) austenite peak superimposed on a broad amorphous peak). No noticeable amorphization was detected in the 40 MeV Ne irradiated sample X-ray spectra, however the martensite was partially austenitized (14 vol.%), as indicated by the small austenite peak ($2\theta = 42.5$) emerging above the surrounding, overlapped martensite peaks. In summary, these low stopping powers produce damage to the extent that phase transformations to metastable states occurs, although it requires high doses and radiation mechanisms in which damage accumulates, driving these phase transformation. The next section will discuss proposed damage mechanisms by which the energy transferred by electronic stopping assists the radiation induced amorphization in the 5 Mev Ni ion irradiated samples.

8.2.4 Linking the high energy ion experiments to the electronic stopping effects in 5 MeV Ni ions

The results of the electronic stopping experiments in the previous section demonstrate that low electronic stopping power ions (~4 keV/nm) produce damage in SMA TiNi films. However, these noticeable changes in the microstructure were only observable after an irradiation to relatively high fluencies, $>10^{15}$ ions cm^{-2}. At these low stopping powers, a single ion does not produce a phase transformation or create an observable track, but during the ion passage through the film, a small number of defects are generated. These defects can accumulate with increasing fluence until a critical threshold value is reached, triggering the transitions. In the 80 MeV Ar study, the amorphous zones were apparent at a fluence of 10^{15} ion cm^{-2} and varied greatly in size and shape. In the 40 MeV Ne ion irradiations, the highest dose specimen was partially transformed into austenite, indicating that a sufficient amount of damage was produced to suppress the martensite below room temperature.

In order that the energy deposited in the electronic system produce damage, it must couple to the atomic system via electron–phonon interactions. For metallic systems, this process is assumed to be much longer than the time it takes for the conduction electrons to dissipate the energy (less than 1 ps), which is why electronic stopping effects are typically neglected. However, as shown by Barbu et al. [17] and by LaGrange et al. [7, 14], electronic stopping effects can produce considerable disorder in TiNi intermetallic alloys. How, then, are the irradiation damage processes influenced by electronic excitations?

If nuclear recoils are assumed to produce all the damage and the extent of this damage is determined by the energy deposited in the electronic system, arguments about electron–phonon coupling time scales become irrelevant. In such a mechanism, the damage scales with the electronic stopping power and the extent of excited atomic bonds in material. Thus, at high electronic stopping powers, where a number of atomic bonds are excited, nuclear recoils can easily trigger a phase transformation, causing the formation of an amorphous track. At lower stopping powers, isolated defects are formed that can accumulate and drive the transform-ation, which are possible mechanisms in 80 MeV Ar and 40 MeV Ne ion irradi-ation. Why, then, do the short time scales of these electronic excitations in metals not prevent such lattice bond softening? In most metals, it does, but TiNi is special; it undergoes displacive phase transformations. The small amplitude modulation in the atomic lattice produced by the martensitic transformation, especially when irradiating near M_s, can couple with electronic excitations and produce metastable configurations that are "locked-in" due to the rapid dissipation of these excitations. Thus, the short time scales of electronic excitations are a benefit and may allow for defects to be quenched-in and accumulate.

It is, therefore, conjectured that the significant amorphization observed in 5 MeV Ni ion irradiated samples at intermediate ion penetration depths (1 μm) results from the synergy between the lattice softening caused by electronic stop-ping effects and the number of primary recoils. However, theoretical modeling and additional experiments are needed to support and conclude that such a damage mechanism exists involving the martensitic transformation. Regardless, electronic stopping effects produce notable microstructural changes in TiNi and need to be taken into account in the damage modeling of materials that undergo rapid displacive phase transformations.

8.3   Using ion beam modification to make novel actuator materials

Amongst the SMAs, TiNi alloys have the greatest potential for applications and actuators requiring large displacements and forces, since they have a high strain energy storage output (exceeding 25 MJ/m^3) and large transformational strains. However, applicability of these materials for large scale devices is limited by the fact that actuation is thermally driven; this can be sluggish and inefficient (Carnot effi-ciencies below 5%) [18]. Furthermore, actuation is non-linear with temperature in most cases, making it difficult to predicatively control motion and integrate into mechanical systems. High cyclic frequency in TiNi SMAs is dependent on two factors, thermal transport and mechanical response time needed to reset the actuator for additional cycles, i.e., the time required to deform the martensite and reset the recoverable shape strains. Thin films, however, have the advantage that they can be quickly heated and cooled by heat transfer through their surfaces, i.e., films have high surface to volume ratios. Mechanical response is governed by the manner in which the martensite is deformed and the actuator is reset on each cycle. In most SMA actuator designs, the repetitive martensite deformation and cyclic motion are accomplished with an external bias spring, e.g. the SMA microvalve design by Krulevitch et al. [19]. However, external bias springs can be awkward to employ;

they are difficult to implement in complex designs with micrometer features and limit actuation geometries and ranges that can be achieved. A more optimized actuator design has both the active SMA film and biasing elements incorporating into a single thin film, e.g., a self-biasing film.

Self-biasing can be accomplished by special processing techniques, many of which use thermomechanical treatments to develop microstructural features that stabilize preferred shapes in both the austenite and martensite, and create a two-way shape memory effect (TWSME), such as "training" techniques (over-deformation of the martensite) [20] and preferential precipitation of Ti_3Ni_4 phase [16, 21]. These treatments develop residual stresses through the nucleation and pinning of dislocations, which act as preferred heterogeneous nucleation sites for the martensite and austenite phases, and thus specific shape strains can be produced on both heating and cooling.

The main drawbacks of these training methods are low transformation forces produced on cooling, the decay and instability (fatigue) of the two-way shape change with continued thermal cycling, and ease in applying these techniques to thin films in complex actuator systems. With the preferred precipitation method (so-called "all around SME"), specific film compositions are required, which is a major drawback, limiting the operational temperature range of the device. The following sections discuss a processing technique that can overcome the afore-mentioned deficiencies and has the major advantage of being able to scale to nanometer sizes. This technique also provides for easy integration into planar photolithography processes common in MEMS device fabrication and, thus, highly flexible, allowing the creation of many different actuator types and a variety of different actuation motions. Most importantly, the actuator material developed from this technique can produce high forces on both heating and cooling. The heart of this technique is to use ion implantation as a means to modify the SME in the preferred location to develop intimately bonded biasing elements.

8.3.1 Actuator design concept

The concept of using ion implantation as a planar processing technique to develop a thermally active bimorph was first introduced by Grummon and Gotthardt [15]. This design is promising for MEMS based actuators since it essentially supplies the same motion as a bimetal but with much lower and discrete[6] temperature changes and with much larger out-of-plane displacements. The novelty of this concept is in the ability to selectively alter the martensitic transformation and the shape memory characteristics through controllable ion implantation parameters, such as the irradiation surface and ion energy. The irradiation defects produced during high energy particle bombardment can depress the martensitic transformation temperatures, since the martensitic transformations are dependent on local atomic

---

[6] Shape memory materials only actuate within a small temperature range (~50–30°C) that depends on the transformation temperatures and opposing stress.

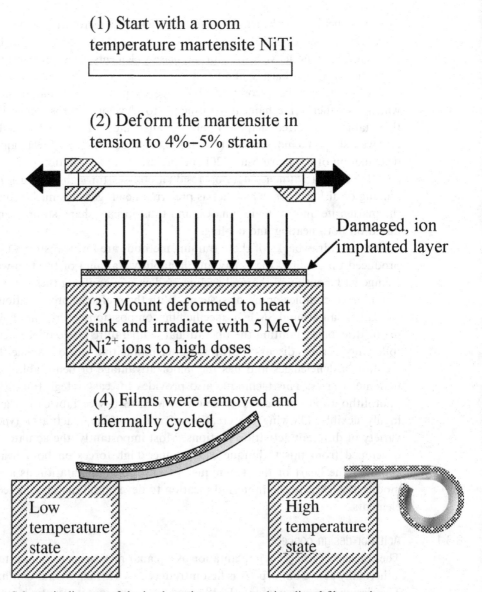

Figure 8.13 Schematic diagram of the implantation process and irradiated film motion.

order, which can be disrupted by the irradiation damage. In addition, irradiation to sufficiently high doses can damage the atomic structure to a degree that it inhibits the martensitic transformation altogether. After ion implantation, this modified ion implanted layer becomes the intimately bonded bias spring used to reset the martensite during cyclic actuation, allowing for rapid actuator speeds.

The approach, which Grummon and Gotthardt [15] took for developing the actuator and was further refined by LaGrange *et al.* [5, 6, 8, 10, 13], is fairly straightforward. It starts (step 1 in Fig. 8.13) with free-standing, 6 μm thick,

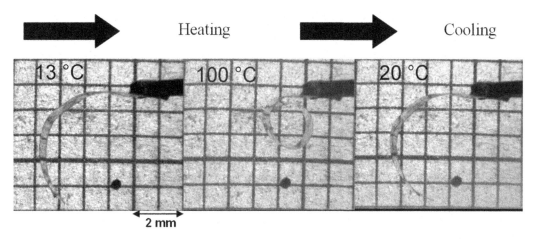

Heating

Cooling

13 °C   100 °C   20 °C

2 mm

Figure 8.14   Experimental photographs of the reversible ion implanted film curling with thermal cycling.

sputtered, near-equiatomic TiNi films with a martensite start temperature is above room temperature. (2) These films are then deformed to tensile strains of 4–5%, using a microtensile device. (3) The deformed films are mounted on a heat sink and irradiated at room temperature with 5 MeV Ni ions to sufficiently high dose to suppress the martensitic transformation in the 2 µm surface layer. (4) The films are removed from the heat sink and exhibit a reversible bending motion with temperature.

The film curls due to the differential strains that occur between the unmodified layer, which contracts during heating to recover the prior martensite deformation, and the implanted layer, for which the martensite transformation is suppressed so that it behaves as a nominally elastic metallic alloy. Upon heating ($T > A_f$), the sharp differential strains make the film curl out-of-plane. On cooling down to room temperature, the stored elastic energy in the implanted layer, acting as an intimately bonded bias spring, is used to redeform the martensite, as shown in images in Fig. 8.14.

The observed film curling motion in Fig. 8.14 exists only if the beam-damaged layer exhibits no athermal transformational behavior and possesses high stiffness and a high yield strength. If the latter condition holds to any degree, cyclic two-way shape-strain behavior is expected (as shown in Figure 8.14). It is obvious that the microstructure of the implanted beam plays a crucial role in the actuator performance. For the optimization of the actuator design, it is important to fundamentally understand how ion implantation changes the microstructure and affects the shape memory and how these changes are related to the observed film motion. This was the main goal of the studies discussed in the previous section. The next sections detail the influence of the irradiation parameters, e.g., the irradiation fluence and temperature, on the implanted film bending and link the microstructural observations discussed in Section 8.2 to the film motion.

8.3.2   Dependence of actuator motion on irradiation dose and temperature
To study the film motion, the ion implanted films were mounted in a precision temperature controlled furnace equipped with an automated optical recording apparatus, and the film curvature was measured as a function of the temperature. The implanted films were cycled either from 0 to 100 °C or 0 to 150 °C, and images were recorded every 1 or 2 °C change in temperature. To normalize the film motion for comparison purposes, a film bending strain (ε) was calculated from the film thickness (t) and ρ, the curvature ($e = t/2\rho$). The following paragraphs describe the film motion for two different fluencies (10^{13} and 10^{15} ions cm^{-2}) implanted at two different temperatures (25 °C and -140 °C). This section is concluded by a study of the cyclic stability of the film motion that compares 10^{14} and 10^{15} fluencies of films irradiated at 25 °C.

8.3.2.1   Bending of films irradiated at 25 °C
Results of the motion experiments for specimens irradiated to a fluence of 1×10^{15} ions cm^{-2} at 25 °C are shown in Fig. 8.15. Bending was observed to occur away from the irradiated face on heating, as theoretically expected. During the first cycle between 0 and 100 °C, the film curled to a strain of 0.2 %, but the motion was not completely reversible upon cooling and reverse transformation, leaving approximately ~0.1 % unrecovered strain. However, during the second cycle, the film showed a completely reversible strain of ~0.1 %. When this film was heated to 150 °C, it curled more tightly to a strain of ~1.4 %, but again failed to recover completely, leaving ~0.6 % unrecovered strain. During the fourth thermal cycle (0↔150 °C), the overall shape change during heating was lower (~0.6 %), but 0.5 % strain was recovered, leaving only 0.1 % unrecovered strain.

The motion between 0↔100 °C on the first and second cycle displays small transformations of the order of 0.1 to 0.5 % that do not reflect a complete transformation and recovery of the deformation strain induced prior to implantation. These small strains are more characteristic of the recovery behavior associated with the R phase and only partial transformation to the austenite phase. The motion at temperatures above 120 °C that resulted in high curvature strains can be related to an increased strain recovery and higher recovery stresses due to a complete reverse martensitic transformation to the austenite. Observing cycle 3, the film motion stopped at a temperature around 140 °C, which is approximately 50–55 °C above the A_f temperature of the unirradiated material observed in DSC (90–95 °C). This shift in the transformation temperatures results from the stress that the implanted layer exerts on the transforming layer during its contractive shape recovery. If the imposed stress is estimated by assuming the stress rate ($d\sigma/dT$) of 10 MPa/ °C, then the stress needed to cause a 45–50 °C change in the A_f would be 450–500 MPa. This is, in fact, close to the measured recovery stress at 4 % tensile deformed film (400 MPa) [12]. The fourth cycle showed a slightly lower A_f temperature, indicating that a small stress relaxation has occurred. This may be linked to the observed incomplete recovery in two-way motion, which is small but indicates an instability in the cyclical motion.

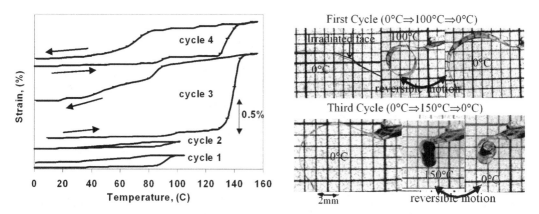

Figure 8.15 Photographs and plot of the irradiated film curvature strain as a function of temperature for 1×10^{15} ions cm^{-2} [6].

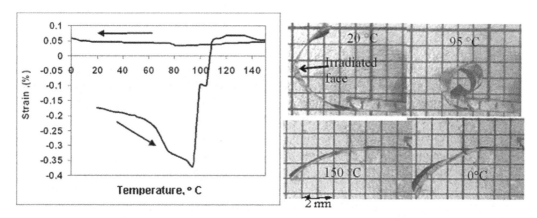

Figure 8.16 Photographs and plot of the irradiated film curvature strain as a function of temperature for 1×10^{13} ions cm^{-2} [6].

The low dose (1×10^{13} ions cm^{-2}) film motion showed a more complex and comparatively different motion, as shown in Fig. 8.16. It should be noted that after removal from the heat sink and before thermal cycling, the film was slightly curved in the direction of the irradiated face. At first and by heating to a temperature of 100 °C, the film curled upward towards the irradiated surface opposite to the expected behavior (curling downward, away from the irradiation damaged layer). With a continued heating to 150 °C, the film uncurled and bent downward in the direction away from the irradiated surface. During cooling and continued thermal cycling, there was no additional film motion, indicating that the irradiation induced defects and microstructure had attained a stable configuration.

This unpredictable motion is related to a multi-layered structure, composed of layers with differing transformation temperatures and recovery properties. The microstructural investigations on 5 MeV Ni ion implantation discussed in the

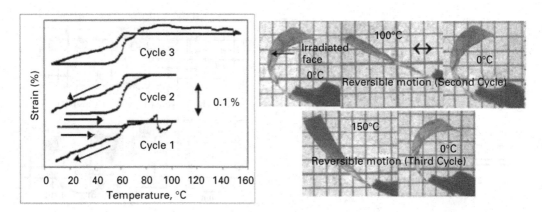

Figure 8.17   Film curvature strain as a function of temperature for specimens irradiated at $<-145\,°C$ to a dose of 1×10^{13} ions cm^{-2} [8].

previous section provides some insight into why this low dose irradiated film behaves differently. Both the X-ray and TEM investigations conclude that the implanted layer of 1×10^{13} ions cm^{-2} specimen was not significantly amorphized, but the layer did contain a large fraction of austenite which was stabilized at room temperature due to radiation effects. The DSC analysis also showed that the irradiated material had similar transformation temperatures and enthalpies as the unirradiated material, thus indicating that the implanted layer partially retained its shape memory properties. However, the transformation peaks were more broad (large hysteresis), suggesting that material within the implanted layer has varying transformation properties with some regions having suppressed martensite start temperatures due to radiation damage. A model will be given at the end of this section, which links the through-thickness variation in transformation temperatures induced by irradiation to the observed motion.

8.3.2.2   Bending of films irradiated at $-145\,°C$

The microstructural analysis of the implanted layer discussed in the previous section clearly showed that the irradiation induced phase transformations were more significant at lower temperatures. Thus, the motion of the irradiated films should follow the same trends. Films with irradiation doses above 1×10^{13} ions cm^{-2} indeed showed similar behavior, i.e. reversible motion with thermal cycling. A detailed description is given in Refs. [10, 8]. Here, it is important to note in films irradiated $\leq 10^{13}$ ions cm^{-2} that the observed motion was opposite to that expected (curling in opposite direction), which is related to the variation in transformation temperatures in the implanted layer (see Fig. 8.17).

8.3.2.3   Comparing film curling of specimens irradiated at the two different temperatures

There was essentially no difference in thermomechanical response between the two irradiation temperatures for the films irradiated at 10^{15} ions cm^{-2}; the response was as predicted, bending motion away from the irradiated face. However, there are some interesting differences between the transformational behavior of films cycled at low temperatures (0–100 °C) and at higher temperature (0–150 °C). During the first heating to 150 °C, the irradiated films curled extensively, six times more than at 100 °C, and the film stopped curling at temperatures that were 50–60 °C above the unirradiated film A_f temperature. This indicates that the differential strains and resulting interfacial stresses between the layers were much higher during high temperature cycling (on the order of several hundred MPa). At such stresses, plastic deformation could occur, resulting in the loss of film bending.

In contrast, the low doses ($\leq 10^{13}$ ions cm^{-2}) irradiated film motion was not as expected and much more complex, the main difference between the samples irradiated at the two different temperatures being the reversible motion observed in the −145 °C sample and the lack of motion in the 25 °C sample on the second thermal cycle. The dissimilarity is related to the difference in the irradiation induced microstructure, the implanted layer of the −145 °C sample had a considerably higher fraction of amorphous material, which, as in the high dose specimens, behaves elastically and resets the actuation stroke on each cycle. Although the motion in the −143 °C sample was reversible, it was in the opposite sense (curling about the implanted layer instead of away from it). In samples irradiated at a temperature of 27 °C, the films curled, at first, in a direction opposite to the expected direction up to 95 °C, and with continued heating to higher temperatures, the film changed the direction of curling in the appropriate sense with no further film movement upon cooling or in a second thermal cycle. This behavior can be explained by the TEM and XRD microstructural observations discussed in Section 8.2.2.

TEM observations of the films showed that the martensite in the implanted layer was partially transformed into austenite and R phase when irradiated to a fluence of 10^{13} ions cm^{-2}. These austenite or R-phase grains extended from the surface to the ion projected range and spanned regions greater than 5 μm. The irradiation induced phases are considered to be a coherent phase product of the martensitic transformation, created by the suppression of the R-phase and austenite start temperatures (A_s) to below room temperature as opposed to the direct nucleation of new phases from ballistic interactions. As "coherent" phase products, upon transformation, these regions recover the martensite deformation strains induced prior to irradiation, causing a partial contraction of the beam-damaged layer. Thus, when the film was released from the heat sink, the film curled about the irradiated face (opposite sense) due to the contractions in the implanted layer and resulting residual stresses. During heating to 95 °C, further shape recovery contraction and curling about the irradiated face occurred due to the transformation of the remaining martensite in the damaged layer, which has a lower austenite start temperature (from radiation defects) than the underlying undamaged martensite. On heating to 150 °C, the undamaged martensite layer transformed to austenite, causing a reversal of the film motion and uncurling away from the irradiated face (bending in the

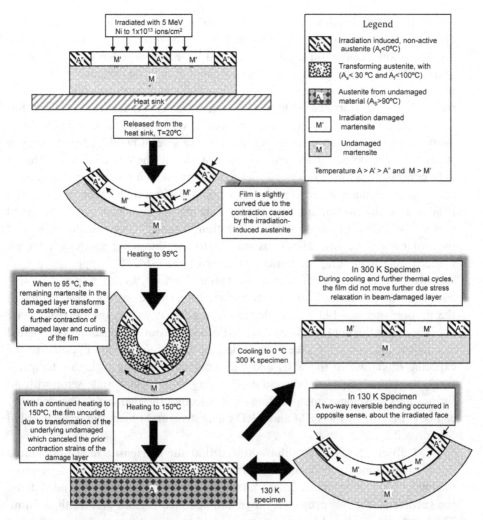

Figure 8.18   Schematic of the film motion for the specimens irradiated to 1×10^{13} ions cm^{-2}.

expected sense), due to the contraction of the undamaged martensite and elimination of the differential strains. Figure 8.18 shows a schematic of the proposed mechanism for these exotic film deflections observed in the specimens irradiated at low doses.

The temperature at which the film irradiated at 25 °C reversed motion and uncurled (95 °C), was attributed to the "stress-induced" increase in the A_s temperature of the undamaged martensite layer and the initiation of the transformation and shape recovery in this layer. This temperature was elevated about 30 °C above the nominal unirradiated film's "zero-stress" A_s temperature (65 °C), and corresponds to about 300 MPa stress (assuming a stress rate of 10 MPa/K) imposed on the undamaged layer by the beam-damaged layer. Partial stress relaxation in the implanted layer may explain the reason for a

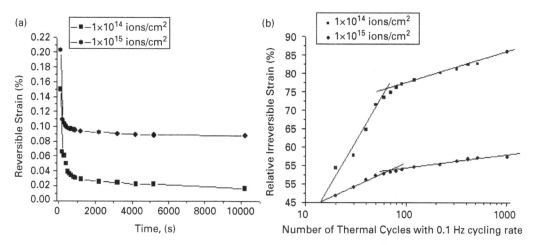

Figure 8.19 (a) Evolution of reversible strain with annealing time at $100\,°C$ for 10^{14} and 10^{15} fluence irradiated at $25\,°C$, (b) the irreversible strain normalized to the maximum curling strain as a function of the number of cycles from 0 to $100\,°C$ with 0.1 Hz cycling rate (lines are not fits but are used to guide the eyes).

reduced bending strain, but it does not explain why it was eliminated altogether, unless the stress was relaxed completely in these films at high temperatures ($150\,°C$).

The films irradiated at $-145\,°C$ also showed a non-conventional behavior, but in a slightly different manner. As in the $25\,°C$ specimen, when the irradiated film was released from the heat sink, the film curled (corresponding to a residual stress of ~50 MPa) as a result of the partial transformation to austenite. In the second cycle, this uncurling during heating and curling during cooling was fully reversible, even at higher temperatures ($150\,°C$). This can be explained by the difference in transformation temperatures between the damaged and undamaged layers, with lower transformation temperatures in irradiation affected zones. The only difference between the -145 and $25\,°C$ specimens is that there was no apparent stress relaxation in the $-145\,°C$ specimen at high temperatures, which led to a loss in the film curling motion. This may be attributed to enhanced radiation hardening and amorphization of the implanted layer at the lower irradiation temperature. The loss in reversible motion with continued cycling may result from relaxation processes in the implanted layer, as will be discussed in the next section.

8.3.3 Cyclic fatigue, decay of two-way shape strains

To explore the cyclic behavior, films irradiated to 1×10^{14} ions cm^{-2} and 1×10^{15} ions cm^{-2} were cycled between 0 and $100\,°C$ at a constant rate of 0.1 Hz, while monitoring the decay in two-way shape strain with time and the number of cycles, as shown in Figs. 8.19(a) and (b). The trend in curves between high and low fluence was similar, both fluencies showing a large decrease in reversible strain initially which then leveled off quickly to a lower constant value. The 10^{14} ion/cm^2 fluence had smaller initial two-way shape strains and showed a larger relaxation in two-way motion with time (steeper slope). These trends are more obvious when the irreversible strain is normalized to the maximum curling strain, as in Fig. 8.19(b).

The evolution of the film motion shows two distinct trends, a steep increase in the irreversible strain in the initial 50 cycles, and a gradual change with increasing number of cycles. If the trend for the 10^{14} ion/cm^2 specimen is extrapolated out to a higher number of cycles, it appears that the film will stabilize to a small reversible bending motion of ~0.01% strain. The trend in the 10^{15} fluence would indicate that the films at this dose will continue to move for an infinite number of cycles with minimal loss in two-way shape strain (less than 0.075%). The dramatic increase during initial times is related to the thermal and structural instability of the implanted layer, while cyclic fatigue at longer times can be attributed to deformation mechanisms at stress within the layer, which indicate the lack of strength at lower fluencies. The thermal instability of the implanted layer may result from the relaxation arising from the deformation processes in the amorphous material near the interface.

Essentially, there are two types of yielding in amorphous metals that can occur, inhomogeneous and homogenous flow. Inhomogeneous flow is the localized yielding of the amorphous metal by slip band propagation, e.g., serrated flow or plastic instability mechanisms, and typically occurs at high stresses ($\sigma = E/50$, where E is Young's modulus), which are on the order of 2 GPa in most glassy metals [22]. The maximum stresses during actuation in the damage layer are far below the implanted (amorphous) layer static yield stress, and, thus, inhomogeneous flow is unlikely to occur. On the other hand, homogenous flow can occur at moderately low stresses ($\sigma = E/100$) within the range of stresses estimated at the interface between the implanted zone and undamaged material. However, the thermally activated homogenous flow should be limited at temperature below 150 °C and cannot account for the sudden drop in the film's reversible motion during the first, short cycles.

Why then does the film's reversible motion decrease rapidly on the onset of cycling? In short, this is due to the fact that glasses formed by irradiation are thermally unstable and, when they are heated, they undergo atomic rearrangements and attain a new equilibrium state associated with a given temperature. These atomic rearrangements structurally relax the crystal which changes the amorphous material's physical properties. Most reports have found that the volume change associated with structural relaxation is relatively small (0.5%) [23, 24]. However, in our case, a volume contraction of this magnitude can significantly relax the internal stresses in the beam-damaged layer (as illustrated by the simple approximation 0.005×110 GPa $= 550$ MPa). In fact, the initial loss in the reversible deflective strain is of the order of 0.2 to 0.5% (see Fig. 8.19).

From a general engineering standpoint, it becomes clear that the nonreversible motion observed is related to the unstable thermodynamic state of the amorphous damage layer, in which stress relaxes due to the high temperature actuation. This is not a desirable aspect for actuator systems, since a mild overheating ($+50\,^\circ$C) can cause serious loss in actuation. This limits the operational temperature of the device considerably and imposes tight restrictions on temperature deviations, complicating the actuation control. If these materials are to be used in devices, it is recommended that the TiNi thin films be irradiated above a fluence of 1×10^{15} ions/cm^2 for two reasons: (1) the beam-damaged layer has a more homogeneous microstructure and uniform mechanical properties, (2) the bimorph is more thermally stable and has lower cyclic fatigue.

8.4   Summary

In this chapter, we have described the influence of ion implantation on the microstructural modifications in TiNi SMA thin films. We focused on investigations involving 5 MeV Ni ion irradiation since it can be used as a means to selectively alter the transformation characteristics and to develop TiNi based thin film actuator material for MEMS devices. The primary effects of ion implantation on the microstructure can be summarized as follows.

- At above 10^{15} ions cm^{-2}, the beam-damaged layer from the surface to ion projected range was largely amorphized with the exception of a small volume fraction (2–5 vol.%) of the irradiation resistant nanocrystalline phases of austenite and Ti$_2$Ni.
- At fluencies of 1×10^{14} ions cm^{-2}, the microstructure was amorphized significantly (~70 vol.%), and a continuous amorphous layer was observed at a depth of 1.2 μm, which was 0.7 μm shallower than the maximum atomic displacements from nuclear collisions predicted by TRIM simulations. This shift in damage and the significant surface amorphization are thought to arise from contributing effects of the electronic stopping processes.
- The ion beam damage is enhanced at cryogenic temperatures due to limited annealing and thermally re-ordering effects as compared to layers irradiated at 25 °C.
- Pre-straining the martensite prior to irradiation also enhances the amount of irradiation induced amorphization.

The role of electronic stopping effects in irradiation induced amorphization observed in TiNi films was also discussed. From electronic stopping experiments using swift ions, it was concluded that ions with electronic stopping powers of 3 keV nm^{-1} produce damage in TiNi films. The synergy that occurs between the lattice softening resulting from electron excitations (electronic stopping effects) and nuclear recoils (nuclear stopping) may explain significant surface amorphization and the formation of a homogenous amorphous implanted layer for 5 MeV Ni ion irradiation. This perspective deviates from the classic view on the ion beam damage in metals, which ignores electronic stopping effects. However, since TiNi possesses unique physical properties that may promote the electronic stopping effect, modeling high energy ion damage must take these effects into account.

The latter part of this chapter described a method for developing an out-of-plane film actuation based on the preferential suppression of SME using ion beam damage, and the motion of this actuator material as a function of the implantation parameters. The main points of this section are summarized in the following.

- Films irradiated at doses $\leq 1 \times 10^{13}$ ion cm^{-2} have complex motion due to variation in microstructure and transformation temperatures within the beam-damaged layer.
- In samples irradiated to high doses, $>10^{14}$ ions/cm^2, significant bending movements can be achieved, but losses in the film's reversible bending motion occur with thermal cycling, especially at high temperatures.
- This loss in the irradiated film's reversible motion was attributed to relaxation of the high elastic stresses in the amorphous ion beam damage layer.
- The use of ion implanted TiNi films as an actuator material in MEMS devices requires that films be irradiated above a fluence of 1×10^{15} ions/cm^2 for two reasons: (1) the beam-damaged layer has more uniform mechanical properties, (2) the bi-morph has lower cyclic fatigue.

References

[1] A. Seeger, On the theory of radiation damage and radiation hardening. In *Proceedings of the Second UN International Conference on the Peaceful Uses of Atomic Energy*, New York: United Nations **6** (1958) p. 250.

[2] M. T. Robinson, in *Proceedings of a Conference on Nuclear Fusion Reactors*, Abingdon, England, 1969. Abingdon, England: UKAEA (1970) p. 364.

[3] J. Biersack and L. G. Haggmark, A Monte Carlo program for the transport of energetic ions in amorphous targets, *Nuclear Instruments & Methods in Physics Research Section B* **174** (1980) 257–269.

[4] J. Ziegler, J. Biersack and U. Littmark, *The Stopping Range of Ions in Matter*, New York, USA: Pergamon Press (1985).

[5] T. B. Lagrange and R. Gotthard, Microstructrual evolution and thermo-mechanical response of Ni ion irradiated TiNiSMA thin films, *J. Optoelectronics Adv. Mat.* **5** (2003) 313–318.

[6] T. LaGrange, R. Schaublin, D. S. Grummon and R. Gotthardt, Nickel ion irradiation of plastically deformed martensitic titanium nickel thin films, *Journal De Physique IV* **112** (2003) 865–868.

[7] T. Lagrange, R. Schaublin, D. S. Grummon, C. Abromeit and R. Gotthardt, Irradiation-induced phase transformation in undeformed and deformed NiTi shape memory thin films by high-energy ion beams, *Phil. Mag.* **85** (2005) 577–587.

[8] T. LaGrange, D. S. Grummon and R. Gotthardt, The influence of irradiation parameters on the behavior of martensitic titanium nickel thin films. In *Materials Research Society Fall Meeting*. Boston, MA: MRS (2003) pp. 495–500.

[9] T. LaGrange and R. Gotthardt, The effect of pre-strain on microstructure of 5 MeV ion irradiated martensitic TiNiSMA thin films, *Scripta Materialia* **50** (2004) 231–236.

[10] T. LaGrange and R. Gotthardt, An ion implantation processing technique used to develop shape memory TiNi thin film micro-actuator devices, *Journal De Physique IV* **115** (2004) 47–56.

[11] T. LaGrange and R. Gotthardt, Post-annealing of ion irradiated TiNiSMA thin films, *Mater. Sci. Eng. A* **378** (2004) 448–452.

[12] T. LaGrange and R. Gotthardt, Bi-metallic model of the free recovery motion of ion irradiated Ti-rich NiTi shape memory alloy thin films, *Mater. Sci. Eng. A* **387–89** (2004) 753–757.

[13] T. LaGrange and R. Gotthardt, Martensitic transformation of partially irradiated Ni-Ti films, resulting in a new technique for designing micro-actuators, *Materials Science Forum*, **476–432** (2003) 2219–2224.

[14] T. LaGrange, C. Abromeit and R. Gotthardt, Microstructural modifications of TiNi shape memory alloy thin films induced by electronic stopping of high-energy heavy ions, *Mater. Sci. Eng. A* **438** (2006) 521–526.

[15] D. S. Grummon and R. Gotthardt, Latent strain in titanium-nickel thin films modified by irradiation of the plastically-deformed martensite phase with 5 MeV Ni2+, *Acta Mater* **48**, (2000) 635–646.

[16] R. Kainuma, M. Matsumoto and T. Honma, The mechanism of the all-round shape memory effect in a Ni-rich TiNi alloy. In *Proceedings of International Conference on Martensitic Transformations*, Naramara, Japan: Japanese Institute of Metals (1986) p. 717.

[17] A. Barbu, A. Dunlop, A. Hardouin Duparc, G. Jaskierowicz and N. Lorenzelli, Microstructural modifications induced by swift ions in the NiTi intermetallic compound 1, *Nuclear Instruments and Methods in Physics Research Section B* **145** (1998) 354–372.

[18] T. W. Duerig, K. N. Melton, D. Stockel and C. M. Wayman, *Engineering Aspects of Shape Memory Alloys*, London: Butterworth-Heinemann (1990).

[19] P. Krulevitch, A. P. Lee, P. B. Ramsey, J. C. Trevino, J. Hamilton and M. A. Northrup, Thin film shape memory alloy microactuators, *J. Microelectromech Syst.* **5** (1996) 270–282.

[20] K. Otsuka and C. M. Wayman, *Shape Memory Materials*, Cambridge: Cambridge University Press (1998).

[21] A. Gyobu, Y. Kawamura, H. Horikawa and T. Saburi, Martensitic transformation and two-way shape memory effect of sputter-deposited Ni-rich Ti-Ni alloy films, *Mater. Sci. Eng.* **A273–275** (1999) 749–753.

[22] F. Spaepen and A. I. Taub, Flow and fracture, in *Amorphous Metallic Alloys*, ed. F. E. Luborsky, London: Butterworth (1983) pp. 231–256.

[23] H. S. Chen. The influence of structural relaxation on the density and Young's modulus of metallic glasses, *J. Appl. Phy.* **49** (1978) 3289.

[24] H. S. Chen. Structural relaxation in metallic glasses, in *Amorphous Metallic Alloys*, ed. F. E. Luborsky, London: Butterworth (1983) pp. 169–186.

9   Laser post-annealing and theory

W. M. Huang and M. H. Hong

Abstract

TiNi shape memory thin films have great potential as an effective actuation material for microsized actuators. However, a high temperature (above approximately 723 K) is required in order to obtain crystalline thin films either during deposition (e.g., sputtering) or in post-annealing. Such a high temperature is not fully compatible with the traditional integrated circuit techniques, and thus brings additional constraint to the fabrication process. Laser annealing provides an ideal solution to this problem, as the high temperature zone can be confined well within a desired small area at a micrometer scale. In this chapter, we demonstrate the feasibility of local laser annealing for crystallization in as-deposited amorphous TiNi thin films and present a systematic study of the theories behind this technique.

9.1   Introduction

TiNi shape memory thin films have great potential as an effective actuation material for microsized actuators [1]. As compared with other materials, such as electrostatic, electromagnetic and piezoelectric thin films, the work output per unit volume of TiNi shape memory thin films is significantly higher, because they are able to provide not only a larger force but also over a longer displacement [2]. In addition, the disadvantage of slow response speed in bulk shape memory alloys (SMAs) can be dramatically improved due to a larger surface-area-to-volume ratio in thin films. 20 Hz frequency has been achieved in a TiNi thin film based microvalve [3]. Furthermore, TiNi shape memory thin films can be fabricated by techniques which are compatible with the well-established batch-processing technology of Si micromachining [1,4]. A number of microdevices have been realized using TiNi shape memory thin films, as presented in Chapters 10 to 19.

    There are several deposition techniques to produce TiNi thin films. Typical ones include sputter deposition [5], laser ablation [6], flash evaporation [7], ion beam

Thin Film Shape Memory Alloys: Fundamentals and Device Applications, eds. Shuichi Miyazaki, Yong Qing Fu and Wei Min Huang. Published by Cambridge University Press. © Cambridge University Press 2009.

sputter deposition [8], and electron beam evaporation [9]. Among them, sputter deposition is more widely used at present due to its better quality in terms of uniformity and strength and, more importantly, it is less problematic in contamination.

There are two methods to produce TiNi shape memory thin films by sputter deposition. One is to deposit TiNi on a high temperature substrate (at around 723 K or above) [3]. The other is that after deposition at room temperature, the as-deposited film, which is amorphous, is post-annealed in a vacuum chamber [5]. Both methods have problems in terms of full compatibility with the traditional integrated circuit techniques, in which such a high temperature may not be applicable. Furthermore, both methods are not applicable for local annealing TiNi films at a micro meter scale.

Laser light can be focused to deliver an enormous amount of energy to a precise location. Hence, it is an ideal tool for micromachining and heat treatment without physically contacting the processed materials [10].

Typically, there are three types of laser for different applications. The q-switched diode-pumped solid-state laser is ideal for microdrilling and other high speed processes. The AVIA laser belongs to this category. The carbon dioxide (CO_2) laser offers the highest average power for materials processing. It emits a $10.6\,\mu m$ wavelength light beam, and the heat effect is significant. The excimer laser is a pulsed gas laser that uses a mixture of gases to provide emissions at a series of discrete wavelengths in the ultraviolet region of the spectrum. The KrF excimer laser emits a 248 nm wavelength light beam.

Various lasers have been widely used for machining [11, 12, 13, 14, 15, 16], welding [17, 18, 19, 20], surface modification [21, 22, 23, 24, 25], and photothermally heating [26] bulk SMAs and their components. On the other hand, they have been used for the deposition of TiNi/TiNiCu shape memory thin films [27, 28, 29, 30], and more recently NiMnGa magnetic shape memory thin films [31].

Laser annealing is a surface treatment method, which utilizes the rapid heating/cooling rate produced on metal surfaces exposed to a laser beam to realize a solid-state transformation [32, 33]. In laser annealing, the fraction of the laser power absorbed by a material depends on the absorptivity of the material surface. In many cases, the absorptivity is relatively low, in particular for materials with a smooth surface and high electrical conductivity. With the increase in wavelength of a laser beam, the absorption normally decreases.

Laser annealing of cold-rolled TiNi shape memory thin sheets has been reported (e.g., [34]). The annealed part exhibits the shape memory effect (SME), while the remaining non-annealed part is still elastic. Since the cold-rolled sheets, which cannot be well-integrated with the modern micromachining process, normally have a much rougher surface than that of the sputter deposited films, the annealing technique for sputter deposited TiNi thin films, which have a typical surface roughness of only about a few nanometers, has some unique features.

In this chapter, the feasibility of laser local annealing will be demonstrated experimentally, followed by a detailed investigation into the theories behind this technique.

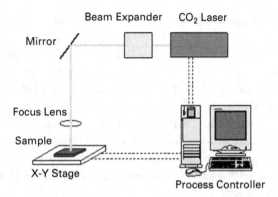

Figure 9.1   Set-up of CO_2 laser annealing [35]. (With permission from Institute of Physics Publishing.)

9.2   Experimental demonstration

CO_2 laser and fiber-injected CW high power near-IR laser diodes have been utilized for local annealing of sputter deposited amorphous TiNi thin films [35, 36].

Figure 9.1 illustrates a typical set-up for CO_2 laser annealing, in which a continuous-wave CO_2 laser (48–2 KWJ, SYNRAD) is used [37]. The laser beam is in the TEM_{00} mode, at 10.6 μm wavelength and 25 W power. The laser beam passes through a beam expander, is reflected by a mirror, passes the focus lens and is finally focused on the sample surface. The diameter of the focused beam spot is about 100 μm. An X–Y stage is used to carry the sample moving horizontally at a speed of 3 mm/s [35].

In [37], the samples for annealing were 3.4 μm thick, amorphous TiNi thin films deposited atop 4-inch (100) Si wafers (450 μm thick) by a DC/RF sputtering machine with argon atmosphere of 1 mtorr at room temperature (about 297 K). Two targets, namely a $Ti_{50}Ni_{50}$ (at%) target (with RF power of 400 W) and a Ti target (with d.c. power of 40 W), were used. The as-deposited films were confirmed to be amorphous by an X-ray diffraction (XRD) test. Before laser annealing, the sample surface has to be cleaned carefully with acetone and deionized water.

As the surface of the as-deposited amorphous TiNi thin films are very smooth (the average surface roughness, R_a, is a few nm only) [35], and TiNi is an alloy with high electrical conductivity (resistivity is about 0.5–1.1 μΩm) [38], most of the infrared laser energy is reflected instead of being absorbed. However, it is proved that the annealing of TiNi thin films can be done from the back side (Si wafer side) [37]. As the laser beam penetrates through the Si substrate, a large portion of the laser energy is reflected by the TiNi film, which causes a localized high temperature region at the Si–TiNi interface, as illustrated in Fig. 9.2. Hence, the amorphous TiNi thin film can be heated up. If the high temperature is over the required threshold energy for crystallization, the amorphous TiNi turns crystalline.

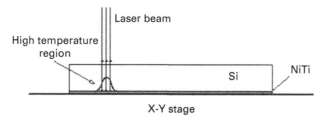

Figure 9.2   Schematic illustration of CO_2 laser annealing from the Si side [35]. (With permission from Institute of Physics Publishing.)

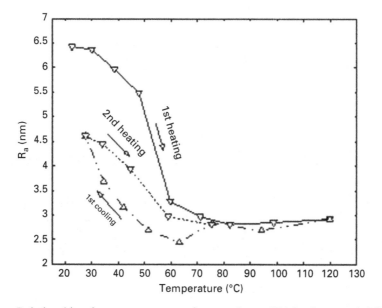

Figure 9.3   Relationship of temperature to surface roughness (R_a) in the annealed film [37].

The highest temperature during the laser scanning process revealed by a thermal image system was 833 K [35], which is over the minimal annealing temperature for TiNi thin films.

Optical images reveal that the surface morphology changes significantly after the laser scanning, since the TiNi surface becomes much rougher at room temperature (about 297 K). Figure 9.3 plots the relationship of temperature to surface roughness (R_a) in an annealed TiNi film upon thermal cycling. XRD test confirms the formation of cubic austenite and monoclinic martensite in the films.

A single annealed line about 100 μm wide is presented in Fig. 9.4. Figure 9.5 shows the edge of an annealed line upon heating. It is clear that the surface roughness of the annealed part is reduced remarkably upon heating, which is the result of the thermally induced phase transformation.

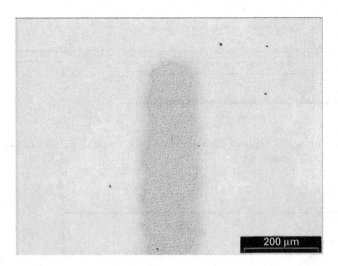

Figure 9.4   A single laser annealed line (optical microscope observation).

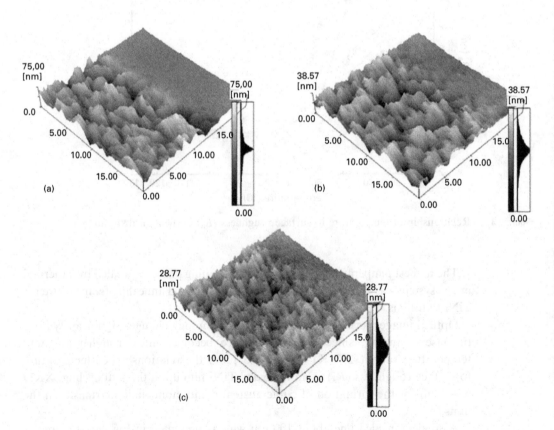

Figure 9.5   Atomic force microscope (AFM) results at the edge of a CO_2 laser annealed line at different temperatures upon heating: (a) at 297.6 K, (b) at 344.2 K, (c) at 367.0 K. The scanned area is $20 \times 20 \, \mu m$. The top-right part is un-annealed. The bar at right indicates the distribution of height (in nm) [37].

It was also observed that the amorphous TiNi films can be easily peeled-off from the Si substrate before the laser annealing. But, after the laser processing, TiNi thin films adhere tightly to the Si substrate [35].

9.3   Theories behind laser annealing

As demonstrated above, the laser is an effective tool for annealing TiNi thin films deposited atop an Si substrate. In order to get a better understanding of the process of laser annealing of TiNi thin films, it is necessary to fully understand the theories behind this technique.

Here, we present a laser irradiation model for multilayer structures and then a detailed investigation on laser annealing of TiNi thin films [39].

9.3.1   Absorption of laser irradiation

9.3.1.1   Fresnel equations

To estimate the amount of laser energy that couples into a film, Fresnel equations,[1] which give the ratio of the amplitude of the reflected electric field over that of the initial electric field for electromagnetic radiation incident on a dielectric material, can be used. In general, when a wave reaches a boundary between two materials with different dielectric constants, one part of the wave energy is reflected and the remaining part is transmitted. Since electromagnetic waves are transverse, there are separated coefficients in the directions perpendicular to and parallel to the surface of the dielectric. We may denote the coefficient for reflection of the transverse electric field (TE) as r_s, and the coefficient for reflection of the transverse magnetic field (TM) as r_p. In addition to the reflection amplitude coefficients, reflection intensity coefficients are often defined as the square of the corresponding amplitude coefficients as

$$R_s = |r_s|^2, \tag{9.1}$$

$$R_p = |r_p|^2. \tag{9.2}$$

For TE radiation shown in Fig. 9.6,

$$r_s = \frac{E_r}{E_i} = \frac{\frac{n_1}{\mu_1}\cos\theta - \frac{n_2}{\mu_2}\cos\theta'}{\frac{n_1}{\mu_1}\cos\theta + \frac{n_2}{\mu_2}\cos\theta'}, \tag{9.3}$$

where n_1 is the refraction index of the first medium, n_2 is that of the second, θ is the angle to the normal of the first medium, θ' is the angle to the normal of the second medium (which is different from θ due to refraction), μ_1 is the magnetic

[1] scienceworld.wolfram.com

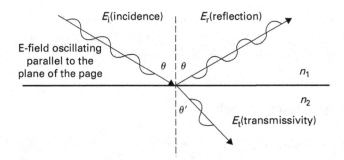

Figure 9.6   Schematic plot of TE radiation [37].

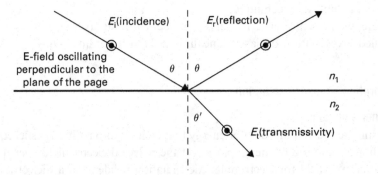

Figure 9.7   Schematic plot of TM radiation [37].

permeability of the first medium, and μ_2 is the permeability of the second medium. In general, $\mu_1 \approx \mu_2 \approx 1$. Hence,

$$r_s = \frac{n_1 \cos\theta - n_2 \cos\theta'}{n_1 \cos\theta + n_2 \cos\theta'}. \qquad (9.4)$$

The relationship between θ and θ' follows Snell's law, i.e.

$$n_1 \sin\theta = n_2 \sin\theta'. \qquad (9.5)$$

For TM radiation as illustrated in Fig. 9.7,

$$r_p = \frac{E_r}{E_i} = \frac{\frac{n_2}{\mu_2}\cos\theta - \frac{n_1}{\mu_1}\cos\theta'}{\frac{n_2}{\mu_2}\cos\theta + \frac{n_1}{\mu_1}\cos\theta'}. \qquad (9.6)$$

If $\mu_1 \approx \mu_2 \approx 1$, it is reduced to

$$r_p = \frac{n_2 \cos\theta - n_1 \cos\theta'}{n_2 \cos\theta + n_1 \cos\theta'}. \qquad (9.7)$$

In the case of normal incidence (i.e., $\theta = 90°$), the reflection amplitude, r_n, can be expressed as

$$|r_n| = |r_s|_{90°}| = |r_p|_{90°}| = \left| \frac{n_2 - n_1}{n_2 + n_1} \right| \qquad (9.8)$$

and the intensity reflectance, R_n, can be expressed as

$$R_n = |r_n|^2 = \left| \frac{n_2 - n_1}{n_2 + n_1} \right|^2. \tag{9.9}$$

For absorbent materials, whose complex indexes of refraction are $m_1 = n_1 - ik_1$ and $m_2 = n_2 - ik_2$, from Eqs. (9.4) and (9.7), it can be deduced that [40]

$$r_s = \frac{(n_1 - ik_1)\cos\theta - (n_2 - ik_2)\cos\theta'}{(n_1 - ik_1)\cos\theta + (n_2 - ik_2)\cos\theta'}, \tag{9.10}$$

$$r_p = \frac{(n_2 - ik_2)\cos\theta - (n_1 - ik_1)\cos\theta'}{(n_2 - ik_2)\cos\theta + (n_1 - ik_1)\cos\theta'}. \tag{9.11}$$

At $\theta = 90°$, the intensity coefficient for reflection can be expressed as

$$R_n = |r_n|^2 = \left| \frac{(n_1 - ik_1) - (n_2 - ik_2)}{(n_1 - ik_1) + (n_2 - ik_2)} \right|^2. \tag{9.12}$$

9.3.1.2 Absorption of laser radiation by metals

When a laser beam strikes a metal surface, the energy is absorbed within a thin layer near the surface. The absorption follows

$$I(z) = I_0 \exp(-az), \tag{9.13}$$

where I_0 is the incident intensity, $I(z)$ is the intensity reaching depth z from the surface, and a is the absorption coefficient. The optical penetration depth is defined as [10]

$$l = \frac{1}{a} = \frac{\lambda}{4\pi k}, \tag{9.14}$$

where λ is the wavelength of incident light. For metals with a value of k of the order of 10, a layer in a thickness of the order of 100 nm may be considered as opaque [10]. In practice, the optical penetration thickness is very small compared to the thickness of metallic films.

For an opaque metal impinged by a laser beam from the air, the fraction of the absorbed incident radiation is

$$A = 1 - R. \tag{9.15}$$

The reflectivity at normal incidence can be calculated from Eq. (9.12) as

$$R = \frac{(n-1)^2 + k^2}{(n+1)^2 + k^2}, \tag{9.16}$$

where $n_1 = 1$, $k_1 = 0$ for air in Eq. (9.12). The complex index of refraction of a metal is $m = n - ik$. Thus, the absorptivity of metals is

$$A = \frac{4n}{(n+1)^2 + k^2}. \tag{9.17}$$

9.3.1.3   Absorption of laser radiation by semiconductors

Now we consider the absorption of laser radiation in an isotropic slab of finite uniform thickness, h, and infinite extension in the xy-plane. For a semi-infinite slab, i.e., the optical penetration depth is smaller than the thickness $(l < h)$, absorptivity can be calculated by Eq. (9.17). In the case of moderate to weak absorption $(l \geq h)$, multiple reflections within the slab cannot be ignored. The absorptivity becomes dependent on the thickness of the slab, i.e., $A = A(h)$.

Due to energy conservation, the absorptivity, A, is related to the reflectivity, R, and the transmittivity, D, by

$$A + R + D = 1. \tag{9.18}$$

For a slab with finite thickness and in the case of weak absorption $(l \gg Z)$ and $k < 1$, one has [41]

$$R = \frac{(1 - n^2)^2 \sin^2(\beta h/2)}{\xi}(1 - A), \tag{9.19}$$

$$A = 2k\frac{(1 + n)\beta h - (1 - n)\sin(\beta h)}{\xi}, \tag{9.20}$$

$$D = \frac{4n^2}{\xi}(1 - A), \tag{9.21}$$

where $\xi = 4n^2 + (1 - n^2)^2 \sin^2(\beta h/2)$, and $\beta = 4\pi n/\lambda$.

9.3.1.4   Absorption of laser energy in multilayer structures

Discontinuity in material properties occurs at the interface of multilayer structures, as shown in Fig. 9.8. Define the complex indexes of refraction of layers 1 and 2 as $m_1 = n_1 - iK_1$ and $m_2 = n_2 - iK_2$. Hereafter, subscripts 1 and 2 refer to layer 1 and layer 2, respectively.

Consider that $h_2 \gg l_2$ (l_2 is the optical penetration depth of layer 2), layer 2 can be considered semi-infinite in terms of its optical behavior. Because $D = 0$, we have

$$A + R = 1, \tag{9.22}$$

where $A (= A_1 + A_2)$ is the total absorptivity of the system. For surface absorption $(h_1 \geq l_1)$, $A = A_1$. For weakly absorbing films $(h_1 < l_1)$, due to interferences, the total absorptivity changes with film thickness, h_1. The heat generated in layers 1 and 2 can be expressed as [41]

$$Q_1 = I_0 f_1(z), \tag{9.23}$$

$$Q_2 = I_0 f_2(z), \tag{9.24}$$

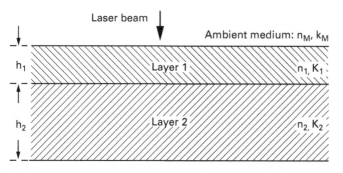

Figure 9.8   Laser irradiation on a multilayer structure [39].

where $f_1(z)$ and $f_2(z)$ describe the attenuation of the laser beam in the z-direction. The functions $f_1(z)$ and $f_2(z)$ can be written as [41]

$$f_1(z) = a_1(1 - R_1)\left|\frac{r_{12}\exp(-i\varphi) + \exp(-i2\psi_1)\exp(i\varphi)}{r_{M1}r_{12} + \exp(-i2\psi_1)}\right|^2 \quad (9.25)$$

and

$$f_2(z) = a_2 n_2 \left|\frac{(1 + r_{M1})(1 + r_{12})}{r_{M1}r_{12} + \exp(-i2\psi_1)}\right|^2 \exp(a_1 h_1)\exp[-a_2(z - h_1)], \quad (9.26)$$

with

$$1 - R_1 = \frac{4n_1}{(1 + n_1)^2 + k_1^2},$$

$$\varphi = \frac{2\pi z(n_2 + ik_2)}{\lambda},$$

$$r_{M1} = \frac{n_M + ik_M - n_1 - ik_1}{n_M + ik_M + n_1 + ik_1},$$

$$r_{12} = \frac{n_1 + ik_1 - n_2 - ik_2}{n_1 + ik_1 + n_2 + ik_2},$$

$$\psi_1 = \frac{2\pi h_1(n_1 + ik_1)}{\lambda}.$$

The energies absorbed within the film and the substrate are

$$A_1 = \int_0^{h_1} f_1(z)\mathrm{d}z \quad (9.27)$$

and

$$A_2 = \int_{h_1}^{\infty} f_2(z)\mathrm{d}z. \quad (9.28)$$

Table 9.1 Selected optical properties of some materials [39, 42, 43]

Material	n	k	l (optical penetration depth, $l = \frac{\lambda}{4\pi k}$)
Air	1	0	N/A
Si	3.4215	1.27×10^{-4}	33 mm [41]
NiTi (estimation)	$7.568 \left(\frac{n_{ni} + n_{ti}}{2} \right)$	$28.339 \left(\frac{k_{ni} + k_{ti}}{2} \right)$	0.03 μm
SiO$_2$	2.184	2.205×10^{-2}	41.15 μm
TiO$_2$	1.445	0.063	13.39 μm

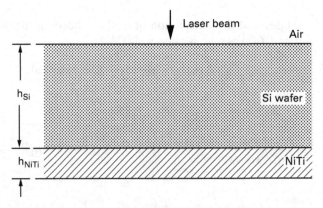

Figure 9.9   Schematic illustration of backside CO_2 laser annealing [39].

The oscillation in the absorptivity shown in Eqs. (9.27) and (9.28) is due to the interference of laser light within layer 1. The period is [41]

$$\Delta h = \frac{\lambda}{2n_1}. \tag{9.29}$$

9.3.2   Backside CO_2 laser annealing

Consider the case in Section 9.2. A continuous wave CO_2 laser is used for annealing TiNi thin films (Fig. 9.9). In real practice, a laser beam scans over a sample surface at a given speed following a pre-determined path by horizontally moving the stage which carries the sample. Normally, the relative speed of the laser beam (i.e., the moving speed of the stage carrying the sample) is not fast. For simplicity, we may consider both laser and sample to be fixed while letting the annealing time be a variable. For example, with a laser beam spot radius of 50 μm, 0.03 second annealing time roughly corresponds to a scanning speed of 3 mm/s.

For optical properties of Si and NiTi, one may refer to Table 9.1.

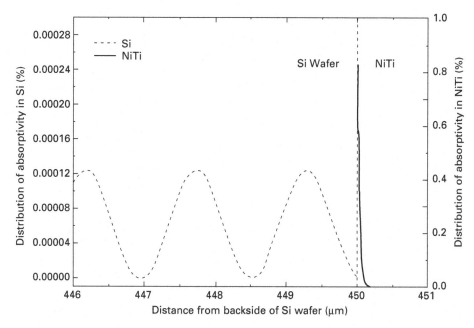

Figure 9.10 Distribution of absorptivity near the Si/NiTi interface ($h_{Si} = 450\,\mu m$) [39].

9.3.2.1 Absorption of Si/TiNi multilayer

TiNi can be considered as a semi-infinite substrate in terms of its optical behavior. Thus, $A + R = 1$. Since $h_{Si} < l_{Si}$ (33 mm), where subscript Si refers to the Si wafer, the absorption behavior of an Si/TiNi multilayer is weak absorption. Both Si wafer and TiNi thin films absorb laser energy. Figure 9.10 shows the distribution of absorptivity near the Si/TiNi interface calculated from Eqs. (9.25) and (9.26).

By integrating absorptivity through the respective material in its thickness direction, we get the absorptivity of Si and TiNi as $A_{Si} = 28.1\%$, $A_{TiNi} = 24.5\%$ and the total absorptivity $A = A_{Si} + A_{TiNi} = 52.6\%$. Here, subscript TiNi or NiTi refers to TiNi. The average absorptivity of Si is 0.06%/μm over its thickness (450 μm). TiNi thin film absorbs laser energy within a thickness of less than 300 nm.

Distribution of absorptivity also changes with h_{Si}. Figures 9.11 and 9.12 show the variation of the absorptivity distributions in Si and TiNi when h_{Si} is gradually reduced from 450 μm to $(450 - \lambda/2n_{Si})\,\mu$m.

The total absorptivity $(A = A_{Si} + A_{TiNi})$ also changes with h_{Si} in a period of 1.55 μm due to the interface (Fig. 9.13).

9.3.2.2 Temperature distribution

A few analytical models have been developed for laser surface treatment. Most of the early works are analytic solutions of one-dimensional heat conduction. Later

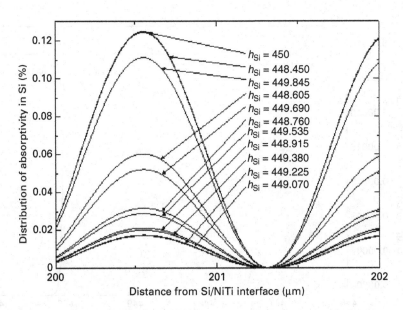

Figure 9.11   Distribution of absorptivity in Si of different thicknesses (only the distance range 200 μm to 202 μm is shown) [39].

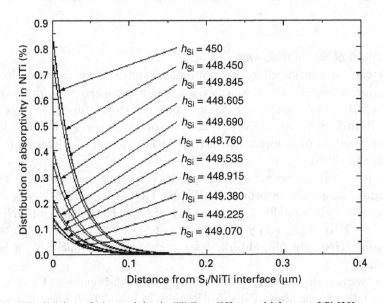

Figure 9.12   Distribution of absorptivity in TiNi at different thickness of Si [39].

on, both analytic and numerical methods were applied to investigate temperature distribution. In the case of multilayer structures discussed here, a few analytical models have been developed (e.g., [44, 45]). However, these models are steady state solutions with many conditions and assumptions. As such, finite element analysis is a more applicable approach.

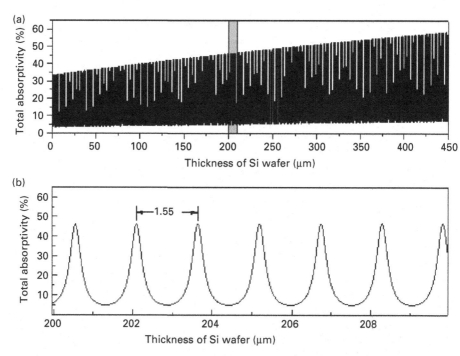

Figure 9.13   Total absorptivity of Si/TiNi system vs. different Si thicknesses. The period of oscillation is $\lambda/2n_{Si}= 1.55$ μm. (a) Overall view; (b) zoom-in view of the highlighted zone in (a) [39].

Obviously, in laser annealing, the density of solid TiNi is almost a constant, but many other material properties are not. Figure 9.14 shows the temperature dependent thermal conductivity and specific heat of Ni, Ti and Si [46]. Since the temperature dependent thermal properties of TiNi are still unknown and, according to Fig. 9.14, the thermal properties of Ni and Ti do not vary tremendously within the temperature range L 300 K $\leq T \leq$ 1500 K, the thermal conductivity and specific heat of TiNi are considered as constants in this study for simplicity.

Physical parameters in laser annealing and properties of TiNi and Si are listed in Tables 9.2–9.4 [39].

ANSYS is used for numerical simulation in this study. The geometry for investigation is illustrated in Figure 9.9. The thickness of Si substrates is a variable, and the thickness of TiNi film is fixed as 4 μm. The sample is chosen in a disk shape with a radius of 8000 μm, and the model is axi-symmetric. The upper and lower surfaces have heat convection with the environment (air). The coefficient of thermal convection is chosen as $H = 20$ W/m²K. Refer to Tables 9.2–9.4 for the physical parameters and thermomechanical properties of TiNi and Si. Note that the properties of Si at a given temperature are determined from the properties of the two nearby temperatures by linear data fitting. A typical mesh for finite element simulation is plotted in Fig. 9.15(a). The energy absorption for crystallization and phase transformation is ignored.

Table 9.2 Physical parameters in the laser annealing experiment [39]

Symbol	Description	Unit	Value
P	laser power	W	22.4
W_0	laser beam spot radius	μm	50
T_0	environment temperature	K	300
V	scanning speed	mm/s	3

Table 9.3 Properties of TiNi at $T = 300$ K [39]

Symbol	Description	Unit	Value
ρ	mass density	kg/m^3	6450
c_p	specific heat	J/(kg'K)	837
K	thermal conductivity	W/(m-K)	13.3
T_m	melting temperature	K	1573

Table 9.4 Properties of Si at $300 \, K < T < 1500 \, K$ [39, 46]

Symbol	Description	Unit	Value
ρ	mass density	kg/m^3	2330
c_p	specific heat	J/(kg K)	712 (300 K), 790 (400 K), 867 (600 K), 913 (800 K), 946 (1000 K), 967 (12003 K)
κ	thermal conductivity	W/(m K)	148 (300 K), 98.9 (400 K), 61.9 (600 K), 42.2 (800 K), 31.2 (1000 K), 25.7 (1200 K)
T_m	melting temperature	K	1685

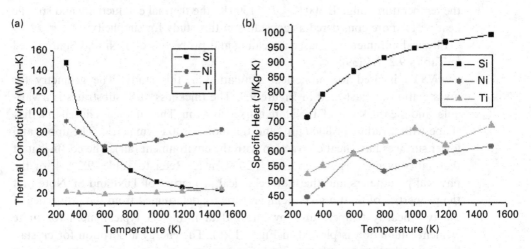

Figure 9.14   Temperature dependent thermal properties of Si, Ni and Ti [39].

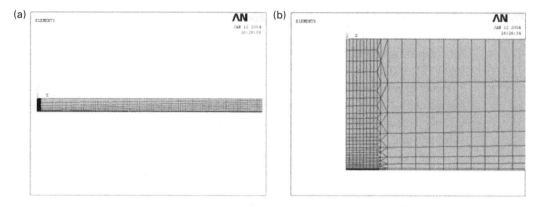

Figure 9.15 Finite element model (*ANSYS*) [39].

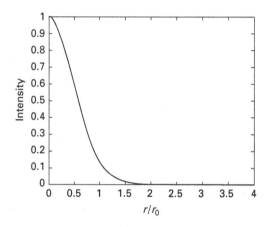

Figure 9.16 Gaussian distribution of laser beam intensity [39].

The intensity of a laser beam generally follows the Gaussian distribution. In the transverse direction of a beam,

$$I(r) = I_0 \exp\left(-\frac{r^2}{w_0^2}\right), \qquad (9.30)$$

where I_0 is the laser intensity at the centre of the beam (maximum laser intensity), r is the distance away from the centre of the laser beam, and w_0 is the focused spot radius of the beam. Figure 9.16 is a graphical expression of the relationship. To get a better approximation of the Gaussian distribution, the area within $2w_0$ is finer meshed [Fig. 9.15(b)].

In this simulation, eight zones are divided equally along the radius of the beam spot according to the following expression:

$$f(r_1, r_2) = \frac{\int_{r_1}^{r_2} 2\pi r I_0 \exp\left(-\frac{r^2}{w_0^2}\right) dr}{I_0(\pi r_2^2 - \pi r_1^2)}, \qquad (9.31)$$

where f is the average laser power distribution ratio between different distances (r_1 and r_2) from the center of the laser beam, the average power distribution ratios for respective zones are 0.969 391, 0.856 599, 0.668 857, 0.461 493, 0.281 365, 0.151 581, 0.072 1583 and 0.030 352.

Heat in Si directly generated by the laser beam is assumed to be uniformly distributed (A_{Si}/h_{Si}) within the thickness in the radiation zone, while the TiNi film is directly heated uniformly at the interface of Si/TiNi. The heat sources in the Si substrate and TiNi film follow a Gaussian distribution within the radiation zone.

If the laser power is 25 W, the thickness of the TiNi film is 4 μm and the thickness of the Si wafer is 450 μm, the highest temperature is 922 K at an annealing time of 0.01 second [Fig. 9.17(a)]. The temperature distribution at slightly different Si thicknesses at 0.01 second annealing time is shown in Fig. 9.17(a)–(l). We can see that the high temperature zone is always located in a small area around the Si/TiNi interface. It is noticed that in Fig. 9.17(a), the radius of the high temperature zone above 773 K, which is higher than the minimal TiNi crystallization temperature [3], is about 56 μm. Hence, the effective annealing zone should be about 112 μm in width. This is consistent with the measured width of the annealed line, which is about 120 μm [35]. In other cases, the high temperature zone is also around 110 μm.

The relationship between the highest temperature and Si thickness is plotted in Fig. 9.18. It is obvious that a slight variation in thickness of the Si wafer has a strong influence on the result of CO_2 laser annealing.

9.3.2.3 Effects of oxidation and interfacial layers

Up to this point, we have assumed that the material is pure and homogeneous, and its surface is isotropic and optically smooth. However, during intense laser irradiation, even an initially ideal material may have its surface composition and quality altered. Laser induced high temperature may cause strong oxidation and/ or other chemical reactions, and produce a surface layer of a material different from the original. As reported in [37], the backside of the Si wafer is rather rough. The root-mean-square roughness, R_q, of the backside (without polishing) of the single-side Si wafer is about 0.85 μm, while R_t (maximum peak-to-valley height of roughness) can be up to 7.6 μm as measured by a WYKO interferometer. It indicates that the thickness of Si wafer may vary around 7.6 μm within a small area. For simplicity, we may only consider the specular loss caused by R_q (< 1.55 μm) and ignore the absorptivity change induced by R_t (> 1.55 μm).

On the other hand, it is well known that, even at room temperature, a thin native oxide layer can be formed on the surface of an Si wafer in air. The native oxide layer is typically 1~10 nm thick (e.g. [41]). During laser processing, laser-enhanced oxidation occurs, which alters the absorption behavior of the Si/TiNi multilayer system. In the following investigation, we ignore oxidation during laser processing, but concentrate only on the influence of different thicknesses in the native oxide layer. In addition, diffusion may occur at the interface between Si and TiNi at a high temperature. The following discussion addresses the

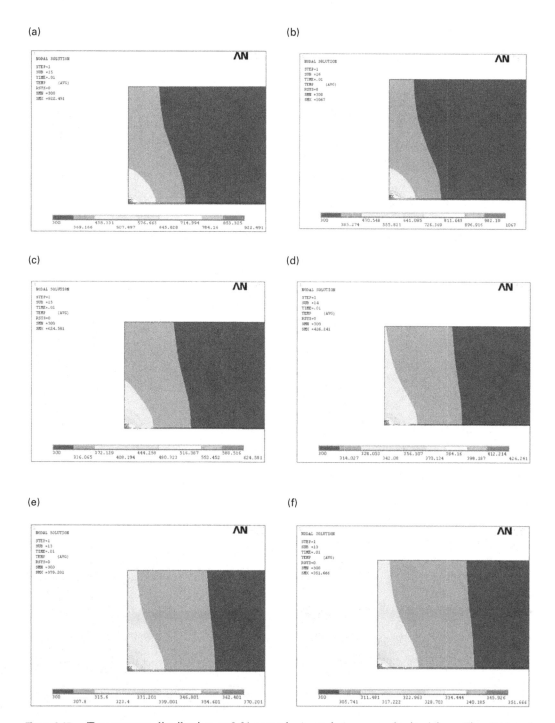

Figure 9.17   Temperature distribution at 0.01 second. A_{Si} and A_{TiNi} are obtained from Figs. 9.13 and 9.18 [39].

(g)

(h)

(i)

(j)

(k)

(l)

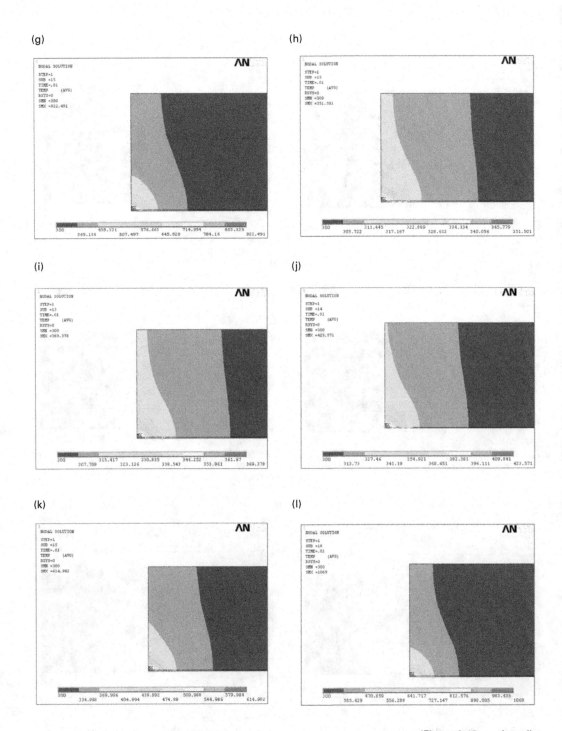

(Figure 9.17 continued)

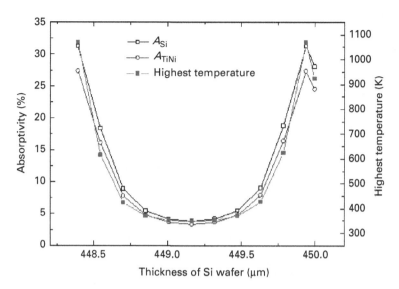

Figure 9.18 Highest temperature and absorptivities of Si and TiNi against thickness of Si [39].

influence of surface roughness, surface oxidation and interfacial layer on the radiative properties.

Effect of surface roughness

A surface is "optically smooth" if the average length scale of the surface is much less than the wavelength of the electromagnetic wave. A common measure of surface roughness is given by R_q. In general, with the increase of roughness, a surface will become less reflective, less specular and more diffusive. This behavior may be explained through geometric optics. In view of this, a laser beam hitting a rough surface may undergo two or more reflections off local peaks and valleys (resulting in increased absorption), after which it leaves the surface into an off-specular direction.

In order to count the loss in specular reflectance due to surface imperfections, one can estimate the effect of surface roughness by, for instance, *IMD* [47], a commercial software developed for calculating specular and non-specular (diffuse/roughness) optical functions of any arbitrary multilayer structure. As a typical result, Fig. 9.19 reveals the change of total absorptivity due to the variation in roughness of the backside Si wafer.

The variation in total absorptivity of the Si/TiNi multilayer appears to be in contradiction to common sense, i.e., a rough surface should result in an increase in absorption. It may be explained as follows: in a single crystalline Si wafer, the absorptivity increases with the increase of roughness due to the increase in reflection. The absorptivity (refer to Fig. 9.19) changes from 3.6% to 35.8% as the roughness is increased by 1 μm. In a Si/TiNi multilayer, a rough surface of Si wafer absorbs more light than a smooth surface and subsequently less light can reach the TiNi thin film. Thus, the interference becomes weaker than that of a smooth Si surface, which results in a decrease in the total absorption.

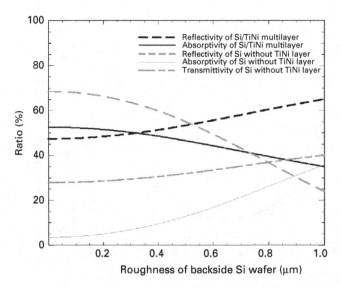

Figure 9.19 Absorptivity ($A = A_{Si} + A_{TiNi}$) and reflectivity of Si/TiNi multilayer vs. roughness of backside Si wafer. $h_{Si} = 450\,\mu m$, $\lambda = 10.6\,\mu m$ [39].

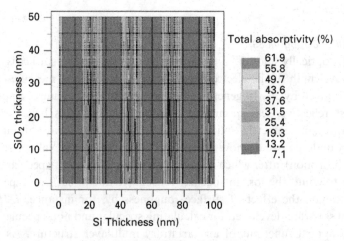

Figure 9.20 Total absorptivity of SiO$_2$/Si/TiNi multilayer. $\lambda = 10.6\,\mu m$ [39].

Effect of an oxide layer

Both Si and TiNi obtain a native oxide layer if one leaves them in air even at room temperature. The oxide layers of Si and TiNi have different optical properties. In the case of backside laser annealing of TiNi thin films (sputter deposited in vacuum), for simplicity, consider only the oxidation of Si and ignore the oxidation of TiNi. To study the influence of an oxide layer in Si, a simple approach is to treat the oxide as a thin SiO$_2$ layer atop of Si. Figure 9.20 presents the total absorptivity of an SiO$_2$/Si/TiNi multilayer (calculated by *Macleod,* a commercial software for computing optical properties of multilayer

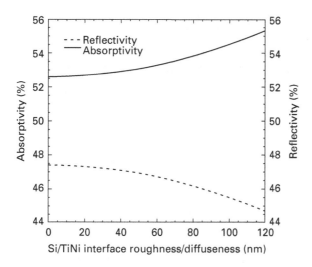

Figure 9.21 Effect of an Si/TiNi interface layer. $h_{Si} = 450\,\mu m$, $\lambda = 10.6\,\mu m$ [39].

systems). It shows that a thin oxidation layer (thickness < 100 nm) does not change the total absorptivity very much.

Interfacial layer
If an Si/TiNi interfacial layer does exist (normally generated in a high temperature process), the total absorptivity of the multilayer may alter. The interfacial layer may be up to about 120 nm thick [48]. This effect can be estimated by *IMD* as shown in Fig. 9.21. It reveals that the total absortivity of the multilayer increases from 52.6% to 55.3% with a 120 nm interface layer, while the reflectivity decreases from 47.4% to 44.7%.

9.3.2.4 Remarks
Now we may conclude that the exact thickness of an Si wafer (in a 1.55 μm period) has significant influence on CO_2 laser annealing of TiNi films. Since the tolerance in the thickness of the Si wafer is about $\pm 25\,\mu m$, special attention is required to be given to the precise control of the wafer thickness. In the annealing process, the high temperature zone is largely localized within a zone with a radius about twice that of the focused beam spot. As demonstrated, the width of effective annealing zone ($T > 773$ K) is about the same size as the diameter of the laser beam spot, which agrees well with the experimental observation reported above. Surface roughness, oxide layer and interface layer may induce instability in laser annealing and should be taken into consideration. A double-side polished wafer is preferred for backside annealing.

9.3.3 Direct laser annealing
Laser annealing of TiNi thin films from the back Si wafer side is not fully compatible with many MEMS fabrication processes. Sputtering a thin Si layer atop of

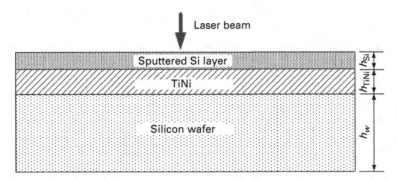

Figure 9.22   Schematic illustration of direct laser annealing [39].

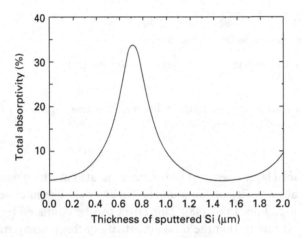

Figure 9.23   Total absorptivity of sputtered Si/TiNi multilayer. $\lambda = 10.6\,\mu m$ (CO_2 laser) [39].

TiNi thin films may be a solution, which makes direct laser annealing possible. After annealing, the sputtered Si layer can be easily etched away by a standard MEMS process. Here, we investigate the required conditions for direct laser annealing following this idea. A typical situation of laser direct annealing is schematically illustrated in Fig. 9.22. Table 9.5 lists the refractive indexes of Si, TiNi, SiO_2 and TiO_2 at some different wavelengths.

9.3.3.1   Absorption of a Si/TiNi/Si multilayer

Similar to backside annealing, TiNi is considered as a semi-infinite substrate in terms of its optical behavior and $A+R=1$ for the Si/TiNi/Si multilayer. Since no light passes to the Si wafer, the Si wafer can be excluded from the calculation of absorption, i.e., only the sputtered Si layer and TiNi thin film absorb laser energy. Figures 9.23 and 9.24 show the total absorptivity[2] of the multilayer impinged by four different types of laser, namely, 10.6 μm CO_2, 1064 nm (HIPPO), 532 nm (Green) and 355 nm (AVIA) Nd:YAG lasers.

---

[2] Absortivity is calculated by *Macleod*.

Table 9.5 Refractive indeces of Si, TiNi, SiO$_2$ and TiO$_2$ at different wavelengths [42, 43]

Wavelength (nm)	n				k			
	10600 (CO$_2$ laser)	1064 (Nd:YAG, HIPPO)	532 (Nd:YAG, Green)	355 (Nd:YAG, AVIA)	10600 (CO$_2$ laser)	1064 (Nd:YAG, HIPPO)	532 (Nd:YAG, Green)	355 (Nd: YAG, AVIA)
Si	3.74	3.597	4.428	3.728	0	0	0.878	2.771
TiNi* (estimation)	7.568 $\left(\frac{n_{mi}+n_{ii}}{2}\right)$	1.03	1.019	2.231	28.339 $\left(\frac{k_{mi}+k_{ii}}{2}\right)$	4.49	1.729	1.049
SiO$_2$	2.184	1.45	1.547	1.476	2.205×10^{-2}	0	0	0
TiO$_2$	1.445	2.74	3	4.023	0.063	0	0	0.173

* *TFCompanion*, Semiconsoft, Inc.

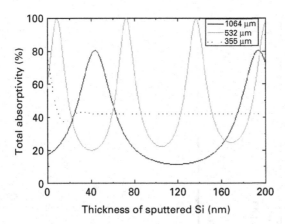

Figure 9.24   Total absorptivity of sputtered Si/TiNi multilayer. λ is 1064 nm, 532 nm and 355 nm (Nd:YAG laser) [39].

Figure 9.23 reveals that the first peak in absorptivity of CO_2 laser ($\lambda = 10.6$ μm) annealing is at around 0.72 μm. The absorptivities of Si and TiNi are calculated as $A_{Si} = 0.07\%$ and $A_{TiNi} = 33.5\%$, respectively. Figure 9.24 shows that the absorptivities of the 1064 nm and 532 nm laser oscillate with a period of $\lambda/2n_{Si}$. For the 355 nm laser, if the thickness of sputtered Si exceeds the optical penetration depth l_{Si} ($l_{Si} = \lambda/4\pi k_{Si} = 9.35$ nm), the absorption of sputtered Si becomes strong and interference disappears. Thus, the absorbed laser power tends to be stable. The absorptivity of different thicknesses of sputtered Si in Table 9.6 proves that the laser power is mainly absorbed by the sputtered Si layer when $h_{Si} > l_{Si}$.

ANSYS results reveal that the highest temperature in the case of a 0.72 μm Si layer on top of a TiNi film is 1032 K for a CO_2 laser ($\lambda = 10.6$ μm) (Fig. 9.25). Hence, we may conclude that the minimal thickness of the sputtered Si layer for CO_2 laser annealing is about 0.72 μm.

The above analysis shows that the thickness of sputtered Si is still a dominant factor. The 355 nm laser may be a good candidate for annealing due to its insensitivity to the thickness of Si as long as $h_{Si} > 45$ nm. However, since the sputtered Si is relatively thin, the effects of oxidation in sputtered Si (SiO_2) and the interfacial layer should not be ignored. Assuming the Si layer is deposited right after the TiNi film in the same vacuum chamber, the oxidation in TiNi may be neglected. The effect of roughness can also be ignored since in general the average roughness is only around a few nanometers in sputtered amorphous TiNi thin films [39].

9.3.3.2   Effects of oxidation and interfacial layer

Using the same software, a study on the effects of oxidation and interfacial layers is carried out. Figures 9.26–9.29 show the dependence of total absorptivity on the thickness of the SiO_2/Si/TiNi multilayer at different wavelengths. We can see that the total absorptivity is mainly dependent on the thickness of sputtered Si.

Table 9.6 Absorptivity of different thicknesses of sputtered Si $\lambda = 355$ nm [39]

Absorptivity (%)	$h_{Si} = 20$ nm	$h_{Si} = 45$ nm	$h_{Si} = 90$ nm
A_{Si}	34.83	41.9	42.35
A_{niti}	3.95	0.28	0.002
$A = A_{Si} + A_{niti}$	38.78	42.18	42.352

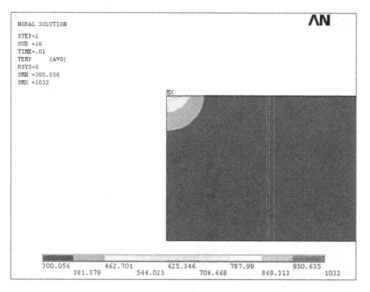

Figure 9.25 Temperature distribution of Si/TiNi/Si multilayer at 0.01. Laser power is 25 W, 0.72 µm sputtered Si layer, 4 µm TiNi layer and 450 µm Si wafer. $A_{Si} = 0.07\%$ and $A_{tini} = 33.5\%$ [39].

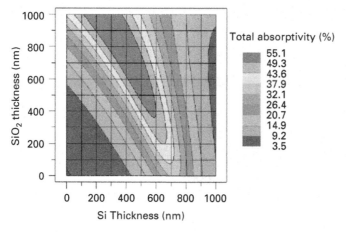

Figure 9.26 Total absorptivity of SiO_2/Si/TiNi multilayer. $\lambda = 10.6$ µm (CO_2 laser) [39].

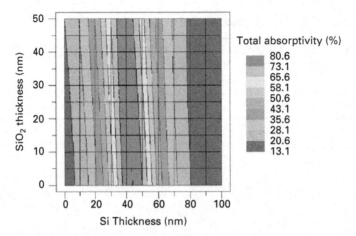

Figure 9.27   Total absorptivity of SiO$_2$/Si/TiNi multilayer. $\lambda = 1.064$ μm (HIPPO Nd:YAG laser) [39].

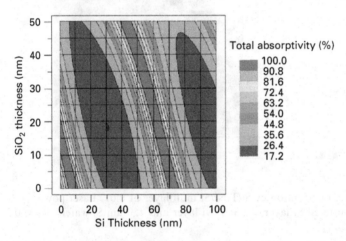

Figure 9.28   Total absorptivity of SiO$_2$/Si/TiNi multilayer. $\lambda = 532$ nm (Green Nd:YAG laser) [39].

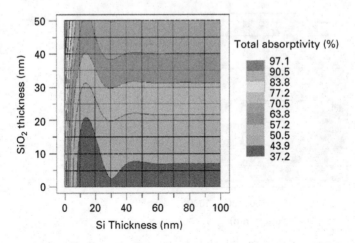

Figure 9.29   Total absorptivity of SiO$_2$/Si/TiNi multilayer. $\lambda = 355$ nm (AVIA Nd:YAG laser) [39].

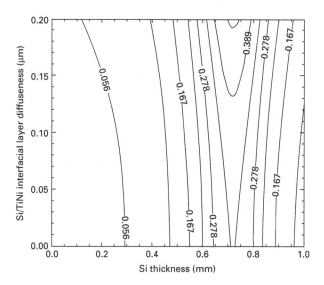

Figure 9.30 Total absorptivity of Si/NiTi multilayer with a Si/TiNi interfacial layer. $\lambda = 10.6$ μm (CO_2 laser) [39].

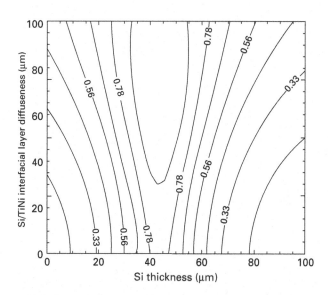

Figure 9.31 Total absorptivity of Si/TiNi multilayer with a Si/TiNi interfacial layer. $\lambda = 1.064$ μm (HIPPO Nd:YAG laser) [39].

Figures 9.30–9.33 present the dependence of total absorptivity on thickness of an Si/TiNi interface layer of four different wavelengths. The total absorptivity increases with the increasing thickness of the Si/TiNi interface layer. As compared with that in backside annealing, the effects of the oxidation layer and interfacial layer are more significant.

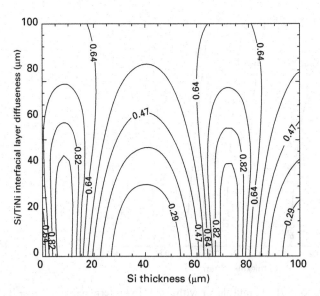

Figure 9.32   Total absorptivity of Si/TiNi multilayer with a Si/TiNi interfacial layer. λ = 532 nm (Nd:YAG laser) [39].

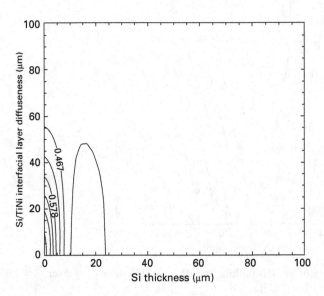

Figure 9.33   Total absorptivity of Si/TiNi multilayer with a Si/TiNi interfacial layer. λ = 355 nm (AIVA Nd:YAG laser) [39].

9.3.3.3   Remarks

The above theoretical study reveals that direct laser annealing of TiNi thin films is possible. The influence of oxidation and an interfacial layer is more significant than that in backside CO_2 annealing.

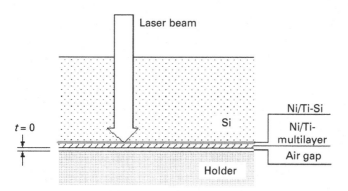

Figure 9.34   Schematic illustration of CO_2 laser annealing of an Ni/Ti multilayer film [50]. (With permission from Institute of Physics Publishing.)

9.3.4   Annealing of Ni/Ti multilayer thin films

As discussed in Section 9.1, sputter deposition is the most popular technique for fabrication of TiNi thin films at present. In this technique, two or more targets, if additional alloying is required, are normally sputtered simultaneously in a vacuum chamber. By controlling the applied voltage on the individual target, in theory, TiNi thin films of different compositions can be obtained [1]. However, in practice, it is not an easy task to control the composition of the resultant thin films precisely, as many other factors, for instance, sputtering yield and angular flux distribution, also play important roles [4]. Since the transformation temperatures of TiNi thin films are very sensitive to the exact composition [38], it turns out to be necessary to find some other approaches which are more suitable for producing high quality TiNi thin films.

As an alternative, furnace anneal sputter deposited Ni/Ti multilayer films, which consist of many alternate thin layers of pure Ni and Ti, has been proposed [49]. As each layer is only a few nanometers in thickness, the composition of the resultant film can be more precisely controlled through the deposition time for each layer.

We can numerically investigate the feasibility of using a CO_2 laser for annealing Ni/Ti multilayer thin films [50]. We assume that the CO_2 laser beam impinges normally on the Ni/Ti thin film from the Si substrate side (Fig. 9.34). As revealed in [48], in an Si-TiNi system, there is a significant inter-diffusion between the Si substrate and TiNi during sputter deposition. The reported diffusion zone can be about 120 nm (60 nm into Si and 60 nm into TiNi). In addition, titanium silicides are observed. As such, the effects of the interfacial layer should be considered in the simulation.

Substituting the optical constants in Table 9.7 into Eq. (9.12), the intensity reflection coefficients of various conditions are obtained and listed in Table 9.8. Note that the wavelength (λ) of the CO_2 laser is 10.6 μm. As the intensity coefficient in the situation of reflection from Si to Ni-Si is higher than that for

Table 9.7 Optical constant at $T \approx 300$ K and $\lambda \approx 10.6$ μm [39, 42, 43]

Material	n	k
Air	1	0
Si	3.4215	1.27×10^{-4}
Ni	7.11	38.3
Ti	3.90	19.79
TiNi (estimation)	5.505 $\left(= \frac{n_{ni}+n_{ti}}{2} \right)$	29.045 $\left(= \frac{k_{ni}+k_{ti}}{2} \right)$
Ni-Si (estimation)	5.266 $\left(= \frac{n_{ni}+n_{si}}{2} \right)$	19.150 $\left(= \frac{k_{ni}+k_{si}}{2} \right)$
Ti-Si (estimated)	3.661 $\left(= \frac{n_{ni}+n_{si}}{2} \right)$	9.895 $\left(= \frac{k_{ni}+k_{si}}{2} \right)$

Table 9.8 Indensity reflection coeficients at $T \approx 300$ K and $\lambda \approx 10.6$ μm [50]

Symbol	First medium	Second medium	Intensity reflection coefficient
R_{air-si}	Air	Si	0.300
$R_{air-niti}$	Air	TiNi	0.975
$R_{si-niti}$	Si	TiNi	0.918
$R_{si-nisi}$	Si	Ni-Si	0.837
$R_{si-tisi}$	Si	Ti-Si	0.662
$R_{tisi-nisi}$	Ti-Si	Ni-Si	0.096
$R_{tisi-nisi}$	Ni-Si	Ti-Si	0.096
$R_{nisi-ti}$	Ni-Si	Ti	0.001
R_{ti-ni}	Ti	Ni	0.101

reflection from Si to Ti-Si, it is obvious that it is better to sputter the Ti layer first onto the Si substrate in order for more absorption of laser energy and better adhesion to the Si substrate [48, 50].

It can be concluded that [50]

(1) at room temperature, the energy loss in a laser beam passing through an ordinary 450 μm thick Si wafer is ignorable;
(2) for whatever case, when a laser beam enters into an Ni/Ti multilayer film, the reflection intensity coefficient is less than 0.1;
(3) the laser energy coupled into the film will be absorbed within a thickness of less than 300 nm.

As such, the exact thickness of the Ti and Ni layers should not affect the laser energy absorption very much. Therefore, given a film thickness of a few μm, for simplicity, we can consider it as a TiNi film instead of an Ni/Ti multilayer film in simulation. Hence, the numerical simulation method described in Section 9.3.2 is directly applicable here. However, according to Figure 9.35, the absorption ratio

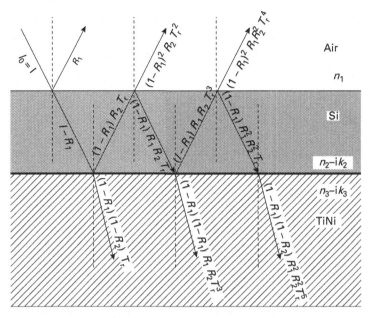

Figure 9.35 Illustration of multiple reflections in laser annealing (where R_1 stands for R_{air-si}, R_2 stands for $R_{si-film}$, and $T_r = Tr_{si} = 2\exp(T/110)$ (where T is temperature in K). The detail of the TiNi-Si interfacial layer is too small to be shown [50]. (With permission from Institute of Physics Publishing.)

of the Ni/Ti film is

$$A_{niti} = \frac{(1 - R_{air-si})Tr_{si}(1 - R_{si-film})}{1 - R_{si-film}Tr_{si}R_{air-si}Tr_{si}} \tag{9.32}$$

and for the Si substrate, it is

$$A_{si} = (1 - R_{air-si})(1 - Tr_{si})$$
$$\times \left(1 + \frac{Tr_{si}R_{si-film} + Tr_{si}R_{si-film}Tr_{si}R_{air-si}}{1 - Tr_{si}R_{si-film}Tr_{si}R_{air-si}}\right). \tag{9.33}$$

The case study in [50] reveals again that the high temperature zone is always located within a small area with a radius about $2w_0$. In about 2 milliseconds, the maximum temperature reaches over 773 K. It demonstrates that a CO_2 laser can be used for local annealing of a Ni/Ti multilayer film on a micron scale.

9.4   Summary

The feasibility of local laser annealing of amorphous TiNi thin films is demonstrated. A systematic study of the theories behind this technique is presented. Backside and direct annealing are investigated, together with the case of annealing of a Ti/Ni multilayer. It is proved that the width of the crystalline TiNi line is about the same as the diameter of the laser beam spot. Surface roughness, oxide layer and interface layer have significant influence on laser annealing, in particular in direct annealing. A double-side polished wafer is preferred in backside annealing.

Acknowledgements

The authors would like to thank Q. He, J.F. Su and Dr X.Y. Gao for their contributions to this chapter.

References

[1] Y. Fu, H. Du, W. Huang, S. Zhang and M. Hu. TiNi-based thin films in MEMS applications: a review. *Sensors and Actuators A*, **112** (2004) 395–408.

[2] P. Krulevitch, A. P. Lee, P. B. Ramsey, J. C. Trevino, J. Hamilton and M. A. Northrup. Thin film shape memory alloy microactuators. *J. Microelectromechanical Systems*, **5** (1996) 270–282.

[3] A. D. Johnson. Vacuum-deposited TiNi shape memory film: characterization and applications in microdevices. *J. Micromech. Microeng.* **1** (1991) 34–41.

[4] A. Ishida and V. Martynov. Sputter-deposited shape-memory alloy thin films: properties and applications. *MRS Bulletin* **27** (2002) 111–114.

[5] J. A. Walker, K. J. Gabriel and M. Mehregany. Thin-film processing of TiNi shape memory alloy. *Sensors and Actuators, A: Physical*, **21** (1990) 243–246.

[6] K. Ikuta, H. Fujishiro, M. Hayashi and T. Matsuura. Laser ablation of Ni-Ti shape memory alloy thin film. In *Proceedings of the First International Conference on Shape Memory and Superelastic Technologies*, eds. A. R. Pelton, D. Hodgson, S. M. Russell and T. Duerig, California: SMST (1994), pp. 13–18.

[7] E. Makino, M. Uenoyama and T. Shibata. Flash evaporation of TiNi shape memory thin film for microactuators. *Sensors and Actuators, A: Physical*, **71** (1998) 187–192.

[8] S. T. Davies and K. Tsuchiya. Ion beam sputter deposition of TiNi shape memory alloy thin films. *Proc. SPIE*, **3874** (1999) 165–172.

[9] H. Y. Noh, K. H. Lee, X. X. Cui and C. S. Choi. Composition and structure of TiNi thin film formed by electron beam evaporation. *Scripta Materialia*, **43** (2000) 847–852.

[10] J. F. Ready and D. F. Farson (eds). *LIA Handbook of Laser Materials Processing*, Berlin: Laser Institute of America /Magnolia Publishing/Springer (2001).

[11] M. Rohde and A. Schussler. On the response-time behaviour of laser micromachined TiNi shape memory actuators. *Sensors and Actuators A*, **61** (1997) 463–468.

[12] M. Kohl, D. Dittmann, E. Quandt, B. Winzek, S. Miyazaki and D. M. Allen. Shape memory microvalves based on thin films or rolled sheets. *Mater. Sci. Eng. A*, **275** (1999) 784–788.

[13] W. M. Huang, Q. Y. Liu, L. M. He and J. H. Yeo. Micro TiNi-Si cantilever with three stable positions. *Sensors and Actuators A*, **114** (2000) 118–122.

[14] K. C. Yung, H. H. Zhu and T. M. Yue. Theoretical and experimental study on the kerf profile of the laser micro-cutting TiNi shape memory alloy using 355 nm Nd:YAG. *Smart Materials and Structures*, **14** (2005) 337–342.

[15] C. Li, S. Nikumb and F. Wong. An optimal process of femtosecond laser cutting of TiNi shape memory alloy for fabrication of miniature devices. *Optics and Lasers in Engineering*, **44** (2006) 1078–1087.

[16] A. T. Tung, B.-H. Park, G. Niemeyer and D. H. Liang. Laser-machined shape memory alloy actuators for active catheters. *IEEE/ASME Transactions on Mechatronics*, **12** (2007) 439–446.

[17] P. Schlobmacher, T. Haas and A. Schubler. Laser welding of Ni-Ti shape memory alloys. In *Proceedings of the First International Conference on Shape Memory and Superelastic Technologies*, eds. A. R. Pelton, D. Hodgson, S. M. Russell and T. Duerig, California: SMST (1994), 85–90.

[18] A. Tuissi, S. Besseghini, T. Ranucci, F. Squatrito and M. Pozzi. Effect of Nd-YAG laser welding on the functional properties of the Ni-49.6at.%Ti. *Mater. Sci. Engi. A*, **275** (1999) 813–817.

[19] A. Falvo, F. M. Furgiuele and C. Maletta. Laser welding of a TiNi alloy: mechanical and shape memory behaviour. *Mater. Sci. Eng. A*, **412** (2005) 235–240.

[20] X. J. Yan, D. Z. Yang and X. P. Liu. Corrosion behaviour of a laser-welded TiNi shape memory alloy. *Materials Characterization*, **58** (2007) 623–628.

[21] F. Villermaux, M. Tabrizian, L. H. Yahia, M. Meunier and D. L. Piron. Excimer laser treatment of TiNi shape memory alloy biomaterials. *Applied Surface Science*, **110** (1997) 62–66.

[22] H. C. Man, Z. D. Cui and T. M. Yue. Corrosion properties of laser surface melted TiNi shape memory alloy. *Scripta Materialia*, **45** (2001) 1447–1453.

[23] H. C. Man, S. Zhang, F. T. Cheng and X. Guo. Laser fabrication of porous surface layer on TiNi shape memory alloy. *Mater. Sci. Eng. A*, **404** (2005) 173–178.

[24] H. C. Man, K. L. Ho and Z. D. Cui. Laser surface alloying of TiNi shape memory alloy with Mo for hardness improvement and reduction of $Ni2^+$ ion release. *Surface & Coating Technology*, **200** (2006) 4612–4618.

[25] M. H. Wong, F. T. Cheng and H. C. Man. Laser oxidation of TiNi for improving corrosion resistance in Hanks' solution. *Mater. Lett.*, **61** (2007) 3391–3394.

[26] S. Inaba and K. Hane. Miniature actuator driven photothermally using a shape-memory alloy. *Review of Scientific Instruments*, **64** (1993) 1633–1635.

[27] S. T. Davies, E. C. Harvey, H. Jin, *et al.* Characterization of micromachining processes during KrF excimer laser ablation of TiNi shape memory alloy thin sheets and films. *Smart Materials and Structures*, **11** (2002) 708–714.

[28] A. Camposeo, N. Puccini, F. Fuso, M. Allegrini, E. Arimondo and A. Tuissi. Laser deposition of shape-memory alloy for MEMS applications. *Applied Surface Science*, **208–209** (2003) 518–521.

[29] A. Camposeo, F. Fuso, M. Allegrini, E. Arimondo and A. Tuissi. Pulsed laser deposition and characterization of TiNi-based MEMS prototypes. *Applied Physics A*, **79** (2004) 1141–1143.

[30] A. Morone. Pulsed laser deposition of the NiTiCu thin film alloys. *Applied Surface Science*, **253** (2007) 8242–8244.

[31] A. Hakola, O. Heczko, A. Jaakkola, T. Kajava and K. Ullakko. Ni-Mn-Ga films on Si, GaAs and Ni-Mn-Ga single crystals by pulsed laser deposition. *Applied Surface Science*, **238** (2004) 155–158.

[32] K. Dai and L. Shaw. Thermal and stress modelling of multi-materials laser processing. *Acta Materialia*, **49** (2001) 4171–4181.

[33] L. Liu, A. Hirose and K. F. Kobayashi. A numerical approach for predicting laser surface annealing process of Inconel 718. *Acta Materialia*, **50** (2002) 1331–1347.

[34] Y. Bellouard, T. Lehnert, J. E. Bidaux, T. Sidler, R. Clavel and R. Gotthardt. Local annealing of complex mechanical devices: a new approach for developing monolithic micro-devices. *Mater. Sci. Engi. A*, **275** (1999) 795–798.

[35] Q. He, M. H. Hong, W. M. Huang, T. C. Chong, Y. Q. Fu and H. J. Du. CO_2 laser annealing of sputtering deposited TiNi shape memory thin films. *J. Micromech. Microengi.*, **14** (2004) 950–956.

[36] X. Wang, Y. Bellouard and J. J. Vlassak. Laser annealing of amorphous TiNi shape memory alloy thin films to locally induced shape memory properties. *Acta Materialia*, **52** (2005) 4955–4961.

[37] Q. He. *CO_2 Laser Annealing of TiNi (Shape Memory) Thin Films for MEMS Applications*. MEng. Thesis, Nanyang Technological University, Singapore (2004).

[38] W. M. Huang. On the selection of shape memory alloys for actuators. *Materials and Design*, **5** (2002) 270–282.

[39] J. F. Su. *Indentation and Laser Annealing of TiNi Shape Memory Alloys*. MEng. Thesis, Nanyang Technological University, Singapore (2005).

[40] L. Ward. *The optical constants of bulk materials and films*, Bristol, Philadelphia: Institute of Physics, (1994).

[41] D. Bauerle. *Laser Processing and Chemistry*, New York: Springer (2000).

[42] E. D. Palik. *Handbook of Optical Constants of Solids*, Orlando: Academic Press (1985).

[43] E. D. Palik. *Handbook of Optical Constants of Solids II*, Boston: Academic Press (1991).

[44] M. L. Burgener and R. E. Reedy. Temperature distributions produced in a two-layer structure by a scanning cw laser or electron beam. *J. Appl. Phys.*, **53** (1982) 4357–4363.

[45] I. D. Calder and R. Sue. Modeling of cw laser annealing of multilayer structures. *J. Appl. Phys.*, **53** (1982) 7545–7550.

[46] F. P. Incropera and D. P. DeWitt. *Fundamentals of Heat and Mass Transfer*, New York: Wiley (1996).

[47] D. Windt. IMD – software for modeling the optical properties of multilayer films. *Computers in Physics*, **12** (1998) 360–370.

[48] Y. Q. Fu, H. J. Du and S. Zhang. Adhesion and interfacial structure of magnetron sputtered TiNi films on Si/SiO_2 substrate. *Thin Solid Films*, **444** (2003) 85–90.

[49] H. Lehnert, H. Grimmer, P. Boni, M. Horisberger and R. Gotthardt. Characterization of shape-memory alloy thin films made up from sputter-deposited Ni/Ti multilayers. *Acta Materialia*, **48** (2000) 4065–4071.

[50] Q. He, W. M. Huang, X. Y. Gao and M. H. Hong. Numerical investigation of CO_2 laser heating for annealing Ni/Ti multilayer thin films. *Smart Materials and Structures*, **14** (2005) 1320–1324.

10 Overview of thin film shape memory alloy applications

A. David Johnson

Abstract

This chapter discusses properties affecting thin film applications and resulting devices that have been developed since sputtered TiNi thin film with shape memory properties was first demonstrated in 1989. As the shape memory alloy technology has matured, the material has gradually gained acceptance. TiNi thin film shape memory alloy (SMA) exhibits intrinsic characteristics similar to bulk nitinol: large stress and strain, long fatigue life, biocompatibility, high resistance to chemical corrosion, and electrical properties that are well matched to joule heating applications. In addition, thin film dissipates heat rapidly so that it can be thermally cycled in milliseconds. These properties make TiNi thin film useful in making microactuators. Microelectromechanical (MEMS) processes – specifically photolithography, chemical etching, and use of sacrificial layers to fabricate complex microstructures – combine TiNi thin film with silicon to provide a versatile platform for fabrication of microdevices. A variety of microdevices have been developed in several laboratories, including valves, pumps, optical and electrical switches and intravascular devices.

Interest in thin film applications is increasing as evidenced by the number of recent publications and patents issued. Intravascular medical devices are currently in clinical trials. The future for thin film devices, especially in medical devices, seems assured despite the fact that to this day no "killer application" has emerged.

10.1 Introduction to TiNi thin film applications

This chapter provides an overview of shape memory alloy thin film applications since TiNi thin film, developed with support from a National Aeronautics and Space Administration Small Business Innovation Research contract, was introduced at the Engineering Aspects of Shape Memory Alloy Conference in Lansing, MI in 1989. Applications of thin film shape memory alloy (SMA) are described,

Thin Film Shape Memory Alloys: Fundamentals and Device Applications, eds. Shuichi Miyazaki, Yong Qing Fu and Wei Min Huang. Published by Cambridge University Press. © Cambridge University Press 2009.

along with experience gained in bringing devices to production readiness. Simple design rules are illustrated for incorporating shape memory microactuators into microelectromechanical (MEMS) devices. Lessons learned and pitfalls to avoid are included in the hope that these will benefit future researchers.

Thin film SMA presents opportunities to extend the roles of nitinol. Research has demonstrated that the fine microstructure of thin films is responsible for superior shape memory characteristics in films compared to that of bulk materials. As with nitinol, TiNi thin film produces a large work output per unit of actuator mass, and thin film may be rapidly cycled thermally due to a large surface to volume ratio. Effort expended to date in the development and fabrication of microdevices actuated by thin film SMA has resulted in limited sales of microvalves and limited expansion to new applications. Thin film shape memory actuators technology is ready for use in MEMS applications, but these have been slow in coming to market.

The most common thin film SMA is binary titanium–nickel (TiNi) of near equiatomic composition in which the transition point is lower than 373 K. High-temperature ternary thin films have also been fabricated. Incorporating a third element provides a means of raising the transformation temperature or decreasing the temperature hysteresis. Besides the ordinary shape memory effect, two-way shape memory effects and superelasticity have been observed in thin film [1].

As discussed in Chapter 3, thin films can be fabricated using several processes. Planar magnetron sputtering is most common. Cylindrical magnetron sputtering is used for making miniature stents. Since the first demonstration, titanium–nickel shape memory alloy thin film, or TiNi thin film, has been recognized by various researchers as a versatile material with many potential applications [2]. Subsequent research and publications emphasize that TiNi thin film exhibits advantageous characteristics for use in miniature devices compared with other technologies [3, 4, 5].

10.2   Properties suitable for applications

Titanium–nickel (TiNi) thin film is the heavy lifter of MEMS technology because the material is able to overcome stiction, make low-ohmic contact switches, operate high pressure self priming pumps and valves, and bend silicon flexures to reposition miniature mirrors and provide motion for biotechnology applications.

Intrinsic characteristics that make bulk nitinol suitable for actuators are also present in thin film. Stress (hundreds of MPa) coupled with large strains (more than three percent) produces large work output (more than a joule per gram per cycle). For comparison, a bulk nitinol rod 1 centimeter in diameter and 1 meter long could lift an automobile 3 centimeters, whereas a TiNi thin film microribbon 1 millimeter long, 5 microns thick, and 100 microns wide will pull with a force of 0.5 newton through a distance of 30 microns to operate a switch or open a valve. Force scales with cross-section and displacement with length, so it is possible to make actuators of almost any size, force, and stroke in sizes from centimeters down to tens of nanometers.

Perhaps most important is the natural connection with MEMS processing, since film is sputtered on silicon and can be patterned using photolithography and selective chemical etching solutions.

The disadvantages of TiNi thin film actuators, as with any heat-driven mechanical device, are low thermodynamic efficiency and limited cycling rate. The shape recovery of SMA is driven by heat energy by the second law of thermodynamics; this fact limits the thermodynamic efficiency to very low numbers, generally one to two percent. Thin film, having a large surface to volume ratio, radiates heat effectively: articles made of thin film cool in milliseconds. However, a millisecond cycle rate simply is not fast enough for electronics applications such as radio frequency relays. For these reasons, devices based on shape memory actuators are better suited to some applications than to others. Since SMA actuators have poor thermodynamic efficiency, applications in which the duty cycle is small are most likely to find commercial appeal.

The characteristic of superelasticity makes TiNi thin film especially suited to intravascular applications. Possibly related to its small grain size, thin film exhibits superelastic behavior without processing such as cold working. As a result, articles for intravascular intervention can be made by MEMS processes and used to augment existing cardiovascular technology.

10.2.1   Mechanical properties

The ultimate tensile properties of TiNi thin films are at least as good as those of bulk nitinol. Stress–strain curves for Ti-50.0 at%Ni thin film shows an elongation of more than 15%, while a Ti-51.5 at%Ni thin film containing a large amount of Ti_3Ni_4 shows a fracture stress as high as 1.6 GPa. In general, the yield stress for plastic deformation of thin films is higher than that of bulk materials, owing to the small grain size compared with that of bulk materials.

TiNi thin film has good ductility. A Ti-48.3 at%Ni thin film annealed at a low temperature (773 K) shows an elongation of almost 20 percent before fracture. It is known that a bulk material with the same composition does not show such plastic deformation. This improvement comes from suppression of the coarse grain-boundary precipitates that cause fracture in bulk nitinol.

A striking difference between TiNi thin film and bulk nitinol is the much smaller crystal domains in film. Whereas nitinol made from melt has crystal domains ranging from tens of microns to a hundred microns, transmission electron microscope images of sputter-deposited TiNi that is crystallized from as-deposited amorphous thin film show crystals of the order of 1–2 microns as pictured in Fig. 10.1.

This characteristic grain size is evident in films as thin as 100 nanometers. With a thickness to diameter ratio of 1 to 10, these crystals are practically two-dimensional (Fig. 10.2).

Despite this difference, the intrinsic stress–strain, resistivity and nano-indentation characteristics of film are similar to bulk material. Nano-indentation experiments have been performed by several research groups [6].

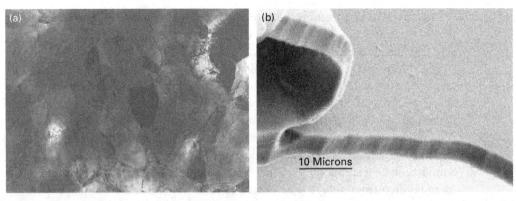

Figure 10.1   Sputter deposited TiNi, a few microns thick after crystallization, consists of densely packed grains 1–3 microns in diameter. (a) Film adhered to substrate. The film is free from inclusions and precipitates. Surface stresses are sufficient to cause wrinkles to appear, which makes the martensite appear hazy while the austenite phase is specularly reflecting (not shown). (b) Free-standing film that has been patterned by photolithography and released from the substrate. The chemically machined (etched) edge of film is smooth so that crack initiation is inhibited, resulting in good fatigue properties.

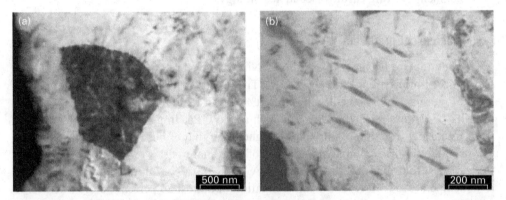

Figure 10.2   Transmission electron microscope image of a single crystal of thin film TiNi less than 100 nanometers thick. Nanoindentation experiments demonstrate shape memory behavior in such films. (a) Crystal diameter is the same as in thicker films, about 2–3 microns, so this crystal is practically two-dimensional. (b) At higher magnification the internal structure can be seen.

Banas *et al.* describe TiNi films with properties that are unattainable in conventional processes: 1200 MPa and 16% strain compared with 1100 MPa and 10% strain in bulk nitinol. Integrating over the volume of a sample, the modulus of toughness is nearly double that of bulk nitinol because both the thin film material's fracture strength and maximum strain levels are higher [7].

These improvements may be attributed to refinement in the sputtering process. Inclusions act as local stress concentrators during mechanical cycling and so can diminish fatigue life. Impurities at any level reduce the recoverable strain. Thin film, compared with bulk nitinol, has fewer impurities and fewer inclusions of foreign particles such as carbon.

Thin film may also exhibit improved fatigue life over conventional bulk material. Conventional wisdom says that lattice imperfections are generated in every thermal cycle through the transformation interval, and therefore cycle lifetime is finite. In film that has thickness about equal to grain size, greater opportunity exists for accommodating the mismatches between martensite and austenite, which can significantly reduce the number of dislocations produced during cycling.

Unpublished data from TiNi Alloy Company show that thin film does exhibit a long fatigue life. Fatigue life is dependent on maximum strain, quality of film, and fabrication technique. As a demonstration of the potential long fatigue life of thin film actuators, one valve in the TiNi Alloy Company's laboratory was operated for more than 50 million cycles at just over one percent deformation and one cycle per second.

The electrical resistivity of TiNi thin film is about 80 micro-ohm per centimeter; so thin film devices using joule heating operate at low voltages. Thin film SMA is therefore safe to use in implanted medical electronic devices. A TiNi actuator is a purely resistive load that easily matches impedance to microelectronics power sources.

10.2.2   Temperature effect of adding a third component to TiNi thin film

Thin film having austenite finish temperature above 373 K can be made as titanium–nickel–palladium, titanium–nickel–hafnium, and titanium–nickel–zirconium. Above about 10 atomic percent hafnium, the transition temperature of TiNi alloys increases. Ductility of the resulting film is maintained if the composition is approximately 50 atomic percent of the elements nickel plus palladium (or platinum) and 50 atomic percent of titanium plus hafnium (or zirconium).

Creation of both high and low transition temperature thin film was accomplished by simultaneous sputtering from two or three sources. Deposition rates for each of the target materials were measured by depositing film and measuring the film thickness as a function of time and power. These were combined in a "dead reckoning" calculation to produce the desired compositions. This procedure worked quite well in producing film with good thermomechanical properties. Unfortunately, there is no practical method of precisely determining the actual composition of the film, so the phase transition temperature must be determined by the electrical and physical behavior of the film.

10.2.3   Corrosion behavior of thin films

Having a very large surface area to volume ratio, TiNi may be expected to be more sensitive to corrosion than bulk material. It has been observed, for example, that thin film TiNi will burn if exposed to flame.

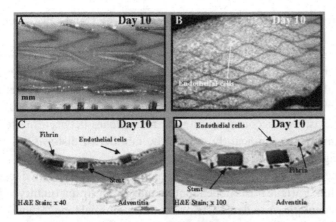

Figure 10.3   Four photographs showing a thin film TiNi stent graft implanted in a swine for 10 days in biocompatibility demonstration by Advanced Bio Prosthetic Surfaces. No irritation of tissues is evident, and endothelial cells have grown inside the vessel in a normal way, covering the struts of the stent graft. Reprinted with permission of ASM International. All rights reserved. Banas *et al.* [7], figure 3 (SMST, 2003).

The corrosion resistance of TiNi is in general due to a thin oxide layer. As long as this layer is undisturbed it protects the underlying material and deters reaction: titanium oxide is extremely inert chemically. But if the layer of titanium oxide is damaged, corrosion by acids and saline solutions such as body fluids is observed. Corrosion resistance in thin film is enhanced by the absence of nickel-rich precipitates that can create nickel oxides [7].

10.2.4   Biocompatibility of thin films

Living tissue generally rejects foreign material. Nickel is among the metals considered toxic and therefore not acceptable for medical implants. However, nitinol, the atoms of which are half nickel, has been demonstrated to be non-cytotoxic along with stainless steel and titanium, and is accepted by the medical community as biocompatible for intravascular uses. TiNi thin film, with a much larger exposed surface than bulk nitinol, might be more prone to toxicity, but experiments have demonstrated that thin film is also biocompatible.

Biocompatibility was demonstrated by Advanced Bio Prosthetic Surfaces who combined thin film stent covers with self-expanding stents, and deployed those stents in the carotid arteries of swine (Fig. 10.3). The thin film stents were inspected immediately after implantation, again at 10 days, and a third time at 28 days. These all-metal stent grafts showed patency in all cases, and no inflammatory foreign body reaction was seen [7], p. 659.

The conclusion is that TiNi thin film stent covers have an advantage over polymer coverings because the TiNi thin film provides biologically friendly scaffolding that promotes tissue growth and endothelialization.

10.3   Thermo mechanical applications of thin films

The remainder of the chapter discusses the combination of TiNi films a few microns thick with silicon to form composite micromachine elements – specifically actuators for microvalves, microrelays and optical switches. The basic MEMS processes, photolithography and chemical etching, are used for fabrication of these devices. Although TiNi is resistant to most chemicals, some acids used in MEMS processes can damage the film. Selection of processes and reagents compatible with TiNi requires care and experimentation.

10.3.1   Microactuators

Successful use of SMA thin films as microactuators brings several challenges. A thorough understanding of stress evolution and relaxation in thin films during film processing and micromachining is essential to prevent detachment of films and to control the actuation of SMA thin films, especially for bimorph type actuators. The stress in SMA thin films consists of intrinsic and extrinsic stresses: the former is caused by deposition and crystallization, while the latter comes from differential thermal expansion. In addition to the stress in films, characteristics of the film/surface interface, such as the presence of reaction products, also affect the film adhesion.

The ability of thin film shape memory alloy to change shape by application of heat can be used to move objects. If recovery is resisted, the SMA generates force that is useful as an actuator. When combined with springs and mechanical linkages, shape memory alloys can be used to convert heat to mechanical energy. Heat energy to the actuator may be supplied by conduction, radiation, electrical joule heating, or even by an incident electron beam for micron-size actuation [8].

The ordinary shape memory effect is one-way, in which a deformed shape in the martensite phase recovers the original shape in the parent phase. To achieve reversible actuation, a bias force is required to deform the film in the martensite phase and complete a thermodynamic cycle.

The stress–strain characteristics of martensite and the parent phase are very different, as illustrated by the high- and low-temperature isotherm curves in Fig. 10.4.

One can create an actuator by utilizing the mechanical energy released during this phase transformation. Physically, this is accomplished by coupling a bias spring to the SMA so that the spring causes deformation of the SMA at room temperature. If the spring constant is chosen appropriately, the SMA element reverses the action as it recovers its shape during heating..

Two stress–strain isotherms are plotted, for austenite (steep curve) and martensite (sloping curve). A "load line" (dashed line in Fig. 10.4) is drawn so that it intersects these curves at a pre-determined stroke and force. The cycle determined by this line represents a predictable, repeatable cycle. The bias force may be supplied by an elastic spring as represented by the dashed line, by another mechanical component, or by electromagnetic or gravity force.

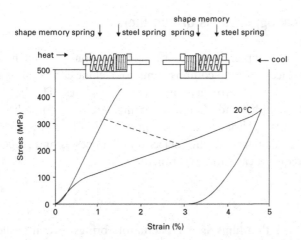

Figure 10.4   SMA actuator incorporating a bias spring to accomplish two-way motion and to establish repeatable cycling at predetermined stroke and force range. The stress–strain curves represent SMA behavior at temperatures above (steep curve) and below the transition temperature (curve labeled 20 °C). The slope of the dashed line connecting the stress–strain curves corresponds to the stiffness of the bias spring, and its position corresponds to the force. Used with permission from A. D. Johnson, Vacuum-deposited TiNi shape memory film: characterization and applications in microdevices, *J. Micromech. Microeng.*, **1**, (1991) figure 4.

Depending on the source of the bias force, thin film SMA microactuators can be classified into bimorph and freestanding types. For bimorphs the bias force is provided by a thermal stress exerted from the substrate. Two-way shape memory in TiNi thin film has been reported [9, 10, 11]. However, for maximum work output in freestanding film an external source of stress such as bias pressure or a bias spring is provided. Bimorph actuators can produce large deflections at low forces, while freestanding films are capable of exerting large forces over a smaller stroke.

10.3.2   Fluid control

Control of fluid flow is essential to the operation of all pneumatic and hydraulic systems from implantable insulin pumps to heating, ventilating, and air conditioning systems. Chemical analyzers such as gas chromatographs are used in many environments, and especially in field measurements where portability is important. The trend to miniaturization is driven by the twin needs for portability and improved performance.

Miniaturization of fluidics systems requires miniaturization of all the components including microvalves and pumps. If these components are large, the internal dead volume of the conduits connecting the components may significantly affect performance. Miniature gas chromatographs work best if the internal dead volume is minimized. Systems for processing extremely expensive solutions such

(a)   (b)   (c)

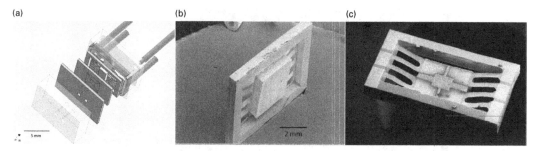

Figure 10.5   Exploded drawing of a TiNi valve assembly. (a) From lower left of drawing: plastic package bottom; silicon orifice die; silicon membrane valve seal; TiNi valve actuator; beryllium–copper bias spring; plastic package top; and spring-loaded electrical contacts. (b) and (c) SEM images of miniature TiNi actuator, 3 mm × 5 mm. Thin film TiNi microribbons connect the central silicon poppet with the frame. Electrical current (100 milliamperes) causes these microribbons to contract against the bias force (a fraction of a newton) and lifts the poppet (tens of microns) from the orifice, thus opening a flow pathway through the valve. Johnson *et al.* [14], figure 3 (SMST, 2003).

as DNA require that all components have small internal dead volume. Quite a few researchers and inventors have contributed to publications and patents regarding miniature fluid handling components [12].

The microvalve manufactured by the TiNi Alloy Company was the first miniature SMA-actuated pneumatic control device to be offered commercially. It consists of an actuator die with a poppet controlled by the eight TiNi thin film strips, 3.5 micrometers thick and 250 micrometers wide, a silicon orifice die, a spacer, and a bias spring. All elements are assembled in a plastic package. The bias spring forces the poppet towards the orifice. Resistive heating of the SMA thin films supporting the poppet causes it to transform from the martensite phase to the parent phase. By this transformation, TiNi strips recover the original length, lifting the poppet against the bias force and opening the valve. This device has a poppet displacement of ~100 micrometers and bias force of 0.5 newton. It is operated with an electric current of 50–100 milliamperes. The response time when operating in air is about 20 milliseconds, and maximum flow is up to 2000 standard cubic centimeters per minute at 1.3 atmospheres.

Figure 10.5(a) is a drawing of a liquid control valve that uses the same actuation principle, and incorporates a membrane to isolate the flowing liquid from the actuator. Figures 10.5(b) and (c) show SEM images of TiNi thin film actuators.

10.3.3   Microswitches

In recent decades the miniaturization of electronic control components, especially transistors, has been dramatic. This has not been matched by a corresponding miniaturization of mechanical switches and relays. Transistors draw power

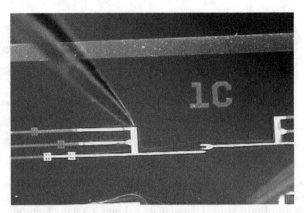

Figure 10.6   Photograph of a miniature latching relay with two beams that conduct current when in contact. Two opposing micromachined TiNi thin film actuators, driven by electric current, engage and disengage the forked latch shown at the center of the photograph. When the latch is engaged, the electrical resistance is small because the two beam-ends are pressed together by the force exerted by the bent beams. The total length is approximately 0.2 millimeters.

continuously while in use, and most electronic components lose information about their state when power is off. A latching miniature relay, a form of non-volatile storage, has low insertion losses and would be applicable in many electronic systems. Thin film TiNi has electrical and mechanical properties suitable for actuating miniature switches.

A micromechanism invented by John Comtois at Wright-Patterson Air Force Base is the basis of the TiNi Alloy Company's latching relay shown in Fig. 10.6.

Two beams are placed end-to-end, one with a concave end and the other convex, close enough that they can be brought into or out of contact by bending one beam or the other. While the outer ends are fixed to the substrate, the two inner ends can be moved by bending so that they latch. In this configuration, the force causing engagement can be removed without disengaging the two contacting ends, and the force pressing the two contacting ends together is sufficient to produce a low-ohmic contact. Power is required only while the switch is changing state – making or breaking contact.

The conducting beams were made of nickel that was electroplated into molds of SU-8 photoresist, which in turn was fabricated by photolithography. Each conducting beam is individually controlled by a TiNi thin film microribbon actuator as shown in the photograph. The microribbon actuators lie parallel to the conducting beams in order to conserve space so that multiple relays can be closely spaced, and are attached by cantilevers to the conducting beams. The TiNi microribbon actuators were about 50 microns wide and 2–5 microns thick. The resulting latching relay measures about 1 mm long by 100 microns wide and conducts up to 1.0 ampere electrical current with about 0.1 ohm resistive impedance.

Figure 10.7   Photograph showing a variety of three-dimensional devices formed from sputtered thin film TiNi. Sizes can range from less than a millimeter to several centimeters. Stiffness depends on film thickness – thin films are very flexible and can be folded for implantation by a microcatheter. Photolithography provides a versatile means for fenestration of thin films.

10.4   Superelastic applications and medical devices

TiNi thin film's superelastic characteristics make it particularly useful for implantable devices such as aneurysm closures, miniature stents, and stent covers. Thin film devices with fenestrations as shown in Fig. 10.7 are intended to be used to retrieve, filter, capture or remove blood clots in small blood vessels. The micro-catheter loaded with the device will be traversed through the blood vessel and deployed to capture and remove a blood clot. Once inside and warmed to body temperature, the TiNi device expands to its original shape and exhibits considerable resistance to deformation, damage or kink after implantation. Sieves and clot retrievers are other potential applications for TiNi thin film, especially in areas of the body distal from the heart where vessels are smaller, stiffer and more muscular, as in the case of intracranial arteries and veins. Thin film stents can be from 1 to 40 microns thick and fenestration feature size can be as small as a micron.

As medical implantable devices made of TiNi alloys become smaller, designers can benefit from the combination of sputter deposition of TiNi and photolithography for manufacturing miniature medical devices. Several United States and world patent publications describe methods for fabricating thin film TiNi stents for use in very small brain vessels – down to one millimeter in size.

Film up to 30 micrometers in thickness, deposited onto a flat substrate and released from it by etching away the sacrificial layer, forms a miniature stent. The film may be photolithographically patterned with overlapping perforations, as illustrated in Fig. 10.8, so it can be stretched.

Currently, superelastic miniature stents and stent covers are the most useful form of thin film SMA in medical devices [13].

Intravascular applications of TiNi (or NiTi) thin film are discussed in a paper by Banas *et al.* [7]. Stent covers were made and tested on swine for physical properties and biocompatibility with living tissue. Banas *et al.* assert that their

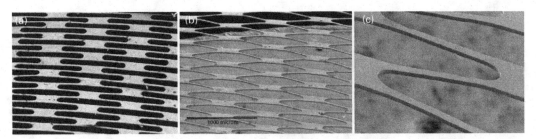

Figure 10.8   Detail of a superlastic thin film TiNi stent cover. SEM images show TiNi film 10 microns thick with overlapping fenestration to provide the ability to stretch in one dimension. Pattern length 1000 microns, strut width 30 microns: (a) unstretched, (b) stretched, (c) higher magnification view showing that the edges of the struts are very regular.

process for manufacturing thin film (sputtering from a cylindrical target) enables them to maintain extremely close compositional and dimensional control. As a result, the material is more ductile and stronger than bulk nitinol, and the product is highly resistant to corrosion.

10.5   Summary

Worldwide research programs in universities and industry have produced many publications, prototypes and patents on SMA thin film. TiNi thin film is a high quality material with excellent physical properties. It has been shown to be highly biocompatible and has a long fatigue life. Intrinsic properties are equivalent to or better than bulk nitinol.

Although practical applications have been slow in coming, it is now safe to predict that TiNi thin film will find its way into more and varied applications. The use of SMA thin films in the fabrication of medical devices will become more common, enabling the performance of intra-cranial procedures using smaller stent devices and providing stent covers for safer angioplasty. The rapid increase in the number of United States and world patents involving SMA thin alloys shows that new applications are being investigated.

Since thin film has some properties that differ markedly from bulk nitinol, the methods used in producing film and manufacturing articles from thin film are also completely different from those of bulk nitinol.

MEMS processes are both capital- and labor-intensive. Marketing of devices utilizing SMA actuators has been held back by the high cost of labor-intensive MEMS processes when producing small quantities. Until large numbers are achieved and the "value-added" is high, MEMS devices will suffer when competing with existing technology.

Currently, a major obstacle to commercialization of thin film actuators is the high cost of development. Most buyers want an off-the-shelf solution at a competitive price. This cannot happen until a very large market is created. Portable and hand-held chemical analyzers such as gas chromatographs require small valves, but the market is relatively small. If miniature fuel cells become a consumer item, the requirement for miniature fluidics control should provide the incentive to make MEMS actuators less expensive.

A particular SMA thin film product, miniature valves, has not been a financial success as yet. In actual applications, it has been possible to circumvent the use of microvalves by separating fluid control functions from on-chip operations or, in special cases, using electrokinesis.

So despite what has seemed like great promise for future applications, this material is not yet in common use. Of the many potential applications, none has reached significance in commercial uses. An explanation has been offered by Kurt Peterson, the first researcher to publish microdevice fabrication by chemical machining of single crystal silicon and founder of NovaSensors and Cepheid, who says three things are needed for a new technology to take root – namely a good technology, robust manufacturing processes, and a starving market.

Following the trend established by bulk nitinol, medical device companies are preparing products for market suggesting that this will be one area to see significant growth. Yet to this day, shape memory alloy thin film has not yet achieved its potential for use in commercial arenas – industry, manufacturing and technology. The viability of shape memory alloy thin film depends on dramatic demand for the benefits of this material in diverse market applications.

Acknowledgements

Support for research and development leading to the results reported here have come from a variety of sources, but mainly from United States government agencies through the Small Business Innovation Research program including National Institutes of Health, National Aeronautics and Space Administration, United States Air Force, United States Navy, Defense Advanced Research Projects Agency, Ballistic Missile Defense Organization and United States Department of Education.

References

[1] A. D. Johnson. Thin film shape-memory technology: a tool for MEMS. *J. Micromachine Devices*, 4:12 (1999) 34–41.
[2] J. A. Walker, K. J. Gabriel and M. Mehregany. Thin-film processing of TiNi shape memory alloy, *Sensors and Actuators*, **A21–A23** (1990) 243.
[3] P. Krulevitch, A. P. Lee, P. B. Ramsey, J. C. Trevino, J. Hamilton and M. A. Northrup. Thin film shape memory alloy microactuators. *J. Microelectromechanical Systems*, **5**:4 (1996) 270–282.
[4] A. D. Johnson. Vacuum-deposited TiNi shape memory film: characterization and applications in microdevices. *J. Micromech. Microeng.*, **1** (1991) 34–41.
[5] A. Ishida and V. Martynov. Sputter-deposited shape-memory alloy thin films: properties and applications. *MRS Bulletin*, **27**:2 (2002) 111–114.

[6] G. A. Shaw, D. S. Stone, A. D. Johnson, A. B. Ellis and W. C. Crone. Shape memory effect in nanoindentation of nickel-titanium thin films. *Appl. Phys. Lett.*, **83**:2 (2003) 257–259.

[7] C. E. Banas, C. P. Mullens, A. G. Sammons, *et al*. New design opportunities in the medical device industry using NiTi thin film technology. In *SMST 2003: Proceedings of the 4th International Conference on Shape Memory and Superelastic Technologies*, eds. S. M. Russell & A. R. Pelton, Asilomar Conference Center, Pacific Grove, USA (2003) 651–660.

[8] K. Clements. Wireless technique for microactivation. United States Patent 6,588,208 (2003). Available from www.freepatentsonline.com/6588208.html

[9] A. Gyobu, Y. Kawamura, T. Saburi and M. Asai. Two-way shape memory effect of sputter-deposited Ti-rich Ti-Ni alloy films. *Mater. Sci. Eng. A*, **312**:1–2 (2001) 227–231.

[10] J. J. Gill, K. Ho and G. P. Carman. Three-dimensional thin-film shape memory alloy microactuator with two-way effect. *J. Microelectromechanical Systems*, **11**:1 (2002) 68–77.

[11] E. Quandt, C. Halene, H. Holleck, *et al*. Sputter deposition of TiNi, TiNiPd and TiPd films displaying the two-way shape-memory effect. *Sensors and Actuators A: Physical*, **53** (1996) 434–439.

[12] R. H. Wolf and A. H. Heuer. TiNi (shape memory) films for MEMS applications. *J. Micromech. Microeng.*, **4** (1995) 206–212.

[13] T. W. Duerig. The use of superelasticity in modern medicine. *Materials Reasearch Society Bulletin*, **27**:2 (2002) 111–114.

[14] A. D. Johnson, V. Gupta, V. Martynov and L. Menchaca. Silicon oxide diaphragm valves and pumps with TiNi thin film actuation. In *SMST 2003: Proceeding of the 4th International Conference on Shape Memory and Superelastic Technologies*, eds. S. M. Russell and A. R. Pelton, Asilomar Conference Center, Pacific Grove, USA (2003) 605–612.

11 Theory of SMA thin films for microactuators and micropumps

Yi-Chung Shu

Abstract

This chapter summarizes several recent theoretical and computational approaches for understanding the behavior of shape memory films from the microstructure to the overall ability for shape recovery. A new framework for visualizing microstructure is presented. Recoverable strains in both single crystal and polycrystalline films are predicted and compared with experiments. Some opportunities for new devices and improvements in existing ones are also pointed out here.

11.1 Introduction

The explosive growth of microsystems has created a great need for the development of suitable microactuators and micropumps. Among these applications, micropumps with large pumping volume per cycle and high pumping pressure are essential to microfluidic devices. This requires a large actuation energy density to transmit a high force through a large stroke. However, common MEMS-integrated actuation schemes can deliver limited stroke and actuation force; specifically, the typical output pressure of these pumps is of the order of several tens of kPa. Therefore, there is an important need for finding suitable materials which are able to deliver a high work output from a small volume. Shape memory alloys show great promise in this aspect since they outperform other actuation material in work to volume ratio; consequently, they are able to recover large strain at high force [1, 2, 3, 4]. A disadvantage of using these alloys is that the frequency of operation is relatively low due to limitations on heat transfer. But this can be improved in thin films because of the increase in the surface area to volume ratio. Currently an operation frequency of 100 Hz has been demonstrated using the R-phase transformation [5]. Preliminary attempts to use thin films of

Thin Film Shape Memory Alloys: Fundamentals and Device Applications, eds. Shuichi Miyazaki, Yong Qing Fu and Wei Min Huang. Published by Cambridge University Press. © Cambridge University Press 2009.

shape memory alloys, while encouraging and promising, have not achieved their exceptional potential (see discussions in [6]). This concern has motivated a number of research efforts for understanding the fundamental behavior of shape memory thin films. This chapter summarizes these recent theoretical and computational models of thin film shape memory alloys from the microstructure to the overall ability for shape recovery.

We present the general framework of pressurized films following the works by Bhattacharya and James [7], James and Rizzoni [8] and Shu [6, 9] in Section 11.2. A typical film has a characteristic geometry where one dimension is much smaller than the other two with large surfaces. As a result, the film thickness appears as a new length-scale comparable to that of the microstructure. Further, shape memory crystals are highly anisotropic and non-linear due to phase transformation, and therefore it was difficult to develop suitable theories to describe their behavior in slender structures until recently. Bhattacharya and James [7] have employed ideas similar to the notion of Γ-convergence to derive a theory of martensitic single crystal film. They showed that the microstructure in thin films can be different from that in the bulk, and this enables a novel strategy that directly uses aspects of this microstructure for building new microactuators [10]. However, thin films of shape memory material TiNi and other closely related alloys are commonly made by magnetron sputtering. Thus, these films are polycrystalline rather than monocrystalline, and the behavior of a polycrystal can be very different from that of a single crystal because of the constraining effect of neighboring grains. In addition, sputtering may produce texture different from that prepared by rolling and drawing of bulk materials in the austenite phase. Thus, it is not clear if TiNi with sputtering texture is the ideal candidate in applications of shape memory films. Shu has discussed it in [11] and concluded that other textures may be favorable from the point of view of large recoverable extention. Shu has further developed the framework of Γ-convergence to study the size effect in polycrystalline films [9]. Indeed, depending on the deposition technique, the size of grains within the film can be larger than, comparable to or smaller than the thickness of the film. Moreover, depending on the material, the length scale of the microstructure can also be larger than, comparable to or smaller than that of grains. Thus, the behavior of the heterogeneous film shows strong size effects, and this can be explained by introducing the *effective theory* in Section 11.3. Two extreme cases including flat and long columnar grains are taken for illustrating the size effect on recoverable strains.

More recently, it has been suggested that the characteristic microstructure of martensite can be exploited as device elements to create micromachines [12]. This requires the design of devices that can take full advantage of the inherent microstructure to achieve this goal, which in turn calls for an appropriate model that not only can capture the spirit of the Bhattacharya–James theory but also can serve as a useful tool for evaluating various conditions in the design process. Shu and Yen [13, 14] have recently developed such a model suitable for simulating microstructure under a variety of boundary conditions. Related problems have been investigated by Khachaturyan, Roytburd, Salje, Saxena, Lookman, and their collaborators [15, 16, 17, 18, 19, 20, 21, 22] for martensite as well as by Chen and his coworkers for ferroelectrics [23, 24]. Other works in this direction include [25, 26, 27, 28, 29]. Basically, they all use the time-dependent Ginzburg–Landua (TDGL) model, and choose the special polynomial expansions of order parameters at high orders for a particular transformation. Instead, Shu and Yen [13] have chosen a non-conventional set of order parameters to represent each martensitic variant. This approach is motivated by the hierarchical structure of multi-rank laminates constructed by Bhattacharya [30] for establishing the rule of mixtures. It has the advantage of expressing the energy-well structure in a unified fashion, irrespective of the different types of transformation. We introduce Shu and Yen's model in Section 11.4 where TiNi in the R-phase state is chosen as the model material. Although the R-phase transformation yields a relatively small shape change, its temperature hysteresis is an order of magnitude smaller than that of the monoclinic phase [31, 32]. Using this idea, Tomozawa et al. [5] have recently developed microactuators using the R-phase transformation of TiNi shape memory films, and found that the working frequency of the microactuators can be improved up to 125 Hz without degradation in displacement. Section 11.4.3 presents various intriguing and fascinating patterns of microstructure predicted by the model. They are found to be in good agreement with those observed in experiments and are consistent with the interface conditions of the Bhattacharya–James thin-film theory [7].

In the last part of Section 11.5, the computational model developed by Shu and Yen [14] is applied to the design of large strain micropumps by targeting the optimal microstructures and film orientations. The R-phase TiNi single crystal is chosen for simulation. It is found that (110) films have the largest principal strains amongst other film orientations. Other computational models for simulating the overall shape of a single crystal film under pressure can be found in [33, 34] where the quasi-convex energy density developed by Bhattacharya and Dolzmann [35] is adopted in their simulation. However, many practical applications use polycrystalline films. In this situation, all of these models need to be modified due to significant computation and lack of texture information. Thus, we introduce the method proposed by Shu [6] for estimating the recoverable deflection of a pressured film in Section 11.5.2. Surprisingly, it is found that the estimation of recoverable deflection is very different from that of recoverable extension. The former is due to the accommodation of biaxial stretch, while the latter is due to the accommodation of one-dimensional tensile strain. For example, TiNi films with {111} texture are much superior to those with {110} texture in view of recoverable extension. However, Shu [6] argued that recoverable deflection is not sensitive to common film textures in TiNi films. The predicted results are compared with several experimental observations of Wolf and Heuer [36], Miyazaki et al. [37] and Makino et al. [38]. Conclusions are drawn in Section 11.6.

Finally, Bhattacharya [39] has reviewed some recent developments of the mechanics of thin films of active materials including works not mentioned above [40, 41, 42, 43, 44].

11.2   A theory of pressurized thin film

Consider a heterogeneous (possibly multilayered) thin film of thickness h. Suppose it is released on a certain region $\omega \subset I\!R^2$ but attached to a substrate outside it, as shown in Fig. 11.1. Let $\mathbf{x} = (x_1, x_2, x_3)$ be the material point of the film relative to an orthonormal basis $\{\mathbf{e}_1, \mathbf{e}_2, \mathbf{e}_3\}$. The deformation of the film is denoted by $\mathbf{y} = (y_1, y_2, y_3)$ which is the function of the material point $\mathbf{x}$. Let d be the period of the in-plane texture (in other words, d is the typical length scale of the representative area element in the plane of the film). Let the in-plane variables x_1 and x_2 be normalized by d and the out-of-plane variable x_3 by h. Thus, the elastic density of this heterogeneous film is

$$W = W\left(\mathbf{F}, \frac{x_1}{d}, \frac{x_2}{d}, \frac{x_3}{h}\right), \tag{11.1}$$

where $\mathbf{F}$ is the deformation gradient defined by $\mathbf{F}(\mathbf{x}) = \nabla \mathbf{y}(\mathbf{x}) = \left(\frac{\partial y_i}{\partial x_j}\right)$ for $i, j = 1, 2, 3$. In the Wechsler–Lieberman–Read (WLR) theory [45], $\mathbf{F}$ is the distortion matrix which is the measure of the crystal deformation.

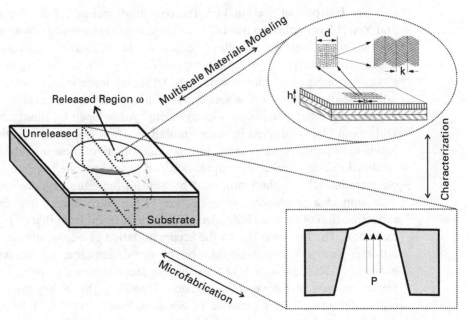

Figure 11.1   Prototype of a micropump using the shape memory film. The film is heterogeneous and contains three different length scales h, d and κ. It is released from the substrate by etching in the chosen region ω, but attached to it outside.

To design a micropump, pressure is usually applied from either above (evacuation type) or below (pressurization type) depending on the actuation method [38]. Suppose a hydrostatic pressure $p^{(h)}$ is applied on the lower surface of the film. The total energy of this pressurized film per unit film thickness is

$$e_1^{(h)}[\mathbf{y}] = \frac{1}{h} \int_{\omega \times (0,h)} \left\{ \kappa^2 |\nabla^2 \mathbf{y}|^2 + W\left(\nabla \mathbf{y}, \frac{x_1}{d}, \frac{x_2}{d}, \frac{x_3}{h}\right) \right\} \mathrm{d}x_1 \mathrm{d}x_2 \mathrm{d}x_3$$

$$- \frac{P}{3} \int_{\omega \times \{0\}} \mathbf{y} \cdot \left(\frac{\partial \mathbf{y}}{\partial x_1} \times \frac{\partial \mathbf{y}}{\partial x_2} \right) \mathrm{d}x_1 \mathrm{d}x_2, \tag{11.2}$$

where $P = p^{(h)}/h$ is assumed to be a constant. Above $\frac{\partial \mathbf{y}}{\partial x_i} = \left(\frac{\partial y_1}{\partial x_i}, \frac{\partial y_2}{\partial x_i}, \frac{\partial y_3}{\partial x_i}\right)$, $\nabla^2 \mathbf{y} = \left(\frac{\partial^2 y_i}{\partial x_j \partial x_k}\right)$ for $i, j, k = 1, 2, 3$, and $|\nabla^2 \mathbf{y}|^2 = \sum_{i=1}^{3} \sum_{j=1}^{3} \sum_{k=1}^{3} \left|\frac{\partial^2 y_i}{\partial x_j \partial x_k}\right|^2$. Further, $\mathbf{a} \cdot \mathbf{b}$ and $\mathbf{a} \times \mathbf{b}$ are the standard notations for the inner and cross products of two vectors $\mathbf{a}$ and $\mathbf{b}$. The term

$$\frac{1}{3} \int_{\omega \times \{0\}} \mathbf{y} \cdot \left(\frac{\partial \mathbf{y}}{\partial x_1} \times \frac{\partial \mathbf{y}}{\partial x_2} \right) \mathrm{d}x_1 \mathrm{d}x_2$$

is the volume enclosed between the plane $\omega \times \{0\}$ and the deformed lower surface of the film $\mathbf{y}(\omega \times \{0\})$ [8].

The interpretation of Eq. (11.2) is as follows. The first term $\kappa^2 |\nabla^2 \mathbf{y}|^2 = \kappa^2 |\nabla \mathbf{F}|^2$ penalizes changes in the deformation gradient and thus is interpreted as the energy of forming an interface separating distinct distortion matrices. Moreover, such a change is supposed to be abrupt if κ is assumed to be very small. Minimizers of the energy in Eq. (11.2) have oscillations on a length scale that scales with κ and hence we call κ the length scale of the microstructure. The second term is the elastic energy with density W as described in Eq. (11.1). The dependence of W on the material point $\mathbf{x}$ reflects the fact that the film is not homogeneous. The final term is interpreted as the energy of a fluid under the film with pressure P.

The behavior of the film at pressure P is determined by minimizing its free energy in Eq. (11.2) amongst all possible deformations $\mathbf{y}$. It is not an easy task for a number of reasons. First, the modeling of phase-transforming materials usually requires that W has a number of energy ground states. Thus, W is not a convex function of deformation gradient $\mathbf{F}$. It has a multi-well structure – one well for each phase. This creates serious problems because of the difficulty of non-convex optimization. Second, if the film is heterogeneous (i.e., W depends on $\mathbf{x}$), it involves the technique of homogenization to find its average behavior. It is also difficult since the information of heterogeneity is typically limited. Finally, as the film thickness is very small, it becomes a new length scale competing with the other two length scales: the typical grain size d and the microstructure length scale governed by κ. The properties of the film are crucially determined by the different ratios of these three length scales. Thus, combining all of these facts indicates that the present problem is much harder than other conventional homogenization problems with convex energy density and fixed geometry.

Since the lateral extent (i.e., the in-plane dimensions) of the film is usually much larger than any of these three length scales, the effective behavior of the film is expected not to depend on every detail of the grains and multilayers, but only on some average features. Indeed, Shu [9] has used the framework of Γ-convergence to show that the average behavior of the film is determined by an effective two-dimensional theory. The limiting theory implies that the overall deformation $\mathbf{y}$ has three components y_1, y_2 and y_3 which depend only on the in-plane variables x_1 and x_2. In addition, the deformation $\mathbf{y}$ of the film is determined by minimizing the *effective potential energy*

$$e_1^{(0)}[\mathbf{y}] = \int_\omega \left\{ \overline{W}\left(\frac{\partial \mathbf{y}}{\partial x_1}, \frac{\partial \mathbf{y}}{\partial x_2}\right) - \frac{P}{3}\, \mathbf{y} \cdot \left(\frac{\partial \mathbf{y}}{\partial x_1} \times \frac{\partial \mathbf{y}}{\partial x_2}\right) \right\}\, \mathrm{d}x_1 \mathrm{d}x_2, \qquad (11.3)$$

where $\overline{W}$ is the *effective, macroscopic* or *overall* energy density of the heterogeneous film. It describes the overall behavior of the thin film after taking into account the martensitic microstructure, grains and multilayers. Note that the density $\overline{W}$ does not explicitly depend on the position; instead, it is dependent only on the reduced deformation gradient $\bar{\mathbf{F}}$ which is a 3×2 distortion matrix

$$\bar{F}_{ia}(x_1, x_2) = \frac{\partial y_i(x_1, x_2)}{\partial x_a}, \qquad \text{for } i = 1, 2, 3 \text{ and } a = 1, 2. \qquad (11.4)$$

The explicit expression of $\overline{W}$ depends on the different limiting ratios of these three length scales, and will be briefly described in the next section. To summarize, the proposed framework establishes a simpler two-dimensional effective energy $e_1^{(0)}$ whose solutions approximate those of the fully three-dimensional theory based on $e_1^{(h)}$. Thus, one can use this effective theory to investigate the microstructure and behavior of thin films.

11.3   Effective behavior

11.3.1   Single crystal film

Consider a single crystal film. The bulk elastic density of it depends only on the deformation gradient: $W = W(\mathbf{F})$ in Eq. (11.1). In this case, Shu [9] has discovered a remarkable result: the effective theory is independent of the ratio of κ/h (it can be zero, finite or infinite). Indeed, define the thin-film elastic density by

$$W_0(\bar{\mathbf{F}}) = \min_{\mathbf{b} \in I\!R^3} W(\bar{\mathbf{F}}|\mathbf{b}), \qquad (11.5)$$

where $\bar{\mathbf{F}}$ is the 3×2 distortion matrix defined in Eq. (11.4), and the notation $(\bar{\mathbf{F}}|\mathbf{b})$ means that the first two columns and final column of the 3×3 matrix are replaced by $\bar{\mathbf{F}}$ and $\mathbf{b}$, respectively. The effective density of the film is determined by $\overline{W}(\bar{\mathbf{F}}) = QW_0(\bar{\mathbf{F}})$, where QW_0 is the relaxation of density W_0. The exact definition of QW_0 can be found in [46, 47] and is not shown here as it requires advanced mathematical analysis which is beyond the purpose of this chapter.

The major difference between bulk materials and thin films is that the relaxation process is associated with the density W_0 given by Eq. (11.5) for films rather than W in Eq. (11.1) for bulk materials. The deformation of the film is determined by two vector fields $\mathbf{y}$ (x_1, x_2) and $\mathbf{b}$ (x_1, x_2) which are independent of the thickness direction. The vector field $\mathbf{y}$ describes the deformation of the middle surface, while the vector field $\mathbf{b}$ describes the transverse shear and normal stretch. The last column of the 3×3 distortion matrix $\mathbf{F}$ originally describing the change of deformation along the thickness direction $\left(\frac{\partial y_1}{\partial x_3}, \frac{\partial y_2}{\partial x_3}, \frac{\partial y_3}{\partial x_3} \right)$ is now replaced by the column vector $\mathbf{b}$ via Eq. (11.5). Therefore, the out-of-place kinematic compatibility becomes insignificant. More precisely, let $\mathbf{e_3}$ be the normal to the film plane, the kinematic compatibility condition for thin films is

$$[[(\overline{\mathbf{F}}|\mathbf{b})]] = \mathbf{a} \otimes \mathbf{n} + \mathbf{c} \otimes \mathbf{e_3}, \qquad (11.6)$$

for some vectors $\mathbf{a}$, $\mathbf{c}$ and some unit vector $\mathbf{n}$ such that $\mathbf{n} \cdot \mathbf{e_3} = 0$. Above in Eq. (11.6), the notation $[[\overline{\mathbf{F}}|\mathbf{b}]]$ denotes the jump in $(\overline{\mathbf{F}}|\mathbf{b})$ across the interface. Another equivalent statement of Eq. (11.6) is

$$[[(\overline{\mathbf{F}}|\mathbf{b})]] \, \mathbf{t} = \mathbf{0}, \qquad (11.7)$$

for some vector $\mathbf{t}$ such that $\mathbf{t} \cdot \mathbf{e_3} = 0$. Thus, Eq. (11.7) denotes an invariant line condition in films in contrast with the invariant plane condition in bulk materials. Next, the vector $\mathbf{c}$ in Eq. (11.6) indicates a mismatch in the thickness direction. The energy cost to overcome this mismatch in the film direction is around an order of h^2 smaller than the effective energy $e_1^{(0)}$; i.e., $e_1^{(h)} = e_1^{(0)} + O(h^2)$ where the term $O(h^2)$ is the energy scaling at an order of h^2. The requirement of coherence is therefore weaker in thin films than in bulk materials. Finally, the energy of stretching and shearing is described by $e_1^{(0)}$, while the energy of bending is contained in the term $O(h^2)$. Thus, the bending energy is insignificant for very thin films.

We now apply these results to shape memory thin films. These materials change their crystal structures under temperature variations. The high temperature *austenite* phase typically has a cubic symmetry, while the low temperature *martensite* phase has less symmetry such as tetragonal, trigonal, orthorhombic or monoclinic symmetry. This gives rise to N symmetry-related variants of martensite. The transformation from the austenite to the ith variant of martensite is described by the distortion (Bain) matrix $\mathbf{U}^{(i)}$. It can be determined from the change of symmetry and lattice parameters, as listed in Table 11.1 [11, 48]. As a result, the bulk elastic density W has a multi-well structure and is minimized on these distortion matrices followed by any proper rotations due to the principle of frame-indifference. Thus, from Eq. (11.5), any energy minimizing deformation takes the form of $(\overline{\mathbf{F}}|\mathbf{b}) = \mathbf{Q}\mathbf{U}^{(i)}$ for some rotation $\mathbf{Q}$ and variant i. In fact, Bhattacharya and James [7] have used the energy-well structure of martensite and the thin-film compatibility Eq. (11.7) to discover many more austenite/martensite and

Table 11.1 List of transformation strains $\varepsilon^{(i)} = \mathbf{U}^{(i)} - \mathbf{I}$ for various symmetries of martensite. Only $\varepsilon^{(1)}$ is shown here; the rest $\varepsilon^{(2)}, \ldots, \varepsilon^{(N)}$ can be obtained from $\varepsilon^{(1)}$ by symmetry by permuting the basis. The symmetry of the austenite is cubic. We choose variant (1) so that $\delta > 0$ (orthorhombic and monoclinic cases) and $\varepsilon > 0$ (monoclinic case). The compositions of all alloys are given in atomic percentages unless otherwise specified. Notice that there are two kinds of cubic to monoclinic transformation. In Monoclinic-I, the axis of monoclinic symmetry corresponds to $\langle 110 \rangle_{cubic}$ while in Monoclinic-II, the axis of monoclinic symmetry corresponds to $\langle 100 \rangle_{cubic}$ [11, 48].

Symmetry of martensite	N	Transformation strain	Examples	Measured parameters
Tetragonal	3	$\begin{pmatrix} \alpha & 0 & 0 \\ 0 & \alpha & 0 \\ 0 & 0 & \beta \end{pmatrix}$	Ni-36.8Al [49]	$\alpha = -0.0608,\quad \beta = 0.1302$
Trigonal	4	$\begin{pmatrix} \beta & \alpha & \alpha \\ \alpha & \beta & \alpha \\ \alpha & \alpha & \beta \end{pmatrix}$	Ti-50.5Ni (R-phase) [31]	$\alpha = 0.0047,\quad \beta = 0$
Orthorhombic	6	$\begin{pmatrix} \alpha & \beta & 0 \\ \beta & \alpha & 0 \\ 0 & 0 & \beta \end{pmatrix}$	γ'_1 Cu-14Al-4Ni (wt%) [50] Ti-40Ni-10Cu (B19) [51]	$\alpha = 0.0425,\quad \beta = -0.0822$ $\delta = 0.0194$ $\alpha = 0.0240,\quad \beta = -0.0420$ $\delta = 0.0310$
Monoclinic-I	12	$\begin{pmatrix} \alpha & \delta & \epsilon \\ \delta & \alpha & \epsilon \\ \epsilon & \epsilon & \beta \end{pmatrix}$	Ti-49.8Ni [52] Ti-45Ni-5Cu [51]	$\alpha = 0.0243,\quad \beta = -0.0437$ $\delta = 0.0580,\quad \epsilon = 0.0427$ $\alpha = 0.0232,\quad \beta = -0.0410$ $\delta = 0.0532,\quad \epsilon = 0.0395$
Monoclinic-II	12	$\begin{pmatrix} \alpha+\epsilon & \delta & 0 \\ \delta & \alpha-\epsilon & 0 \\ 0 & 0 & \beta \end{pmatrix}$	Cu-15Zn-17Al [53] β'_1 Cu-14Al-4Ni (wt%) [54]	$\alpha = 0.0483,\quad \beta = -0.0907$ $\delta = 0.0249,\quad \epsilon = 0.0383$ $\alpha = 0.0442,\quad \beta = -0.0822$ $\delta = 0.0160,\quad \epsilon = 0.06$

martensite/martensite interfaces in thin films. They have shown that there are interfaces that are possible in thin films which are not possible in bulk. They have also proposed a strategy to take advantage of such interfaces in designing tent and tunnel based micropumps [10].

While various exact interfaces can be found using this geometrically non-linear framework, many applications require the use of linearized kinematics to simplify the problems. If only the in-plane deformation is concerned, the linearized kinematics require the in-plane displacement $\mathbf{u}^P = (u_1, u_2) = (y_1 - x_1, y_2 - x_2)$ and the in-plane strain ε^P to be related by

$$\varepsilon^P[\mathbf{u}^P] = \frac{1}{2}\left[\nabla_p\mathbf{u}^P + \left(\nabla_p\mathbf{u}^P\right)^T\right], \tag{11.8}$$

where $\nabla_p = \mathbf{e}_1\frac{\partial}{\partial x_1} + \mathbf{e}_2\frac{\partial}{\partial x_2}$ is the in-plane gradient with respect to the planar variables $\mathbf{x}_p = (x_1, x_2) \in \omega$. In this case, the energy-well structure of the thin-film elastic density W_0 can be described by

$$W_0(\varepsilon^P)\begin{cases} = 0 & \text{if } \varepsilon^P \in \mathbf{Z} = \varepsilon_p^{(1)} \cup \varepsilon_p^{(2)} \cup \cdots \varepsilon_p^{(N)}, \\ > 0 & \text{otherwise}, \end{cases} \tag{11.9}$$

where $\varepsilon_p^{(i)}$ is the in-plane transformation strain of the ith martensitic variant. They are obtained from the distortion matrices $\mathbf{U}^{(i)}$ by

$$\varepsilon_p^{(i)} = \Pi\left(\mathbf{R}\varepsilon^{(i)}\mathbf{R}^T\right)\Pi^T, \quad \varepsilon^{(i)} = \mathbf{U}^{(i)} - \mathbf{I}, \quad \Pi = \begin{pmatrix} 1 & 0 & 0 \\ 0 & 1 & 0 \end{pmatrix}, \tag{11.10}$$

where $\mathbf{R}$ is the crystal orientation relative to the reference basis, $\mathbf{I}$ is the identity matrix, and the transformation strain $\varepsilon^{(i)}$ is the linearized version of the distortion matrix $\mathbf{U}^{(i)}$.

The compatibility criterion between variants of martensite needs adjustment for this linearized kinematics. Two variants $\varepsilon^{(i)}$ and $\varepsilon^{(j)}$ of bulk martensite are compatible if there exist $\mathbf{a} \in IR^3$ and $\mathbf{n} \in IR^3$ such that

$$\varepsilon^{(i)} - \varepsilon^{(j)} = \mathbf{a} \otimes \mathbf{n} + \mathbf{n} \otimes \mathbf{a} \quad \text{(bulk compatibility)}. \tag{11.11}$$

Similarly, two variants $\varepsilon_p^{(i)}$ and $\varepsilon_p^{(j)}$ of thin film martensite are compatible if they satisfy

$$\varepsilon_p^{(i)} - \varepsilon_p^{(j)} = \mathbf{a}^P \otimes \mathbf{n}^P + \mathbf{n}^P \otimes \mathbf{a}^P \quad \text{(thin-film compatibility)}, \tag{11.12}$$

for some $\mathbf{a}^P \in IR^2$ and $\mathbf{n}^P \in IR^2$. Thus, if two variants are compatible in bulk, their projections through Eq. (11.10) are also compatible in thin films. However, the converse is not true. It can be shown that certain variants of monoclinic martensite are not compatible in the bulk sense as in Eq. (11.11), while they may satisfy the thin-film compatibility as in Eq. (11.12) for some oriented films. This also confirms that martensitic materials can form many more interfaces in a thin film than in bulk.

The ability that shape memory alloys have to form compatible patterns of microstructure is the key to their shape recovery on heating. However, only certain strains can be recovered: those that can be accommodated by coherent mixtures of martensitic variants. Larger strains will introduce stress, giving rise to lattice defects and non-recoverability. Thus, $\mathcal{S}_f$, the set of recoverable strains in a single crystal film, is exactly the set of strains one can make by the coherent rearrangement of microstructures of martensite. Bhattacharya [30] has proposed the rule of mixture to estimate this set. Simply speaking, the strain compatibility in Eq. (11.12) implies that the two variants can form a laminate structure by alternating layers of these two, and the overall strain of this laminate is the average of these two variants. Therefore, if each pair of variants is twin-related or satisfies Eq. (11.12), this set $\mathcal{S}$ is the convex hull of all in-plane transformation strains; i.e.,

$$\mathcal{S}_f = \left\{ \varepsilon^P : \varepsilon^P = \sum_{i=1}^{N} \lambda_i \varepsilon_p^{(i)}, \lambda_i \geq 0, \sum_{i=1}^{N} \lambda_i = 1 \right\}. \tag{11.13}$$

The hypothesis of pairwise compatibility is valid for transformations with tetragonal, trigonal or orthorhombic martensite and we can obtain $\mathcal{S}_f$ in these cases. However, it is valid for monoclinic martensite only for some oriented films, and consequently, the set $\mathcal{S}_f^{mono}$ is not always the convex hull of in-plane transformation strains. Indeed, it can be shown that all pairs of monoclinic variants are thin-film compatible for $\{100\}$ and $\{110\}$ films in TiNi, and $\{110\}$ and $\{111\}$ films in Cu-Zn-Al. Thus, for these films, $\mathcal{S}_f^{mono}$ is the convex hull of all in-plane transformation strains. However, not all pairs are compatible for $\{111\}$ films in TiNi and $\{100\}$ films in Cu-Zn-Al. In these two cases, the set $\mathcal{S}_f^{mono}$ is more complicated and needs further analysis [55].

11.3.2   Polycrystalline film

Shape memory thin films are usually made by sputtering [2, 56, 57, 58, 59], and therefore they are polycrystalline in general. A polycrystal is an aggregate of a great number of single crystal grains with different orientations. The texture (i.e., the size and orientations of the grains) may be described by a rotation-valued function $\mathbf{R}(\mathbf{x})$ which provides the orientation of the crystal at the point $\mathbf{x}$ relative to some fixed reference crystal. In a typical polycrystal, $\mathbf{R}(\mathbf{x})$ is piecewise constant, though we shall not assume any such restriction here. As the grains in sputtered films are typically columnar (for example, see Figure 2 of [57]), thus, the texture is independent of the thickness direction. In other words, the orientation function $\mathbf{R} = \mathbf{R}(x_1, x_2)$ and the elastic energy density $W = W(\mathbf{F}, x_1, x_2)$. Further, the microstructure is usually smaller than the grains (e.g., see Figure 5 in [3]). So we may assume $d \gg \kappa$. In this case, the elastic energy dominates the interfacial energy and materials can form microstructures freely. As a result, the effective energy density $\overline{W}$ in Eq. (11.3) is *impervious to the presence of interfacial energy*. But the behavior of the film depends on the ratio of the film thickness h to the typical size of grains d, and we now discuss it.

11.3.2.1 Flat columnar grains

Consider a columnar film with flat grains ($d \gg h$). The grains are flat and thin and have "pancake" shape. As a result, the intergranular constraints are now only in-plane, and any out-of-plane incompatibility is easily overcome with very small elastic energy. Further, within each grain the interface condition is an "invariant line" rather than an "invariant plane" condition, as discussed in Section 11.3.1. Therefore, the effective behavior of the film is obtained by passing to the two-dimensional limit first and then homogenizing in the plane of the film. Mathematically, the effective energy density $\overline{W}$ is obtained by homogenizing the thin-film elastic energy density W_0.

Each grain has its own set of recoverable strains, and thus, $\mathcal{S}_f = \mathcal{S}_f(\mathbf{x_p})$. For example, if the rule of mixtures can be applied or Eq. (11.13) is valid,

$$\mathcal{S}_f(\mathbf{x_p}) = \Pi \mathbf{R}(\mathbf{x_p}) \mathcal{S}_b \mathbf{R}^T(\mathbf{x_p}) \Pi^T,$$
$$\mathcal{S}_b = \left\{ \varepsilon : \varepsilon = \Sigma_{i=1}^N \lambda_i \varepsilon^{(i)}, \ \lambda_i \geq 0, \ \Sigma_{i=1}^N \lambda_i = 1 \right\}. \tag{11.14}$$

Now a constant strain ε_0^f is recoverable in a polycrystalline film if it is able to find a non-uniform compatible in-plane strain field $\varepsilon^f(\mathbf{x_p})$ that satisfies different constraints in different grains ($\varepsilon^f(\mathbf{x_p}) \in \mathcal{S}_f(\mathbf{x_p})$ for each $\mathbf{x_p}$) and whose average is this strain ε_0^f. However, the determination of recoverable strains is in general very difficult for both bulk and thin film polycrystals. A way to estimate it is to use the Taylor bound assuming each grain undergoes identical deformation to avoid intergranular incoherence [60]. Thus, the overall strain is recoverable if it is recoverable for every grain of the polycrystal. Precisely, the Taylor bound $\mathcal{T}_f^0$ of this case is defined by

$$\mathcal{T}_f^0 = \bigcap_{\forall \mathbf{x_p}} \mathcal{S}_f(\mathbf{x_p}). \tag{11.15}$$

Shu and Bhattacharya [11] have demonstrated that the results obtained from the Taylor bound are surprisingly good in estimating recoverable strain and agree very well with experiment. Bhattacharya and Kohn [61] have derived rigorous results to support this argument.

Finally, Table 11.2 lists the results for predicted uniaxial recoverable extension for different textured TiNi and Cu-Zn-Al films. Note that they are larger for flat grains than for long grains, which will be explained later. We also note here that neither the random nor {110} texture which is common for BCC materials [56, 59] are ideal textures for large recoverable extension. The ideal textures appear to be {100} for Cu-Zn-Al (this texture can be produced by melt-spinning) and {111} for TiNi.

11.3.2.2 Long columnar grains

We now turn to another extreme case $h \gg d$. The grains are now long and rod-like; and it is no longer possible to overcome out-of-plane constraints. Therefore, the intergranular constraints are fully three-dimensional. Consequently, the effective behavior is obtained by homogenizing in three dimensions and then

Table 11.2 Predicted uniaxial recoverable extension for various textures in TiNi and Cu-based thin films [11]

| | Uniaxial recoverable strains (%) | | | |
| | Ti-Ni | | Cu-Zn-Al | |
Texture	long grains	flat grains	long grains	flat grains
{100} film	2.3	2.3	7.1	7.1
{110} film	2.3	2.3	1.7	1.7
{111} film	5.3	8.1	1.9	5.9
random	2.3	2.3	1.7	1.7

passing to the two-dimensional limit. In other words, we first consider a bulk material with density W and texture $\mathbf{R}(\mathbf{x}_p)$. Then, we homogenize this heterogeneous material and obtain an effective density QW which has no position dependence. Finally, the thin-film behavior of this homogenized material is obtained by relaxing the deformation gradients along the thickness direction such as that in Eq. (11.5).

One consequence of this result is that the estimation of recoverable strains for such films is basically similar to bulk materials. Indeed, the Taylor bound $\mathcal{T}_f^\infty$ of this case is

$$\mathcal{T}_f^\infty = \Pi \mathcal{T} \Pi^{\mathrm{T}},$$
$$\mathcal{T} = \bigcap_{\forall \mathbf{x}_p} \mathcal{S}(\mathbf{x}_p), \qquad (11.16)$$

where $\mathcal{S}(\mathbf{x}_p)$ is the set of recoverable strains of bulk martensite with texture $\mathbf{R}(\mathbf{x}_p)$ and $\mathcal{T}$ is the Taylor bound of this bulk martensite. Table 11.2 lists the predicted recoverable extension for TiNi and Cu-Zn-Al films with various textures. It contrasts the behavior of films with long or rod-like ($h \gg d$) grains and films with flat or pancake shaped ($h \ll d$) grains. Recoverable extension is smaller than that for the same films with flat grains, since the intergranular constraints are weakened in films with flat grains.

11.4   Simulation of thin film microstructure

This section describes how the "rule of mixtures" motivates the development of a multi-variant framework suitable for microstructure simulation. This approach was first proposed by Shu and Yen [13, 14] and we briefly describe their ideas here. The linearized kinematic variables in Eq. (11.8) are adopted here.

11.4.1   Free energy

Consider a single crystal film of martensite. Let ε_p^* be a macroscopically homogeneous in-plane strain. It is recoverable if it can be obtained by a coherent mixture of martensitic variants. If each pair of variants is twin-related, any recoverable strain in this case must belong to the set $\mathcal{S}_f$ given by Eq. (11.13), as described in Section 11.3.1. A deeper analysis carried out by Bhattacharya [30] indicates that this strain can be achieved by a rank-$(N-1)$ compatible laminate; i.e.,

$$\varepsilon_p^*(\gamma_i) = \sum_{j=1}^{N} \gamma_j \varepsilon_p^{(j)}, \tag{11.17}$$

where γ_i is the global volume fraction of the ith variant and can be expressed in terms of

$$
\begin{aligned}
\gamma_1 &= \mu_1, \\
\gamma_2 &= (1 - \mu_1)\mu_2, \\
\gamma_3 &= (1 - \mu_1)(1 - \mu_2)\mu_3, \\
&\cdots \\
\gamma_{N-1} &= (1 - \mu_1)\cdots(1 - \mu_{N-2})\mu_{N-1}, \\
\gamma_N &= (1 - \mu_1)\cdots(1 - \mu_{N-2})(1 - \mu_{N-1}),
\end{aligned} \tag{11.18}
$$

and μ_i is the local volume fraction of the ith rank laminate. Now suppose each value of μ_i at each point $\mathbf{x}_p$ is restricted to take on only two: 0 or 1. Then, when $\mu_i = 1$ and $\mu_1 = \mu_2 = \cdots = \mu_{i-1} = 0$, $\gamma_i = 1$ and $\gamma_j = 0$ for $j \neq i$. If, however, all $\mu_i = \mu_2 = \cdots = \mu_{N-1} = 0$, then $\gamma_N = 1$ and $\gamma_j = 0$ for $j \neq N$. Thus, assigning discrete values to each of the μ_i guarantees that at each point $\mathbf{x}_p$, only one of the γ_i is equal to 1 and the rest of them vanish. As a result, $\varepsilon_p^*(\mathbf{x}_p)$ here can also be interpreted as a locally inhomogeneous strain such that it is equal to one of $\varepsilon_p^{(i)}$ at each point $\mathbf{x}$. An advantage of using Eq. (11.18) is that it provides a unified way for specifying the energy-well structure of martensite, as described next.

Let $\mu_j(\mathbf{x})$ be relaxed to continuously vary across the interfaces at the boundaries of martensitic variants. The free energy of a martensitic thin film per unit film thickness at some fixed temperature below the critical temperature is described by

$$
\begin{aligned}
\mathcal{I}(\boldsymbol{\mu}) &= \int_\omega \{W^{\text{int}}(\boldsymbol{\mu}) + W^{\text{ani}}(\boldsymbol{\mu}) + W^{\text{elas}}(\boldsymbol{\mu}) - \sigma^{p0} \cdot \varepsilon^p\} \mathrm{d}\mathbf{x}_p + O(h^2), \\
W^{\text{int}}(\boldsymbol{\mu}) &= \kappa^2 |\nabla_p \boldsymbol{\mu}|^2, \\
W^{\text{ani}}(\boldsymbol{\mu}) &= K \sum_{j=1}^{N-1} \mu_j^2 (1 - \mu_j)^2, \\
W^{\text{elas}}(\boldsymbol{\mu}) &= \tfrac{1}{2}\left[\varepsilon^p - \varepsilon_p^*(\boldsymbol{\mu})\right] \cdot \mathbf{C}^p\left[\varepsilon^p - \varepsilon_p^*(\boldsymbol{\mu})\right],
\end{aligned} \tag{11.19}
$$

where the in-plane strain ε^p is determined by a solution to

$$\nabla \cdot \sigma^p = 0, \quad \sigma^p = \mathbf{C}^p\left[\varepsilon^p - \varepsilon_p^*(\boldsymbol{\mu})\right], \tag{11.20}$$

$\varepsilon_p^*(\boldsymbol{\mu})$ is given by Eq. (11.17), $\boldsymbol{\mu} = (\mu_1, \mu_2, \ldots, \mu_{N-1})$, σ^p is the in-plane stress, and $\mathbf{C}^p$ is the plane-stress elastic modulus and is approximated to be the same for all phases.

Above, in Eq. (11.19), σ^{p0} is the in-plane auxiliary stress state which is divergence-free and consistent with the applied traction at the boundary [62].

Each of the terms in Eq. (11.19) has a physical interpretation. The first term W^{int}, called the *interfacial* energy density, penalizes changes in the field variables and thus is interpreted as the energy of forming a martensitic interface. It is similar to the term $\kappa^2|\nabla \mathbf{y}|^2$ in Eq. (11.2) which uses the change in deformation gradient to estimate the interfacial energy. The second and third terms, with $K > 0$, are the *anisotropy* and *elastic* energy densities. The sum of these two denotes the energy cost that the crystal must pay if the field variables and strain deviate from the preferred states; thus, this builds in the information that the crystal prefers a certain spontaneous strain. If the sum of these two energy densities is minimized with respect to μ for fixed ε^p, the resulting term is similar to the non-linear elastic energy density W in Eq. (11.2).

11.4.2   Evolution of microstructure under driving forces

We postulate that the martensitic microstructure is obtained by minimizing the total free energy in Eq. (11.19). However, it is not an easy task due to non-linear optimization. Alternatively, the energy is decreasing if it follows the path [62]

$$\frac{\partial \mu}{\partial t} = -M\frac{\delta \mathcal{I}}{\delta \mu} = M(\mathbf{F}_p^{int} + \mathbf{F}_p^{ani} + \mathbf{F}_p^{elas}), \qquad (11.21)$$

where $M > 0$ is the mobility and $-\delta \mathcal{I}/\delta \mu$ is the total *thermodynamic driving force* which is the sum of the following three forces:

$$\mathbf{F}_p^{int} = 2\kappa^2\nabla_p^2\mu,$$

$$\mathbf{F}_p^{ani} = -\frac{\partial W^{ani}(\mu)}{\partial \mu},$$

$$\mathbf{F}_p^{elas} = \mathbf{C}^p[\varepsilon^p - \varepsilon_p^*(\mu)] \cdot \frac{\partial \varepsilon_p^*(\mu)}{\partial \mu}. \qquad (11.22)$$

Physically, $\mathbf{F}_p^{int}$ is the driving force for coarsening microstructure, $\mathbf{F}_p^{ani}$ is the driving force for selecting variants, and $\mathbf{F}_p^{elas}$ is the driving force for refining microstructure to accommodate the specified boundary constraints.

As the length scale of martensitic microstructure is typically much smaller than the lateral boundary size of the film, it is reasonable to assume the periodic boundary condition in Eq. (11.21). Besides, the in-plane stress σ^p can be decomposed as the sum of homogeneous $\langle\sigma^p\rangle$ and perturbed $\sigma^{p'}$ states, where $\langle\cdots\rangle$ denotes the average. The perturbed inhomogeneous stress $\sigma^{p'}$ can be explicitly obtained in the Fourier reciprocal space. The homogeneous in-plane stress $\langle\sigma^p\rangle$, however, depends on the imposed boundary conditions, such as specifying the overall strain or the overall stress.

11.4.3 Self-accommodation patterns

The chosen material for simulation is TiNi in the trigonal R-phase state. Thus, from Table 11.1, the number of variants is $N=4$ and the corresponding transformation strains are

$$\varepsilon^{(1)} = \begin{pmatrix} a & \delta & \delta \\ \delta & a & \delta \\ \delta & \delta & a \end{pmatrix}, \qquad \varepsilon^{(3)} = \begin{pmatrix} a & -\delta & \delta \\ -\delta & a & -\delta \\ \delta & -\delta & a \end{pmatrix},$$

$$\varepsilon^{(2)} = \begin{pmatrix} a & -\delta & -\delta \\ -\delta & a & \delta \\ -\delta & \delta & a \end{pmatrix}, \qquad \varepsilon^{(4)} = \begin{pmatrix} a & \delta & -\delta \\ \delta & a & -\delta \\ -\delta & -\delta & a \end{pmatrix}, \qquad (11.23)$$

where a and δ are material parameters. From Table 11.1, $a=0$, $\delta=0.0047$ in Eq. (11.23). The elastic moduli of TiNi single crystals are not available; therefore, we take $C_{11}=C_{22}=80$ GPa, $C_{12}=20$ GPa, $C_{66}=30$ GPa, and $C_{16}=C_{26}=0$ (Voigt notation), which are typical parameters for TiNi polycrystals [63]. Besides, there are two dimensionless parameters in the evolution Eq. (11.21). The first one is $D=\kappa^2/(Kl_0^2)$, where l_0 is the size of simulation. It is related to the length scale of the microstructure and is taken to be $D=0.0001$. Another parameter, K, is chosen such that the energy densities W^{ani} and W^{elas} are of the same order. As described by Eq. (11.10) in Section 11.3.1, the criterion for compatibility in thin films depends on film normals, so we consider three common crystallographic orientations: (001), (110) and (111) films, given by

$$\mathbf{R}_{(001)} = \begin{pmatrix} \frac{1}{\sqrt{2}} & \frac{1}{\sqrt{2}} & 0 \\ \frac{-1}{\sqrt{2}} & \frac{1}{\sqrt{2}} & 0 \\ 0 & 0 & 1 \end{pmatrix}, \quad \mathbf{R}_{(110)} = \begin{pmatrix} \frac{1}{\sqrt{2}} & \frac{1}{\sqrt{2}} & 0 \\ 0 & 0 & 1 \\ \frac{1}{\sqrt{2}} & \frac{-1}{\sqrt{2}} & 0 \end{pmatrix},$$

$$\mathbf{R}_{(111)} = \begin{pmatrix} \frac{1}{\sqrt{2}} & \frac{-1}{\sqrt{2}} & 0 \\ \frac{1}{\sqrt{6}} & \frac{1}{\sqrt{6}} & \frac{-2}{\sqrt{2}} \\ \frac{1}{\sqrt{3}} & \frac{1}{\sqrt{3}} & \frac{1}{\sqrt{3}} \end{pmatrix},$$

$$(11.24)$$

in Eq. (11.10). Suppose the film is unstressed as deposited and is released in a chosen region ω. Since the rest part of the film is still attached to the substrate in the surrounding region, as illustrated in Fig. 11.1, it is reasonable to assume the clamped boundary condition for simulation; i.e., $\langle\varepsilon_{\text{p}}\rangle = \mathbf{0}$. In addition, the fast Fourier transform (FFT) is employed to enhance the speed of computation. Finally the nucleation problem, while important in general, is not the central issue in the present study. Thus, we take the random initial conditions for all simulations.

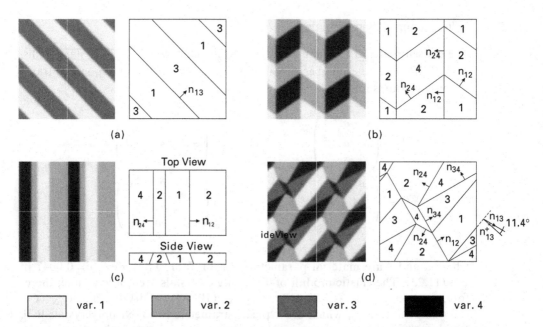

| | var. 1 | | var. 2 | | var. 3 | | var. 4 |

Figure 11.2   Self-accommodation patterns for various film orientations. (a) is for (001) films, (b) and (c) for (110) films, and (d) for (111) films. All of the interfaces are compatible except those with normals $\mathbf{n}_{13}^{+}$ in (d). Notice that four identical patterns are packed together to obtain a better image. Each variant is presented by a different gray level as listed at the bottom [13].

11.4.3.1   (001) film

According to Eq. (11.10), the in-plane transformation strains under the rotation $\mathbf{R}_{(001)}$ given by Eq. (11.24) are

$$\varepsilon_{\mathrm{p}}^{(1)} = \varepsilon_{\mathrm{p}}^{(4)} = \begin{pmatrix} a+\delta & 0 \\ 0 & a-\delta \end{pmatrix},$$

$$\varepsilon_{\mathrm{p}}^{(2)} = \varepsilon_{\mathrm{p}}^{(3)} = \begin{pmatrix} a-\delta & 0 \\ 0 & a+\delta \end{pmatrix}. \tag{11.25}$$

Thus, variants 1 and 4 and variants 2 and 3 are indistinguishable and therefore there are only two different variants. The simulation result shows that the only self-accommodation pattern is the lamellar type, as shown in Fig. 11.2(a). Notice that different variants are presented by different gray levels, as listed on the bottom of Fig. 11.2. While such a lamellar pattern is not a basic unit for self-accommodation in bulk trigonal martensite [64], it is self-accommodated in thin films due to the zero average of the in-plane transformation strains; i.e., $0.5\,\varepsilon_{\mathrm{p}}^{(1)} + 0.5\,\varepsilon_{\mathrm{p}}^{(3)} = \mathbf{0}$. Hence, this pattern is commonly observed in many (001) films with trigonal symmetry [65].

Table 11.3 Compatible interfacial normals in (110) films [13]

	(110) film		
Variants	1, 2 or 1, 3	1, 4	2, 4 or 3, 4
{100} type	(1, 0)	(0, 1)	(1, 0)
{110} type	$(1,\sqrt{2})$	(1, 0)	$(-1,\sqrt{2})$

11.4.3.2 (110) film

According to Eq.(11.10), the in-plane transformation strains under the rotation $R_{(110)}$ given by Eq. (11.24) are

$$\varepsilon_p^{(1)} = \begin{pmatrix} a+\delta & \sqrt{2}\delta \\ \sqrt{2}\delta & a \end{pmatrix}, \quad \varepsilon_p^{(2)} = \varepsilon_p^{(3)} = \begin{pmatrix} a-\delta & 0 \\ 0 & a \end{pmatrix},$$

$$\varepsilon_p^{(4)} = \begin{pmatrix} a+\delta & -\sqrt{2}\delta \\ -\sqrt{2}\delta & a \end{pmatrix}.$$

(11.26)

Thus, variants 2 and 3 are indistinguishable and there are three distinct variants. The simulation results give two distinct patterns. The first one, in Fig. 11.2(b), is similar to the commonly observed "herring-bone" patterns in trigonal martensite [64]. Another simpler pattern is shown in Fig. 11.2(c), which is not an allowable pattern in bulk martensites, since the third components in the interfacial normals are different as can be seen in the right of Fig. 11.2(c). However, it is a legitimate one in thin films, and this confirms that martensitic materials can form many more interfaces in a thin film than in bulk.

Finally, in a bulk trigonal martensite, there are two typical interfaces: one is {100} type, and the other is {110} type. Table 11.3 contains all possible compatible interfaces in (110) films based on the thin-film compatibility. The simulation results confirm that all the interfacial normals in Figs. 11.2(b) and (c) agree very well with those listed in Table 11.3.

11.4.3.3 (111) film

According to Eq. (11.10), the in-plane transformation strains under the rotation $R_{(111)}$ given by Eq. (11.24) are

$$\varepsilon_p^{(1)} = \begin{pmatrix} a-\delta & 0 \\ 0 & a-\delta \end{pmatrix}, \quad \varepsilon_p^{(2)} = \begin{pmatrix} a+\delta & \frac{2}{\sqrt{3}}\delta \\ \frac{2}{\sqrt{3}}\delta & a-\frac{1}{3}\delta \end{pmatrix},$$

$$\varepsilon_p^{(3)} = \begin{pmatrix} a+\delta & \frac{-2}{\sqrt{3}}\delta \\ \frac{-2}{\sqrt{3}}\delta & a-\frac{1}{3}\delta \end{pmatrix}, \quad \varepsilon_p^{(4)} = \begin{pmatrix} a-\delta & 0 \\ 0 & a+\frac{5}{3}\delta \end{pmatrix}.$$

(11.27)

All of the in-plane transformation strains are different in this case, and a self-accommodation pattern containing all of these four martensitic variants is

Table 11.4   Compatible interfacial normals in (111) films [13]

Variants	\(1, 2\)	\(1, 3\)	\(1, 4\)	\(2, 3\)	\(2, 4\)	\(3, 4\)
			(111) film			
{100} type	$(\sqrt{3}, 1)$	$(-\sqrt{3}, 1)$	$(0, 1)$	$(0, 1)$	$(-\sqrt{3}, 1)$	$(\sqrt{3}, 1)$
{110} type	$(\sqrt{3}, 1)$	$(-\sqrt{3}, 1)$	$(0, 1)$	$(1, 0)$	$(1, \sqrt{3})$	$(1, -\sqrt{3})$

shown in Fig. 11.2(d). Moreover, the theoretical values of volume fractions for self-accommodation are indentical for each variant, since $0.25\varepsilon_p^{(1)} + 0.25\varepsilon_p^{(2)} + 0.25\varepsilon_p^{(3)} + 0.25\varepsilon_p^{(4)} = \mathbf{0}$. The simulated volume fractions are $\gamma_1 = 0.247$, $\gamma_2 = 0.250$, $\gamma_3 = 0.252$ and $\gamma_4 = 0.251$. Each of them is close to the theoretical value (0.25). Finally, Table 11.4 contains all possible compatible interfaces in (111) films. All of the interfacial normals shown in Fig. 11.2(d) are found to be in good agreement with those listed in Table 11.4 except $\mathbf{n}_{13}^+$. A further investigation reveals that the discrepancy between $\mathbf{n}_{13}$ and $\mathbf{n}_{13}^+$ is around 11.4°.

11.5   Application to micropumps

We now apply the framework to the design of shape memory micropumps with pumping volume as large as possible. We consider the case where a film is deposited on the substrate in the austenite state. Assume that the film is unstressed as deposited and is released in some chosen region ω by etching. It is flat and taut at high temperatures, while it bulges up in the martensite phase at low temperatures, under perhaps some back pressure. Thus, the released portion of the film swells and shrinks, functioning as an actuator by thermal cycling. In the case of single crystal film, the task is to determine the optimal microstructure and orientation of the film to maximize the deformed volume and, therefore, the computational framework developed in Section 11.4 is chosen for analysis. But for the practical case such as polycrystalline films, the computational model needs significant improvements for microstructure simulation. Thus, we resort to the estimation proposed by Shu [6] for analyzing the out-of-plane recoverable deflection.

11.5.1   Single crystal micropumps

Let the single crystal martensitic film under consideration be the R-phase TiNi. Suppose the released region ω of film, shown in the left of Fig. 11.3, is a circle. It is expected to bulge out like a dome in the martensite phase. Thus, a small element of the film, taken out away from the lateral boundary, may sustain $\langle \sigma_{11}^p \rangle = \sigma^0 > 0$, $\langle \sigma_{22}^p \rangle = \sigma^0 > 0$ and $\langle \varepsilon_{12}^p \rangle = 0$, as illustrated in the right of Fig. 11.3. The tensile stress $\sigma^0 > 0$ is induced due to some back pressure, while its magnitude may be different at distinct locations.

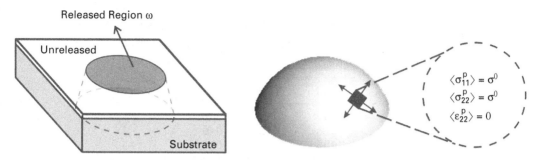

Figure 11.3   A prototype of a micropump with dome-shaped deformation.

To obtain a large pumping volume, we seek the optimal microstructure such that the trace of $\langle \varepsilon^P \rangle \left(= \langle \varepsilon_{11}^P + \varepsilon_{22}^P \rangle \right)$ is maximized [6]. The simulation results via Eq. (11.21) under $\sigma^0 = 5$ MPa are shown in Fig. 11.4(a) for (110) films and Fig. 11.4(b) for (111) films. The pattern for (001) films is similar to the self accommodated pattern shown in Fig. 11.2(a) due to symmetry. Thus, it is not shown here. To explain these stressed patterns for (110) and (111) films, an energy argument is employed. The auxiliary stress state is $\sigma^{P0} = \sigma^0 \mathbf{I}$ in Eq. (11.19), where $\mathbf{I}$ is the identity tensor. For (110) films, the potential energy due to this biaxial stress is positive for variant (2): $-\left(\sigma^0 \mathbf{I} \cdot \varepsilon_p^{(2)} \right) = \delta\sigma^0 > 0$. However, it is negative and identical for both variants (1) and (4): $-\left(\sigma^0 \mathbf{I} \cdot \varepsilon_p^{(1)} \right) = -\left(\sigma^0 \mathbf{I} \cdot \varepsilon_p^{(4)} \right) = -\delta\sigma^0 < 0$. Thus, variant (2) disappears in Fig. 11.4(a). Moreover, from Table 11.3, the only compatible interface separating variants (1) and (4) is $\mathbf{n}_{14} = (1,0)$ or $\mathbf{n}_{14} = (0,1)$, giving rise to the final pattern shown in Fig. 11.4(a). Similarly, this energy argument shows that variants (1) is unfavorable in stressed (111) films, while other variants are energetically equally favorable, leading to a pattern shown in Fig. 11.4(b).

Finally, the sums of the principal strains for these patterns are almost vanishing for (001) films, 0.48 % for (110) films and 0.34 % for (111) films. It shows that (110) films provide the largest biaxial stretch under the same stress state.

11.5.2   Polycrystalline micropumps

We turn to the practical case – polycrystalline thin films with common {100}, {110} or {111} textures. In this situation, the properties of the film are basically transversely isotropic. So we consider a circular diaphragm and let m be the ratio of the maximum central deflection to the radius of the circle. Besides the in-plane displacement u_1 and u_2, the deflection of the film is included as the third kinematic variable. Thus, the strain ε^P is modified as

$$\varepsilon^P[u_1, u_2, \eta] = \varepsilon^1[u_1, u_2] + \varepsilon^2[\eta],$$
$$\varepsilon_{\alpha\beta}^1[u_1, u_2] = \tfrac{1}{2}\left(\frac{\partial u_\alpha}{\partial x_\beta} + \frac{\partial u_\beta}{\partial x_\alpha} \right), \quad \varepsilon_{\alpha\beta}^2[\eta] = \tfrac{1}{2}\frac{\partial \eta}{\partial x_\alpha}\frac{\partial \eta}{\partial x_\beta}, \tag{11.28}$$

Table 11.5 The predicted ratio of the maximum recoverable deflection to the radius of the circular diaphragm for various films with different textures [6].

| Texture | m: maximum deflection / radius | | |
	Cu-Al-Ni	TiNi	Cu-Zn-Al
{100} film	0.19	0.15	0.20
{111} film	0.08	0.13	0.09
{110} sputtered film	0.10	0.15	0.09

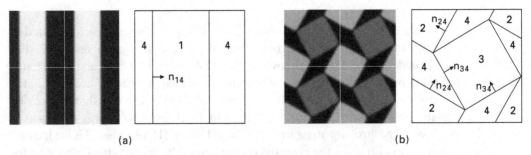

(a)   (b)

Figure 4: Patterns of microstructure in dome-shaped deformations: (a) for (110) films and (b) for (111) films. Notice that these two stressed patterns remain compatible during evolution [13, 14]

for a, $\beta = 1$ and 2. The estimation of maximum deflection under pressure is not easy, and we refer to [6] for details and briefly describe the results here. First, the grains are assumed to be columnar and their sizes are much larger than the film thickness ($d \gg h$). This common case has been discussed in Section 11.3.2 although the effective density $\overline{W}$ is in general not available. Therefore, we have to resort to the Taylor bound for estimating recoverable deflection. Shu [6] has proposed the following rule of thumb:

$$m = \frac{\eta_0}{r_0} \propto \max_i \sqrt{\varepsilon_{p_{11}}^{(i)} + \varepsilon_{p_{22}}^{(i)} - \varepsilon_p^I} \,, \qquad (11.29)$$

where η_0 and r_0 are the central deflection and radius of the diaphragm, $\varepsilon_p^{(i)}$ is given by Eq. (11.10) depending on the texture of the film, ε_p^I is the internal or misfit tensile strain exerted from the remaining part of the film adhered to the substrate. Equation (11.29) is maximized over all possible variants.

Table 11.5 lists the ratio m for a variety of shape memory films with several common textures. The pre-stress ε_p^I is set to be zero in Eq.(11.29) It can be seen that the ratio m for {110} sputtered TiNi film is about 0.15, and is almost the same for other textured TiNi films. So recoverable deflection is insensitive to texture for TiNi films. It is also suprising to see relatively small recoverable deflection for {111} TiNi films, since these films are able to recover large uniaxial tensile strain as shown in Table 11.2. This suggests that the estimation of recoverable deflection

Table 11.6 Comparisons of the prediction of recoverable deflection with several experimental observations for TiNi films [6]

	Prediction	Wolf & Heuer [36]	Miyazaki *et al.* [37]	Markino *et al.* [38]
$m = \frac{\text{deflection}}{\text{radius}}$	0.15	0.12	0.07	0.04

is very different from that of recoverable extension. The former is due to accommodation of biaxial stretch, while the latter is due to accommodation of one-dimensional tensile strain. Next the ratio m is large for {100} Cu-based shape memory film and is sensitive to texture for other Cu-based films. It follows that {100} Cu-based film can have better behavior that TiNi film in view of large recoverable deflection.

Table 11.6 lists several observed deflections of TiNi films under pressure. Obviously, they are all smaller than the predicted value, and there are various reasons for explaining it. First, we assume that the film has a perfect {110} sputtered texture with grain size much larger than the film thickness. Whether this assumption holds or not for their experiments is not clear. In particular, the TiNi films made by Makino *et al.* [38, 66] are not produced by sputtering, instead, they are produced by a flash evaporation method. So the texture we use in the calculation may not be the one in their experiments. Next, the experiment performed by Miyazaki *et al.* [37] used the multilayer with TiNi on the top and SiO_2 on the bottom. They have cleverly created a mechanical two-way shape memory diaphragm by taking advantage of different thermal expansion coefficients among TiNi, SiO_2 and Si. Therefore, the bending effect is a dominant mechanism, leading to a smaller recoverable deflection. Further, there is a shape anisotropy in the experiment of Wolf and Heuer [36] who have used a square diaphragm, while we assumed a circular diaphragm for the prediction.

Finally, we assume that the film is unstressed as deposited in our calculation. However, an internal tensile stress may exist during deposition (for example, see [3, 36]). If that happened, the TiNi diaphragm is subject to biaxial pre-stretch resulting from the remaining part of the film adhered to the Si substrate. In that case, the central recoverable deflection will decrease significantly due to Eq. (11.29).

11.6   Summary

This chapter presents the theory of a pressurized SMA film following the works by Bhattacharya and James [7], James and Rizzoni [8] and Shu [6, 9]. It shows that the behavior of a thin film is different from that of the identical bulk material. It also points out that a heterogeneous film shows strong size effects and its behavior depends crucially on the different ratios of length scales including film thickness, grain size and microstructure length scale. Moreover, this theory is able to predict recoverable strains in both single crystal and polycrystalline films. One crucial result by this theory is that the estimation of recoverable deflection is very different from that of recoverable extension. The consequences of it are summarized as follows.

- Common sputtering {110} texture may not be ideal for recoverable deflection and extension in both TiNi and Cu-based shape memory films.
- Recoverable deflection is not sensitive to common film textures in TiNi films while it is sensitive in Cu-based shape memory films.
- It turns out that {100} texture is ideal for both recoverable deflection and extension in Cu-based films.

Finally, this chapter also presents a framework by Shu and Yen [13, 14] for visualizing microstructure of martensitic thin films. This computational model can serve as a useful tool for evaluating various conditions in the design consideration.

Acknowledgements

The author is grateful to K. Bhattacharya and J. H. Yen for many pleasant collaborations and very helpful comments on this chapter. The author is glad to acknowledge the financial support of the National Science Council of Taiwan under the Grant No. 96–2221-E-002–014 and 97–2221-E-002-125-MY3.

References

[1] Y. Q. Fu, H. J. Du, W. M. Huang, S. Zhang and M. Hu. TiNi-based thin films in MEMS applications: a review. *Sensors and Actuators* A, **112**: 395–408 (2004).

[2] P. Krulevitch, P. B. Ramsey, D. M. Makowiecki, A. P. Lee, M. A. Northrup and G. C. Johnson. Mixed-sputter deposition of Ni-Ti-Cu shape memory films. *Thin Solid Films*, **274** (1996) 101–105.

[3] P. Krulevitch, A. P. Lee, P. B. Ramsey, J. C. Trevino, J. Hamilton and M. A. Northrup. Thin film shape memory alloy microactuators. *J Microelectromechanical Systems*, **5** (1996) 270–282.

[4] S. K. Wu and H. C. Lin. Recent development of TiNi-based shape memory alloys in Taiwan. *Mater. Chem. Phys.*, **64** (2000) 81–92.

[5] M. Tomozawa, H. Y. Kim and S. Miyazaki. Microactuators using R-phase transformation of sputter-deposited Ti-47.3Ni shape memory alloy thin films. *J. Intelligent Material Systems and Structures*, **17** (2006) 1049–1058.

[6] Y. C. Shu. Shape-memory micropumps. *Mater. Trans.*, **43** (2002) 1037–1044.

[7] K. Bhattacharya and R. D. James. A theory of thin films of martensitic materials with applications to microactuators. *J. Mech. Phys. of Solids* **47** (1999) 531–576.

[8] R. D. James and R. Rizzoni. Pressurized shape memory thin films. *J. Elasticity*, **59** (2000) 399–436.

[9] Y. C. Shu. Heterogeneous thin films of martensitic materials. *Archive for Rational Mechanics and Analysis*, **153** (2000) 39–90.

[10] K. Bhattacharya, A. DeSimone, K. F. Hane, R. D. James and C. J. Palmstrm. Tents and tunnels on martensitic films. *Mater. Sci. Eng.* A, **273–275** (1999) 685–689.

[11] Y. C. Shu and K. Bhattacharya. The influence of texture on the shape-memory effect in polycrystals. *Acta Materialia*, **46** (1998) 5457–5473.

[12] K. Bhattacharya and R. D. James. The material is the machine. *Science*, **307** (2005) 53–54.

[13] Y. C. Shu and J. H. Yen. Pattern formation in martensitic thin films. *Appl. Phys. Lett.*, **91** (2007) 021908.

[14] Y. C. Shu and J. H. Yen. Multivariant model of martensitic microstructure in thin films. *Acta Materialia*, **56** (2008) 3969–3981.

[15] Y. Wang and A. G. Khachaturyan. Three-dimensional field model and computer modeling of martensitic transformations. *Acta Materialia*, **45** (1997) 759–773.

[16] A. Artemev, Y. Jin and A. G. Khachaturyan. Three-dimensional phase field model of proper martensitic transformation. *Acta Materialia*, **49** (2001) 1165–1177.

[17] Y. M. Jin, A. Artemev and A. G. Khachaturyan. Three-dimensional phase field model of low-symmetry martensitic transformation in polycrystal: simulation of ζ'_2 martensite in AuCd alloys. *Acta Materialia*, **49** (2001) 2309–2320.

[18] J. Slutsker, A. Artemev and A. L. Roytburd. Morphological transitions of elastic domain structures in constrained layers. *J. Appl. Phys.*, **91** (2002) 9049–9058.

[19] A. Artemev, J. Slutsker and A. L. Roytburd. Phase field modeling of self-assembling nanostructures in constrained films. *Acta Materialia*, **53** (2005) 3425–3432.

[20] E. K. H. Salje. *Phase Transitions in Ferroelastic and Co-Elastic Crystals*. Cambridge: Cambridge University Press (1990).

[21] T. Lookman, S. R. Shenoy, K. Ä. Rasmussen, A. Saxena and A. R. Bishop. Ferroelastic dynamics and strain compatibility. *Phys. Rev.* B, **67** (2003) 024114.

[22] R. Ahluwalia, T. Lookman, A. Saxena and R. C. Albers. Landau theory for shape memory polycrystals. *Acta Materialia*, **52** (2004) 209–218.

[23] L. Q. Chen. Phase-field models for microstructure evolution. *Ann. Rev. Mater. Res.*, **32** (2002) 113–140.

[24] J. Wang, S. Q. Shi, L. Q. Chen, Y. Li and T. Y. Zhang. Phase field simulations of ferroelectric/ferroelastic polarization switching. *Acta Materialia*, **52** (2004) 749–764.

[25] Y. F. Gao and Z. Suo. Domain dynamics in a ferroelastic epilayer on a paraelastic substrate. *ASME – J. Appl. Mech.*, **69** (2002) 419–424.

[26] D. J. Seol, S. Y. Hu, Y. L. Li, L. Q. Chen and K. H. Oh. Computer simulation of martensitic transformation in constrained films. *Mater. Sci. Forum*, **408–412** (2002) 1645–1650.

[27] A. E. Jacobs, S. H. Curnoe and R. C. Desai. Simulations of cubic-tetragonal ferroelastics. *Phys. Rev.* B, **68** (2003) 224104.

[28] V. I. Levitas, A. V. Idesman and D. L. Preston. Microscale simulation of martensitic microstructure evolution. *Phys. Rev. Lett.*, **93** (2004) 105701.

[29] K. Dayal and K. Bhattacharya. A real-space non-local phase-field model of ferroelectric domain patterns in complex geometries. *Acta Materialia*, **55** (2007) 1907–1917.

[30] K. Bhattacharya. Comparison of the geometrically nonlinear and linear theories of martensitic transformation. *Continuum, Mechanics and Thermodynamics*, **5** (1993) 205–242.

[31] S. Miyazaki, S. Kimura and K. Otsuka. Shape-memory effect and pseudoelasticity associated with the R-phase transition in Ti-50.5 at.% Ni single crystals. *Phil. Mag. A*, **57** (1988) 467–478.

[32] S. Miyazaki and A. Ishida. Martensitic transformation and shape memory behavior in sputter-deposited TiNi-base thin films. *Mate. Sci. Eng. A*, **273–275** (1999) 106–133.

[33] P. Belik and M. Luskin. A computational model for the indentation and phase transformation of a martensitic thin film. *J. Mech. Phys. Solids*, **50** (2002) 1789–1815.

[34] P. W. Dondl, C. P. Shen and K. Bhattacharya. Computational analysis of martensitic thin films using subdivision surfaces. *Int. J. Num. Meth. Eng.*, **72** (2007) 72–94.

[35] K. Bhattacharya and G. Dolzmann. Relaxed constitute relations for phase transformation materials. *J. Mech. Phy. Solids*, **48** (2000) 1493–1517.

[36] R. H. Wolf and A. H. Heuer. TiNi (shape memory) films on silicon for MEMS applications. *J. Microelectromech. Sys*, **4** (1995) 206–212.

[37] S. Miyazaki, M. Hirano and T. Yamamoto. Dynamic characteristics of Ti-Ni SMA thin film microactuators. In *IUTAM Symposium on Mechanics of Martensitic Phase Transformation in Solids*, ed. Q. P. Sun *Kluwer Academic Publishers (2002)*, p. 189–196.

[38] E. Makino, T. Shibata and K. Kato. Fabrication of TiNi shape memory micropump. *Sensors and Actuators A*, **88** (2001) 256–262.

[39] K. Bhattacharya. Thin films of active materials. In *Nonlinear Homogenization and its Applications to Composite, Polycrystals and Smart Materials*, ed, P. Ponte Castañeda *et al*. Kluwer Academic Publishers (2004), p. 15–44.

[40] Y. C. Shu and K. Bhattacharya. Domain patterns and macroscopic behavior of ferroelectric materials. *Phil. Mag. B*, **81** (2001) 2021–2054.

[41] Y. C. Shu. Strain relaxation in an alloy film with a rough free surface. *J. Elasticity*, **66** (2002) 63–92.

[42] G. Gioia and R. D. James. Micromagnetics of very thin films. *Proceedings of the Royal Society of London Series A*, **453** (1997) 213–223.

[43] R. D. James and M. Wuttig. Magnetostriction of martensite. *Phil. Mag. A*, **77** (1998) 1273–1299.

[44] A. Desimone, R. V. Kohn, S. Miller and F. Otto. A reduced theory for thin-film micromagnetics. *Comm. Pure Appl. Math.*, **55** (2002) 1408–1460.

[45] M. S. Wechsler, D. S. Lieberman and T. A. Read. On the theory of the formation of martensite. *Transactions AIME Journal of Metals*, **197** (1953) 1503–1515.

[46] H. Le Dret and A. Raoult. The nonlinear membrane model as variational limit of nonlinear three-dimensional elasticity. *Journal de Mathematiques Pures et Appliques*, **74** (1995) 519–578.

[47] A. Braides and A. Defranceschi. *Homogenization of Multiple Integrals*. Oxford: Oxford University Press (1998).

[48] K. Bhattacharya. *Microstructure of Martensite*. Oxford: Oxford University Press (2003).

[49] T. Saburi and C. M. Wayman. Crystallographic similarities in shape memory martensites. *Acta Metallurgica*, **27** (1979) 979–995.

[50] K. Otsuka and K. Shimizu. Morphology and crystallography of thermoelastic Cu-Al-Ni martensite analyzed by the phenomenological theory. *Trans. Japan Institute of Metals*, **15** (1974) 103–108.

[51] T. H. Nam, T. Saburi, Y. Nakata and K. Shimizu. Shape memory characteristics and lattice deformation in TiNi-Cu alloys. *Mater. Trans. JIM*, **31** (1990) 1050–1056.

[52] K. M. Knowles and D. A. Smith. The crystallography of the martensitic transformation in equiatomic nickel-titanium. *Acta Metallurgica*, **29** (1981) 101–110.

[53] S. Chakravorty and C. M. Wayman. Electron microscopy of internally faulted Cu-Zn-Al martensite. *Acta Metallurgical* **25** (1977) 989–1000.

[54] K. Otsuka, T. Nakamura and K. Shimizu. Electron microscopy study of stress-induced acicular β'_1 martensite in Cu-Al-Ni alloy. *Transactions of the Japan Institute of Metals*, **15** (1974) 200–210.

[55] Y. C. Shu. Shape-memory effect in bulk and thin-film polycrystals. Ph.D. Thesis. California Institute of Technology, 1998.

[56] K. R. C. Gisser, J. D. Busch, A. D. Johnson and A. B. Ellis. Oriented nickel-titanium shape memory alloy films prepared by annealing during deposition. *Appl. Phys. Lett.*, **61** (1992) 1632–1634.

[57] A. Ishida, A. Takei and S. Miyazaki. Shape memory thin film of TiNi formed by sputtering. *Thin Solid Films*, **228** (1993) 210–214.

[58] S. Miyazaki and A. Ishida. Shape memory characteristics of sputter-deposited TiNi thin films. *Materials Transactions JIM*, **35** (1994) 14–19.

[59] Q. Su, S. Z. Hua and M. Wuttig. Martensitic transformation in $Ni_{50}Ti_{50}$ films. *J. Alloys and Compounds*, **211/212** (1994) 460–463.

[60] K. Bhattacharya and R. V. Kohn. Symmetry, texture and the recoverable strain of shape-memory polycrystals. *Acta Materialia*, **44** (1996) 529–542.

[61] K. Bhattacharya and R. V. Kohn. Elastic energy minimization and the recoverable strains of polycrystalline shape-memory materials. *Archive for Rational Mechanics and Analysis*, **139** (1997) 99–180.

[62] Y. C. Shu, M. P. Lin and K. C. Wu. Micromagnetic modeling of magnetostrictive materials under intrinsic stress. *Mechanics of Materials*, **36** (2004) 975–997.

[63] K. Otsuka and C. M. Wayman. *Shape Memory Materials*. Cambridge: Cambridge University Press (1998).

[64] T. Fukuda, T. Saburi, K. Doi and S. Nenno. Nucleation and self-accommodation of the R-phase in TiNi alloys. *Materi. Trans., JIM*, **33** (1992) 271–277.

[65] S. K. Streiffer, C. B. Parker, A. E. Romanov, *et al.* Domain patterns in epitaxial rhombohedral ferroelectric films. Geometry and experiments. *J. Appl. Phys.*, **83** (1998) 2742–2753.

[66] E. Makino, T. Shibata and K. Kato. Dynamic thermo-mechanical properties of evaporated TiNi shape memory thin film. *Sensors and Actuators A*, **78** (1999) 163–167.

12 Binary and ternary alloy film diaphragm microactuators

S. Miyazaki, M. Tomozawa and H. Y. Kim

Abstract

TiNi based shape memory alloy (SMA) thin films including TiNi, TiNiPd and TiNiCu have been used to develop diaphragm microactuators. The TiNi film is a standard material and the ternary TiNiPd and TiNiCu alloy films have their own attractive characteristics when compared with TiNi films. The TiNiPd alloy is characterized by high transformation temperatures so that it is expected to show quick response due to a higher cooling rate: the cooling rate increases with increasing the temperature difference between the transformation temperature and room temperature which is the minimum temperature in conventional circumstances. The martensitic transformation of the TiNiCu and the R-phase transformation of the TiNi are characterized by narrow transformation hystereses which are one-fourth and one-tenth of the hysteresis of the martensitic transformation in the TiNi film. Thus, these transformations with a narrow hysteresis are also attractive for high response microactuators. The working frequencies of two types of microactuators utilizing the TiNiPd thin film and the TiNiCu film reached 100 Hz, while the working frequency of the microactuator using the R-phase transformation reached 125 Hz.

12.1 Introduction

The demand for the development of powerful microactuators has stimulated the research to develop sputter-deposited TiNi shape memory alloy (SMA) thin films, because they possess attractive characteristics useful for microactuators. The most prominent characteristics of SMA thin films are small dimensions and high response speed in addition to conventional ones, such as a large recoverable strain of 6–8 % and recoverable stress up to 600 MPa [1, 2, 3]. These values

Thin Film Shape Memory Alloys: Fundamentals and Device Applications, eds. Shuichi Miyazaki, Yong Qing Fu and Wei Min Huang. Published by Cambridge University Press. © Cambridge University Press 2009.

are extremely large when compared with other actuator materials or mechanisms, such as piezoelectric materials and the electrostatic force.

Recently, several types of microactuators utilizing TiNi SMA thin films such as microcantilevers [4, 5, 6], microgrippers [7, 8] and micropumps [8, 9, 10, 11, 12, 13, 14, 15] were successfully fabricated and their actuation properties were characterized. All the microactuators utilized the martensitic transformation and its reverse transformation for generating actuation force. The working frequencies of these microactuators were mostly from 0.2 to 50 Hz. However, it is important to note that the displacement of these microactuators decreased with increasing working frequency up to 50 Hz and no effective displacement was detected over 50 Hz [8].

In order to increase the actuation speed of the microactuator utilizing SMA thin films, it is necessary to reduce the time for the martensitic transformation during cooling. Two different approaches can be suggested to reduce the time as follows. One is to increase the cooling speed of the SMA thin film microactuator. The cooling speed of the microactuator can be raised by increasing the temperature difference between the transformation and room temperatures. Therefore, utilizing high temperature SMA thin films is effective to improve the actuation speed. Microactuators utilizing the TiNiPd high temperature SMA thin films were fabricated [16, 17, 18] and the working frequency reached 100 Hz when the transformation temperature was raised by about 70 K with the addition of 22 at% Pd [16]. However, the microactuators utilizing high temperature SMA thin films require larger electrical power to actuate when compared with the microactuators utilizing the TiNi binary alloy thin films, because they need to be heated up to higher temperatures.

Another way to increase the response speed is to reduce the temperature range of heating and cooling for completing the forward and reverse transformation. Hence, reducing transformation temperature hysteresis is effective in improving the actuation speed. It is well known that the R-phase (rhombohedral phase) transformation of TiNi SMAs show an extremely narrow transformation temperature hysteresis when compared with the M-phase transformation of TiNi SMAs [19, 20]. In fact, microactuators using the R-phase transformation of TiNi thin films worked at a frequency of 125 Hz without any decrease of displacement [16, 21]. Another way is to use TiNiCu films, which reveal the martensitic transformation but have a transformation temperature hysteresis of about a quarter of the martensitic transformation in the TiNi films. In this chapter, the characteristics of microactuators utilizing TiNi and ternary alloy thin films are reviewed, mainly based on the work of the present authors.

12.2   Shape memory behaviour of TiNi thin films

As a result of extensive research [1, 2, 5, 22, 23, 24, 25, 26, 27, 28, 29], it is well established that sputter-deposition is the most successful method for making TiNi thin films with a thickness of less than 10 μm. The TiNi thin films exhibit almost the same shape memory effect as that of a bulk specimen due to the formation of Ti-rich thin plate precipitates. It is important to evaluate the shape memory characteristics such as the transformation temperatures and the transformation and recovery strains of the TiNi SMA thin films in order to apply them to microactuators. The shape memory behaviour of the thin films was evaluated quantitatively by thermal cycling under various constant stresses using a microtensile testing machine.

Figure 12.1 shows three types of schematic strain–temperature curves obtained by thermal cycling under a constant tensile stress. Figure 12.1(a) represents a two-step shape memory effect associated with the R-phase and M-phase transformations and their reversion to the parent B2 phase, i.e., M↔R↔B2. On cooling, the thin film starts to transform from the B2 phase to the R phase at temperature R_s, and the transformation finishes at temperature R_f. The transformation strain associated with the R-phase transformation is denoted ε_R. Further cooling results in the transformation from the R phase to the M phase between temperatures M_s and M_f exhibiting the transformation strain ε_M associated with the M-phase transformation. On heating, the corresponding reverse transformations take place inducing a two-stage shape recovery. The reverse transformation from the M phase to the R phase starts at temperature A_s and finishes at temperature A_f. The R phase starts to reverse transform to the B2 phase at temperature RA_s and the transformation finishes at temperature RA_f. The total recovery strain in the two reverse transformations is denoted ε_A. The remained strain after heating is defined as plastic strain ε_P. ΔH_R and ΔH_M represent the transformation temperature hysteresis associated with the R-phase transformation and the M-phase transformation, respectively.

If the specimen is thermally cycled above M_s, only the R-phase transformation and its reversion take place, as shown in Fig. 12.1(b). Using the R-phase transformation, a quick response speed is expected in operating microactuators although the transformation strain is small. On the other hand, if M_s is higher than R_s, only the M-phase transformation and its reversion take place, as shown in Fig. 12.1(c). Both M_s and R_s temperatures are dependent on the composition and microstructure of the thin films. Experimental data revealing the above transformation behavior are shown in Fig. 12.2.

Figure 12.2(a) shows strain–temperature curves of a Ti-47.3Ni thin film heat-treated at 823 K for 0.6 ks [21]. Two-step shape changes associated with the R-phase and M-phase transformations are observed in the Ti-47.3Ni thin film at lower stresses, the former and latter transformations inducing small and large shape changes, respectively. The R-phase transformation disappears at higher stress levels, because M_s becomes higher than R_s. This is attributed to the fact that M_s increases more rapidly with increasing applied stress when compared with R_s.

In order to investigate the characteristics of the R-phase transformation in the Ti-47.3at%Ni thin film, partial thermal cycling tests between the R phase and the parent phase were conducted under low constant stresses. The strain–temperature curves associated with only the R-phase transformation and its reversion are shown in Fig. 12.2(b) [21]. It reveals that the R-phase transformation

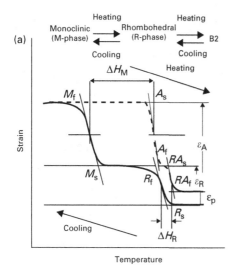

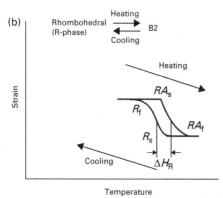

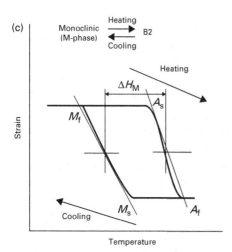

Figure 12.1 Schematic illustration of strain–temperature curves representing (a) M–R–B2 transformation, (b) R–B2 transformation and (c) M–B2 transformation.

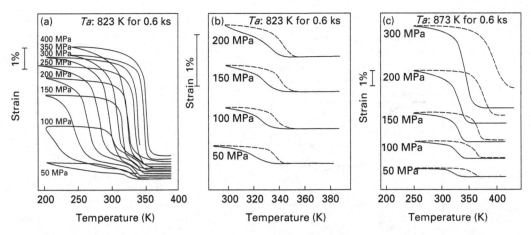

Figure 12.2   Strain–temperature curves associated with the (a) B2–R–M phase transformation in Ti-47.3Ni thin film, (b) B2–R phase transformation in Ti-47.3Ni thin film and (c) B2–M phase transformation in Ti-48.2Ni thin film.

in the Ti-47.3Ni thin film exhibits a small transformation temperature hysteresis when compared with the M-phase transformation. It is clear that the transformation temperatures, transformation temperature hysteresis and transformation strain of the R-phase transformation are less affected by the applied stress when compared with those associated with the M-phase transformation. It means that the R-phase transformation is relatively insensitive to applied stress. The small transformation temperature hysteresis and the stability against applied stress are great advantages for applying the films for microactuators.

Figure 12.2(c) shows strain–temperature curves under various constant stresses for a Ti-48.2Ni thin film heat-treated at 873 K for 0.6 ks. The curves show a one-step transformation, i.e. B2↔M, because M_s is higher than R_s. As seen in Fig. 12.2 (c), the M-phase transformation in the Ti-48.2Ni thin film shows large recoverable strains although plastic strain is introduced under stresses above 200 MPa. The M_s of the Ti-48.2Ni thin film heat-treated at 873 K is higher than that of the Ti-47.3Ni thin film heat-treated at 823 K. M_s and R_s under zero stress are evaluated by extrapolating the Clausius–Clapeyron relationship: M_s and R_s of the Ti-47.3Ni and M_s of the Ti-48.2Ni under zero stress were 220 K, 331 K and 335 K, respectively. The difference in M_s is explained as follows. The Ti-47.3Ni thin films heat-treated below 973 K for 3.6 ks contain a large amount of fine Ti_2Ni precipitates when compared with the Ti-48.2Ni thin films heat-treated at the same temperature [23]. The fine precipitates suppress the M-phase transformation resulting in the decrease of M_s, while R_s was less affected by these precipitates. Moreover, the heat-treatment temperature of the Ti-48.2Ni thin film is 50 K higher than that of Ti-47.3Ni thin film. The precipitates become coarser and larger in Ti-rich TiNi thin films with increasing heat-treatment temperature, resulting in the increase of both M_s and R_s. It is noted that M_s increases more rapidly with increasing heat-treatment temperature when compared with R_s. As a result, it is concluded that microactuators using only the R-phase or M-phase transformation can be fabricated by changing the Ti-content and heat-treatment condition.

12.3   Fabrication and characterization methods

TiNi-base thin films were fabricated by using an r.f. magnetron sputtering apparatus. For characterizing the shape memory behaviour, free standing films were fabricated in the following process. A thin film was deposited onto a Cu foil substrate. The substrate temperature was kept at 473 K during deposition. The thickness of the thin film was 6 μm. The Cu-foil substrate was dissolved in a solution of 70 % HNO_3 and 30 % H_2O by volume to extract the thin film. Since the substrate temperature during deposition was not higher than the crystallization temperature of the TiNi thin film, the as-deposited thin film was amorphous. Crystallization of the thin film was achieved by heat-treatment at different temperatures. The internal structures of the thin films were investigated by using a transmission electron microscope (TEM) and a high-resolution TEM (HRTEM). The samples for TEM and HRTEM observations were prepared by electrolytic polishing using a twin-jet polishing apparatus. The shape memory behavior (strain versus temperature relationship) of the heat-treated specimens was investigated by thermal cycling tests under various constant stresses. The dimension of the specimens was 5 mm in gauge length, 1 mm in width and 6 μm in thickness.

For making diaphragm-type microactuators, TiNi-base thin films were deposited onto SiO_2/Si substrates using an RF magnetron sputtering apparatus. The substrate temperature was kept at 473 K during deposition, similarly to the deposition onto a Cu substrate. The thicknesses of the thin films were determined to be 1.5–2.5 μm by observing their cross-sections using a scanning electron microscope (SEM). In order to crystallize the thin film on the flat SiO_2/Si substrate, it was heat-treated at 973 K for 0.6 ks. At the same time, the flat shape was memorized in the TiNi-base thin film.

Diaphragm-type microactuators with a multilayer diaphragm were fabricated by using a conventional Si micromachining technique. The fabrication procedure is shown in Fig. 12.3 and it is briefly explained as follows [30]. (1) A negative photoresist layer is coated on each of the SMA film and SiO_2 surfaces. (2) The photoresist layers are patterned by the photolithography process including UV exposure and development. (3) The exposed SMA film and SiO_2 layers are wet etched by an HF/HNO_3 solution and an HF/NH_4F buffer solution, respectively. Other unexposed parts of the SMA film and SiO_2 layers are protected against the etching solutions by the remaining photoresist layers. After etching, a stripe of SMA film with a width of 2.0 mm is remaining on the SiO_2/Si substrate. (4) Anisotropic etching for Si is carried out using an etchant containing ethylene diamine and pyrocatechol (EDP) to make an SMA/SiO_2 dual-layer diaphragm consisting of a 1.5 μm thick SMA film and a 1.0 μm thick SiO_2 layer. The diaphragm is square with

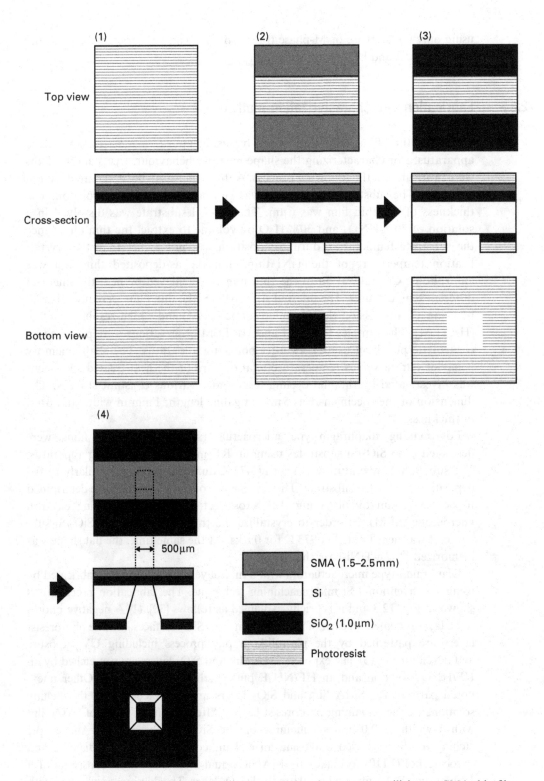

Figure 12.3   Procedure for fabricating a diaphragm-type microactuator utilizing an SMA thin film.

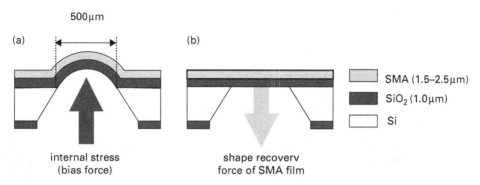

500 μm

(a)   (b)

SMA (1.5–2.5 μm)
SiO₂ (1.0 μm)
Si

internal stress
(bias force)

shape recovery
force of SMA film

Figure 12.4   Schematic illustration of the cross-section of a microactuator utilizing a SMA thin film deposited on an SiO₂/Si substrate at room temperature (a) and at high temperature (b).

a width of 500 μm. Electric current is directly passed through the SMA stripe for Joule heating.

Two types of microactuators were fabricated, i.e., a multilayer diaphragm consisting of an SMA film and a SiO₂ layer, as shown in Fig. 12.4, and a single layer diaphragm consisting of an SMA film, as shown in Fig. 12.5. The multilayer diaphragm microactuators were operated by the martensitic transformation induced in the TiNi, TiNiCu and TiNiPd films, while the single layer diaphragm microactuators were operated by the M-phase or R-phase transformation induced in the TiNi films. Figure 12.4 illustrates the cross-sections of the microactuator at room temperature and high temperature, respectively. Since the diaphragm consists of two layers with different thermal expansion coefficients, an internal stress is generated in the diaphragm after heat-treatment, i.e. compression in the SiO₂ layer and tension in the SMA film layer. At room temperature, the TiNi film layer is of martensite and can be easily deformed. The crystal lattice of the martensite (low temperature phase) is an orthorhombic or monoclinic structure, while that of the parent phase (high temperature phase) is a B2 structure [4, 23]. Therefore, the diaphragm becomes convex as shown in Fig. 12.4(a) to relax the internal stress. By heating to a temperature above the reverse transformation temperature of the SMA film layer, the diaphragm reverts to the initially memorized flat shape due to the shape memory effect. By cooling to a temperature below the martensitic transformation temperature, the diaphragm shape becomes convex again.

As shown in Fig. 12.4, the microactuator operates due to temperature variation. Therefore, the temperature dependence of the height at the centre of the diaphragm was measured to investigate the actuation process. The height at the center of the diaphragm is notated by h. The parameter h was measured at each fixed temperature during cooling and heating in a step-by-step way to characterize a quasi-static actuation. Dynamic actuation was investigated by using a three-dimensional shape analyzer equipped with a laser scanner. The microactuator was dynamically operated by thermal cycling. The thermal cycling was conducted by applying a pulse current to the TiNi film layer in the microactuator, i.e., by means

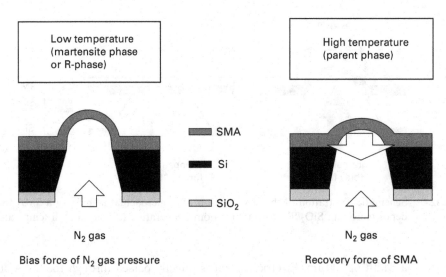

Figure 12.5   Schematic illustration of the cross-section of a microactuator consisting of an SMA thin film deposited on an Si substrate at room temperature (a) and a high temperature (b).

of joule heating and natural cooling. The height h was continuously measured during dynamic actuation. The displacement was estimated by measuring the difference between the maximum and minimum values of h, and it was used as one of the measures of dynamic actuation characteristics. The temperature of the microactuator was measured by a thermocouple microwelded to a part of the SMA film attached to the SiO_2/Si substrate. The working frequency and temperature of the microactuator were adjusted by changing the frequency and amplitude of the pulse current, respectively. The ratio of heating time to cooling time, i.e. a duty ratio, was fixed at 5:95 for each working frequency. The measurement was conducted at an ambient atmosphere (296–298 K).

As for the second type of microactuator with a single layer diaphragm, TiNi films were deposited on Si substrates by r.f. magnetron sputtering. The thicknesses of the TiNi films were 2.0 µm. The TiNi films on the substrates were heat-treated at 873 K and 823 K for 0.6 ks, respectively, in order to memorize the initial flat shape. Diaphragm-type microactuators were fabricated again by using the conventional Si micromachining technique.

Figure 12.5 illustrates the cross-sections of a microactuator at low and high temperatures. The shape of the diaphragm is square with a width of 500 µm. N_2 gas pressure of 40 kPa is used as a bias force for the microactuator. The microactuator is convex at room temperature, because the TiNi film is deformed by the N_2 gas pressure in a low temperature phase such as the M phase or R phase.

In order to investigate the displacement and the transformation temperatures of the microactuator, the height at the center of the diaphragm was measured as a function of temperature during heating and cooling at each fixed temperature in a step-by-step way. Figure 12.6 reveals the temperature dependence of the height at

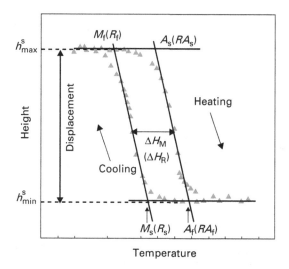

Figure 12.6 Schematic illustration of the temperature dependence of the height at the center of the diaphragm.

the center of the diaphragm. Upon cooling, the height starts to increase at $M_s(R_s)$ due to the start of the M-phase (R-phase) transformation and the increase in height finishes at $M_f(R_f)$ because the M-phase (R-phase) transformation finishes. Upon heating, the height decreases with increasing temperature between $A_s(RA_s)$ and $A_f(RA_f)$ due to the progress of the reverse M-phase (R-phase) transformation. Displacement is defined as the difference between the maximum height and the minimum height ($h^s_{max} - h^s_{min}$), where the superscript "s" stands for "static" because each height is measured at each fixed temperature. ΔH_M and ΔH_R represent the transformation temperature hysteresis associated with the M-phase and R-phase transformations, respectively.

12.4 Microactuators using the R phase of TiNi and the M phase of TiNiPd

12.4.1 Microactuators using the R-phase of TiNi

Figure 12.7 shows the change of height at the center of the diaphragm during heating and cooling in the microactuator using the M-phase transformation of the Ti-48.2Ni thin film or in that using the R-phase transformation of the Ti-47.3Ni under an N_2 gas pressure of 40 kPa [16]. The microactuator using the R-phase transformation of the Ti-47.3Ni thin film is abbreviated to "TiNi R-phase microactuator". The microactuator using the M-phase transformation of the Ti-48.2Ni thin film is abbreviated to "TiNi M-phase microactuator". Open squares and circles respectively denote the results of the cooling and heating processes in the TiNi M-phase microactuator, while closed squares and circles respectively denote the cooling and heating processes in the TiNi R-phase microactuator. Arrows

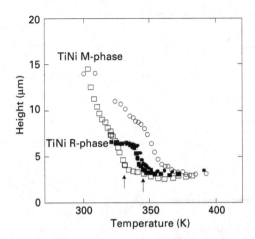

Figure 12.7   Temperature dependence of the height at the center of the diaphragm in the TiNi M-phase microactuator and the TiNi R-phase microactuator under an N_2 gas pressure of 40 kPa.

indicate the M-phase and R-phase transformation start temperatures (M_s and R_s). The heights of the TiNi M-phase and TiNi R-phase microactuators reach almost the minimum values at the A_f temperature of 360 K and 355 K, respectively. The minimum heights of these microactuators do not reach zero due to elastic deformation of the parent phase by applied gas pressure. It can be seen that the displacement of the TiNi R-phase microactuator is smaller than that of the TiNi M-phase microactuator. However, the transformation temperature hysteresis of the TiNi R-phase microactuator is only 3.2 K. This is significantly small when compared with the TiNi M-phase microactuator.

12.4.2   Microactuators using the M phase of TiNiPd

High transformation temperature is also effective in increasing actuation speed, like a narrow transformation hysteresis. Figure 12.8 shows the effect of Pd-content on M_s in TiNiPd thin films heat-treated at 973 K for 3.6 ks [31]. As shown in Fig. 12.8, M_s decreases slightly with increasing Pd-content up to 9 at%. Then, M_s increases up to 560 K with further increasing Pd-content to 36 at%.

Ti-48.2Ni and Ti-26.5Ni-22.7Pd SMA thin films were deposited on SiO_2/Si substrates by r.f. magnetron sputtering. The thicknesses of these thin films were 2.5 μm and 2.0 μm, respectively. The Ti-48.2Ni and Ti-26.5Ni-22.7Pd thin films on the substrates were heat-treated at 873 K for 0.6 ks in order to memorize the initial flat shape. Diaphragm-type microactuators were fabricated by using the Si micromachining technique.

Figure 12.9 shows the change of the height at the center of the diaphragm during heating and cooling in the TiNi M-phase microactuator and the TiNiPd microactuator [16]. The height at the center of the diaphragm decreases with increasing temperature due to the progress of the reverse transformation. The

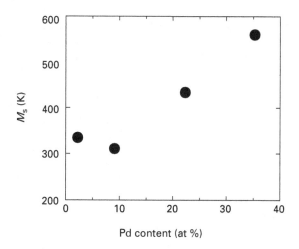

Figure 12.8 Pd content dependence of the M_s in TiNiPd thin films heat-treated at 973 K for 3.6 ks.

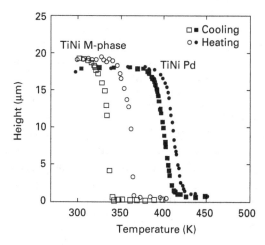

Figure 12.9 Temperature dependence of height at the center of the diaphragm in the TiNi M-phase microactuator and the TiNi-Pd microactuator deposited on SiO_2/Si substrates.

height reached almost zero at the A_f temperature of 370 K and 430 K for the TiNi M-phase microactuator and the TiNiPd microactuator, respectively. The TiNiPd microactuator exhibits almost the same displacement as the TiNi M-phase microactuator. M_s of the TiNiPd microactuator is about 70 K higher than that of the TiNi M-phase microactuator. Also it is noted that the transformation temperature hysteresis of the TiNiPd microactuator is smaller than that of the TiNi M-phase microactuator.

12.4.3 Dynamic actuation of microactuators

Table 12.1 shows the displacement, $M_s(R_s)$ and transformation temperature hysteresis of the fabricated microactuators measured by a quasi-static thermal

Table 12.1 Properties of the TiNi and TiNiPd microactuators measured by quasi-static thermal cycling

	Substrate	Displacement	$M_s(R_s)$	Hysteresis
TiNi M-phase microactuator	SiO_2/Si	20 μm	341 K	25 K
TiNiPd microactuator	SiO_2/Si	18 μm	410 K	11 K
TiNi R-phase microactuator	Si	3.6 μm	347 K	3.2 K

cycling test. The dynamic properties of these microactuators were investigated by a three-dimensional shape analyzer equipped with a laser scanner. The experiment was conducted in the atmosphere at room temperature (296–298 K). Since the temperature of the microactuator was adjusted by controlling pulse current, the microactuator was heated by joule heat when the current was turned on, and it was cooled by natural cooling when the current was turned off. The actuation speed and displacement of the microactuator were adjusted by changing the amplitude and frequency of the pulse current. The ratio of heating and cooling times was fixed at each frequency.

Figure 12.10(a) shows the displacement of the TiNi M-phase microactuator as a function of current amplitude at various constant frequencies [16]. When the frequency is 1 Hz, the displacement of the microactuator increases with increasing current amplitude reaching up to the maximum value, and then decreases with further increasing current amplitude. This relationship can be explained by considering the temperature of the diaphragm during heating and cooling as follows. When the applied current is small, the temperature of the diaphragm does not increase up to A_f during heating. Therefore, a complete shape recovery can not be obtained. When the current is too large, the displacement becomes small, because the temperature of the diaphragm does not easily decrease down to M_f during cooling. The largest displacement is obtained when the temperature variation during cooling and heating in the diaphragm becomes most efficient for inducing both the martensitic and reverse transformations. The current amplitude corresponding to the maximum displacement increases with increasing working frequency, because heating time decreases with increasing working frequency. The maximum displacement decreases with increasing frequency, and finally the TiNi M-phase microactuator does not show effective displacement at the frequency of 50 Hz.

Figure 12.10(b) shows the results of the TiNiPd microactuator [16]. It is important to note that the maximum displacement of the TiNiPd actuator was almost constant irrespective of frequency, while the maximum displacement of the TiNi M-phase microactuator decreased with increasing frequency. The TiNiPd microactuator worked effectively even at the frequency of 100 Hz. This result confirms that increasing transformation temperatures is effective in increasing the actuation speed of the microactuator utilizing SMA thin films.

Figure 12.10(c) shows the results of the TiNi R-phase microactuator [16, 21]. The TiNi R-phase microactuator could also work at the frequency of 100 Hz. This

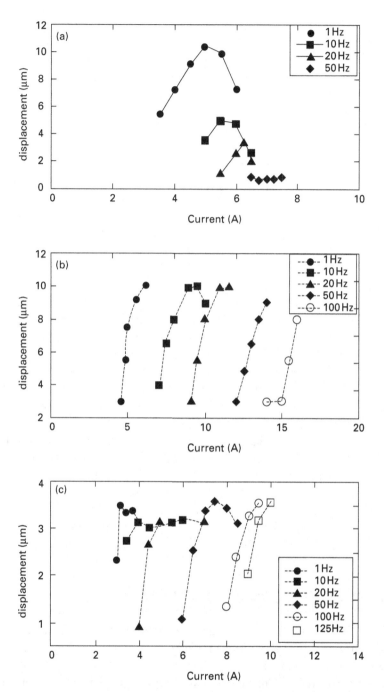

Figure 12.10 Displacement as a function of current amplitude for various constant frequencies in the (a) TiNi M-phase microactuator, (b) TiNiPd microactuator and (c) TiNi R-phase microactuator.

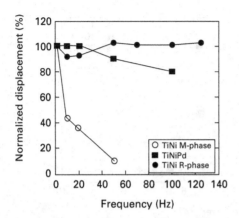

Figure 12.11 Comparison of the normalized displacement as a function of frequency for the TiNi M-phase, TiNiPd and TiNi R-phase microactuators.

is attributed to the decrease in the temperature range between A_f and M_f by reducing the transformation temperature hysteresis. It is also noted that the TiNi R-phase microactuator worked by a smaller current when compared with the TiNiPd microactuator due to the small temperature variation necessary for the full and reverse transformations. It is noted that the maximum displacement was almost constant irrespective of the working frequency up to 125 Hz.

Normalized displacements as a function of working frequency for all micro-actuators are shown in Fig. 12.11. The displacement is normalized by the displacement obtained at a working frequency of 1 Hz. The normalized displacement of the TiNi M-phase microactuator decreases with increasing frequency. However, the normalized displacement of the TiNiPd microactuator does not decrease up to 50 Hz, and the microactuator works effectively at the frequency of 100 Hz. Furthermore, the displacement of the TiNi R-phase microactuator hardly decreases though the frequency varies between 1 Hz and 125 Hz. These results confirm that both increasing transformation temperatures and decreasing the transformation temperature hysteresis are effective to increase the actuation response.

12.5   Microactuators using the M phase of TiNiCu

Figure 12.12 shows the temperature dependence of h (height at the center of the diaphragm) during cooling and heating in the microactuator utilizing a Ti-38.0Ni-10.0Cu thin film which was heat-treated at 973 K for 0.6 ks. Data obtained during the cooling and heating processes are denoted by open and closed circles, respectively. Upon cooling, h starts to increase at M_s and finishes at M_f, while h starts to decrease at A_s and finishes at A_f upon heating. The increase and decrease in h are due to the forward and reverse transformations, respectively, in the TiNiCu layer of the microactuator. The transformation temperature hysteresis ΔH_M is measured to be 6 K, as shown in Fig. 12.12.

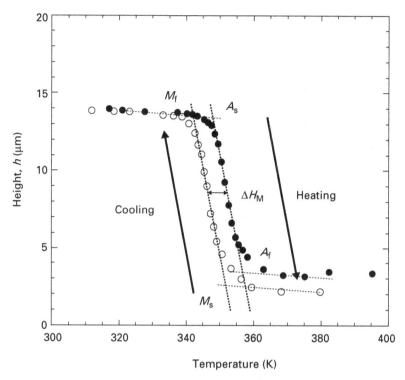

Figure 12.12 Temperature dependence of the height of a microactuator utilizing the Ti-38.0Ni-10.0Cu thin film which was heat-treated at 973 K for 0.6 ks.

The ΔH_M of the present microactuator is approximately a quarter of that of the microactuator Ti-Ni M-phase [8, 21]. Therefore, the microactuator utilizing the TiNiCu thin film is expected to exhibit a higher actuation speed than that using the M-phase transformation of the TiNi binary thin film.

The dynamic actuation behavior of the microactuator is investigated as follows. The microactuator is continuously operated by applying a pulse current to the TiNiCu layer. When the current is applied, the TiNiCu layer is heated, while the layer is cooled naturally by cutting off the current. Thus, the repeated cycling of joule heating and natural cooling enables the microactuator to work continuously. Figure 12.13 shows the relationship between the height at the center of the diaphragm h^d and the voltage generated between both ends of the TiNiCu layer. The superscript d stands for dynamic actuation process in order to distinguish h^d from the height h in Fig. 12.12, because the latter h is measured statically. The frequency and the amplitude of the applied pulse current are 10 Hz and 1.25 A, respectively. The heating and cooling times in Fig. 12.13 are 5 ms and 95 ms, respectively. The h^d immediately decreases when the current is applied, while it immediately increases to the highest original position when the current is cut off. The displacement is defined as the difference between the highest and lowest h^d.

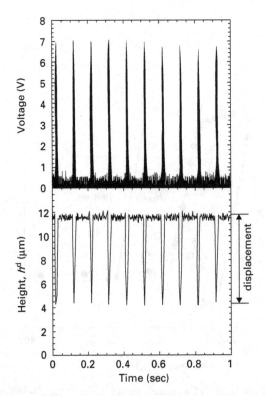

Figure 12.13 Variation of the applied voltage and the height of a microactuator subjected to pulsed current with an amplitude of 1.25 A and a frequency of 10 Hz.

Figure 12.14 shows the current amplitude dependence of the displacement at each working frequency. As shown in the displacement–current curve for the working frequency of 10 Hz, the displacement increases with increasing current amplitude reaching a maximum, then becomes almost constant. This is explained as follows. When the current amplitude is small, the temperature of the TiNiCu layer does not increase up to A_f, resulting in a small displacement due to incomplete reverse transformation. By increasing the current amplitude, the temperature approaches A_f, leading to an increase of the displacement. The current amplitude necessary for generating a maximum displacement increases with increasing working frequency. This is because each heating time decreases with increasing working frequency under a fixed duty ratio condition.

The maximum displacement at each working frequency is plotted in Fig. 12.15. For comparison, the result of the TiNi M-phase microactuator is also included. The M_s and ΔH_M of both microactuators are listed in Table 12.2. The microactuators utilizing the TiNiCu thin film are abbreviated to TiNiCu microactuator.

As shown in Fig. 12.15, the displacement of the TiNi M-phase microactuator is almost the same as that of the TiNiCu microactuator at working frequency below

Table. 12.2 M_s and ΔH_M of the TiNi and TiNiCu microactuators

	TiNi microactuator	TiNiCu microactuator
M_s (K)	341	352
ΔH_M (K)	27	6

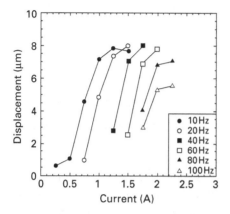

Figure 12.14 Displacement as a function of current amplitude at each working frequency.

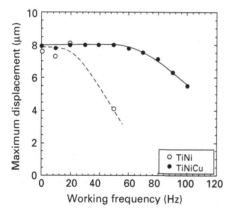

Figure 12.15 Comparison of the maximum displacements expressed as a function of working frequency in two types of microactuator.

20 Hz. However, it decreases by increasing working frequency up to 50 Hz. The displacement of the TiNiCu microactuator does not decrease until the working frequency reaches 60 Hz. Above 60 Hz of working frequency, it gradually decreases. However, the microactuator operates even at 100 Hz. As revealed in Table 12.2, The M_s of the TiNiCu microactuator is slightly higher than that of the TiNi micro-actuator. Moreover, the ΔH_M of the TiNiCu microactuator is less than one-forth of that of the TiNi microactuator. Both higher M_s and narrower ΔH_M result in high-speed actuation for the TiNiCu microactuator.

The TiNi R-phase microactuator operated at a working frequency of 125 Hz without decreasing displacement The R-phase transformation temperature R_s and ΔH_M are 347 K and 3 K, respectively. However, the displacement of the microactuator was only 3.6 µm. As shown in Fig. 12.15, the displacement of the TiNiCu microactuator is even larger than that of the TiNi M-phase microactuator within the working frequency range from 1 Hz to 100 Hz. Thus a high-speed microactuator with large displacement was successfully fabricated by utilizing the TiNiCu thin film.

12.6   Summary

It is confirmed that microactuators have been successfully fabricated by sputter-deposition. The working frequency of the microactuators utilizing TiNiPd and TiNiCu thin films reached 100 Hz, and that of the microactuator using the R-phase transformation reached 125 Hz. Since the TiNiPd and TiNiCu show the shape memory behaviour associated with the martensitic transformation, the displacements of the microactuators utilizing these two alloy films are almost equal to that of the TiNi M-phase microactuator. However, the displacement of the TiNi R-phase microactuator is one third of that of the TiNi M-phase microactuator. We can choose either of the microactuators depending on the requirements of response speed and magnitude of displacement. Further systematic researches on micromachining processes, fatigue properties during actuation and the interface reaction between an SMA thin film and a substrate are still needed to apply the TiNi-based microactuators various fields.

References

[1] S. Miyazaki and A. Ishida, Martensitic transformation and shape memory behavior in sputter-deposited TiNi-base thin films, *Mater. Sci. Eng. A*, **273–275** (1999) 106–133.

[2] A. Ishida and S. Miyazaki, Microstructure and mechanical properties of sputter-deposited Ti-Ni alloy thin films, *J. Eng. Mater. Tech.*, **121** (1999), 2–7.

[3] S. Miyazaki, Thermal and stress cycling effects and fatigue properties in Ti-Ni alloy. In *Engineering Aspects of Shape Memory Alloys*, eds., T. W. Duerig, K. N. Melton, D. Stockel and C. M. Wayman. Guildford, UK: Butterworth-Heinmann (1990), 394.

[4] M. Kohl, K. D. Skrobanek, E. Quandt, P. Schloßmacher, A. Schlüßler and D. M. Allen, Development of microactuators based on the shape memory effect, *Journal de Physique IV C8*, **5** (1995) 1187–1192.

[5] Y. Q. Fu, W. M. Huang, H. Du, X. Huang, J. Tan and X. Gao, Characterization of TiNi shape-memory alloy thin films for MEMS application, *Surf. Coat. Technol.*, **145** (2001) 107–112.

[6] W. M. Haung, Q. Y. Liu, L. M. He and J. H. Yeo, Micro NiTi-Si cantilever with three stable positions, *Sensors and Actuators A*, **114** (2004) 118–122.

[7] M. Kohl, E. Just, W. Pfleging and S. Miyazaki, SMA microgripper with integrated antagonism, *Sensors and Actuators A*, **83** (2000) 208–213.

[8] S. Miyazaki, M. Hirano and V.H. No, Dynamic characteristics of diaphragm microactuators utilizing sputter-deposited Ti-Ni shape memory alloy thin films, *Materials Science Forum*, **394–395** (2002) 467–474.

[9] M. Hirano, H. Hosada, H. Suzuki and S. Miyazaki, Effect of cooling efficiency on dynamic characteristics of diaphragm-type microactuators utilizing sputter-deposited Ti-Ni shape memory alloy thin films, *Trans. Materials Research Society of Japan*, **26** (2001) 133–135.

[10] E. Makino, K. Kato and T. Shibata, Thermo-mechanical properties of TiNi shape memory thin film formed by flash evaporation, *Sensors and Actuators*, **75** (1999) 156–161.

[11] E. Makino, T. Mitsuya and T. Shidata, Dynamic thermo-mechanical properties of evaporated TiNi shape memory thin film, *Sensors and Actuators*, **79** (1999) 163–167.

[12] E. Makino, T. Mitsuya and T. Shidata, Micromachining of TiNi shape memory thin film for fabrication of micropump, *Sensors and Actuators*, **79** (2000) 251–259.

[13] D. Xu, L. Wang, G. Ding, Y. Zhou, A. Yu and B. Caiet, Characteristics and fabrication of TiNi/Si diaphragm, *Sensors and Actuators A*, **93** (2001) 87–92.

[14] D. D. Shin, P. Kotekar, P. Mohanchandra and P. Carman, High frequency actuation of thin film NiTi, *Sensors and Actuators A* **111** (2004) 166–171.

[15] D. D. Shin, P. Kotekar, P. Mohanchandra and P. Carman, Development of hydraulic linear actuation using thin film SMA, *Sensors and Actuators A* **119** (2005) 151–156.

[16] M. Tomozawa, K. Okutsu, H. Y. Kim and S. Miyazaki, Characterization of high-speed microactuator utilizing shape memory alloy thin films, *Materials Science Forum*, **475–479** (2005) 2037–2042.

[17] M. Kohl, D. Dittamann, E. Quandt and B. Winzek, Thin film shape memory microvalves with adjustable operation temperature, *Sensors and Actuators A*, **83** (2000) 214–219.

[18] M. Kohl, D. Dittamann, E. Quandt, B. Winzek, S. Miyazaki and D. M. Allen, Shape memory microvalves based on thin films or rolled sheets, *Mater. Sci. Eng. A*, **273–275** (1999) 784–788.

[19] S. Miyazaki and C.M. Wayman,. The R-phase transition and associated shape memory mechanism in Ti-Ni single crystal, *Acta Metallurgica*, **36** (1988) 181–192.

[20] Miyazaki and K. Otsuka,. Deformation and transition behavior associated with the R-phase in Ti-Ni alloys, *Metallurgical Transactions A*, **17A** (1986) 53–63.

[21] M. Tomozawa, H. Y. Kim and S. Miyazaki, Microactuator using R-phase transformation of sputter-deposited Ti-47.3Ni shape memory alloy thin films *J. Intelligent Material Systems and Structure*, **17** (2006) 1049–1058.

[22] J. Sakurai, H. Hosoda, S. Kajiwara and S. Miyazaki, Nanoscale microstructure formed during crystallization and shape memory behavior in sputter-deposited Ti-rich Ti-Ni thin films, *Materials Science Forum*, **327–328** (2000) 175–178.

[23] J. Sakurai, J. I. Kim and S. Miyazaki, Shape memory behavior and microstructure in sputter-deposited Ti-47.3at%Ni thin films, *The Fourth Pacific Conference on Advanced Materials and Processing (PRICM-4)* (2001) 1509–1512.

[24] J. Sakurai, J. I. Kim, H. Hosoda and S. Miyazaki, Microstructre and tensile deformation behavior of sputter-deposited Ti-51.3at%Ni thin films crystallized at various temperatures, *Transactions of the Materials Research Society of Japan*, **26** (2001) 319–322.

[25] J. Sakurai, J. I. Kim, H. Hosoda and S. Miyazaki, Shape memory characteristics of sputter-deposited Ti-51.3at%Ni thin films aged at various temperatures, *Transactions of the Materials Research Society of Japan*, **26** (2001) 315–318.

[26] A. Ishida and M. Sato, Thickness effect on shape memory behavior of Ti-50.0at%Ni thin film, *Acta Materialia*, **51** (2003) 5571–5578.

[27] A. Ishida, M. Sato and S. Miyazaki, Mechanical properties of Ti-Ni shape memory thin films formed by sputtering, *Mater. Sci. Eng. A* **273–275** (2003) 754–757.

[28] J. Zhang, M. Sato and A. Ishida, On the Ti2Ni precipitates and Guinier–Preston zones in Ti-rich Ti-Ni thin films, *Acta Materialia*, **51** (2003) 3121–3130.

[29] J. Zhang, M. Sato and A. Ishida, Structure of martensite in sputter-deposited Ti-Ni thin films containing Guinier–Preston zones, *Acta Materialia*, **49** (2001) 3001–3010.

[30] M. Tomozawa, H. Y. Kim and S. Miyazaki, Shape memory behavior and internal structure of Ti-Ni-Cu shape memory alloy thin films and their application for microactuators, *Acta Materialia*, doi: 10.1016/j.actamat.2008.09.026.

[31] K. Okutsu, H. Hosoda and S. Miyazaki, Shape memory behavior of sputter-deposited Ti-Ni-Pd thin films, *The Fourth Pacific Rim International Conference on Advanced Materials and Processing (PRICM-4)* (2001) 1521–1523.

13 TiNi thin film devices

K. P. Mohanchandra and G. P. Carman

Abstract

This chapter provides a brief review of TiNi thin film devices, both mechanical and biomedical, that have been studied during the last decade. Prior to reviewing devices, we first provide a description of physical features critical in these devices, including deposition, residual stresses and fabrication. In general, this chapter concludes that the main obstacle for implementing devices today remains control of TiNi properties during manufacturing. If this can be overcome, the next issue is to develop acceptablemicro machining techniques. While several techniques have been studied, considerable work remains on fully developing processes that would be required in mass manufacturing processes. Finally, if these issues can be resolved, the area with the most promise is biomedical devices, the argument here is that biomedicine remains one of the major applications areas for macroscopic TiNi structures today.

13.1 Introduction

Shape memory alloys (SMAs) have fascinated researchers for the last few decades in a range of "macroscopic" industrial and medical applications due to their intrinsic properties such as large stress output, recoverable strain, excellent damping and biocompatibility. More recently the research community has begun to investigate research opportunities in the microscale with the use of thin film SMAs integrated into microdevices [1, 2, 3, 4, 5]. SMAs (e.g. TiNi) exhibit a thermally induced crystalline transformation between the martensitic phase, a low temperature phase, and the austenitic phase, a high temperatures phase. Upon cooling below the martensitic finish (M_f) temperature, undeformed SMAs have a twinned martensitic structure. Under a relatively low applied stress, the twinned martensitic structure is rearranged along the stress direction and produces deformation. When heated above the austenitic finish (A_f) temperature the SMAs recover their highly ordered austenite phase and regain their original crystallized

Thin Film Shape Memory Alloys: Fundamentals and Device Applications, eds. Shuichi Miyazaki, Yong Qing Fu and Wei Min Huang. Published by Cambridge University Press. © Cambridge University Press 2009.

shape, i.e. a shape memory effect (SME). In addition to the SME, the material provides a superelastic effect wherein the material exhibits large recoverable deformations through a stress-induced phase transformation. The SME effect was first reported in a gold-cadmium alloy in 1932 [6]. Prior to the discovery of equiatomic nickel-titanium by Buehler and his coworkers at the US Naval Ordnance Laboratory in 1962 [7, 8], research on the shape memory effect was somewhat limited. The discovery of TiNi with large recoverable strain and actuation stress, corrosion resistance, excellent damping and biocompatibility led to extensive research and development efforts on macroscopic applications over the next four decades. Recently that research focus has turned to developing novel micro electro mechanical systems (MEMS) utilizing fundamental properties inherent in thin film TiNi and other SMA materials.

Shape memory and superelastic properties of TiNi have been studied extensively in its "macroscopic" form since its initial discovery [9, 10, 11]. These studies reveal that TiNi has excellent shape memory and superelastic properties compared with other SMAs. In addition, researchers report the material to have exceptional corrosion or wear resistance [12, 13], damping and fatigue [14, 15], and biocompatibility [16, 17] properties. The large mechanical stress (500 MPa) and strain output (10%) of TiNi produces energy densities on the order of 10^7 J/m^3 which is substantially larger than achievable with other actuators (e.g. PZT is 40 MPa 0.1% strain producing 10^5 J/m^3) [18]. These large stress, strain and energy values make SMA an excellent candidate for microscale actuator design where stroke and force limitations are intrinsic problems. In addition, thin films have substantially larger surface to volume ratios when compared with macroscopic materials. This larger surface to volume ratio dissipates heat more rapidly with a corresponding increase in the material's cyclic response (e.g. 10 to 100 times). Researchers have long been aware of the limited cyclic response of macroscopic SMA (e.g. 1 Hz) and this feature represents the Achilles' heel in many SMA actuation systems. Therefore, microscale actuation systems based on thin film TiNi are expected to have all the beneficial attributes of macroscopic material with a substantially faster cyclic response. That is, the material can be heated and cooled more quickly. The first published attempt at fabricating a thin film TiNi actuator was in 1990 by Walker *et al.* [19] and in the same year Busch *et al.* [20] demonstrated shape memory behavior in sputter-deposited TiNi thin film. Since those initial papers, studies into deposition, characterization and MEMS fabrication processes have proceeded at a rapid rate. However, several issues still inhibit the widespread adoption of thin film TiNi in MEMS structures such as materials availability, repeatability and developing standard fabrication/ deposition processes to construct small-scale structures. Without advancements along these three fronts it is doubtful that commercially viable TiNi MEMS structures will become commonplace in society. Below we provide a brief review of attributes related to thin film SMA materials followed by a description of several mechanical and biomedical devices and associated fabrication processes.

13.2   Fabrication of TiNi thin films

Deposition remains a key issue in the development of future TiNi microscale devices. It should be pointed out that as the deposition processes for TiNi film evolve, they influence fabrication processes to construct a MEMS component. In this light, various types of physical vapor deposition (PVD) techniques have been employed to fabricate thin film TiNi [21, 22, 23, 24, 25, 26] i.e. sputter deposition, pulse laser deposition (PLD) [24, 25], molecular beam epitaxial (MBE) [27], etc. Currently, the vast majority of microscale devices rely on sputter-deposition techniques because of high deposition rates, large area depositions, and compatibility with the MEMS processes. However, even this method presents problems related to film composition with various attempts to resolve the issue [25, 28, 29, 30, 31]. One method that has been reported in the literature to produce fairly uniform compositions is the hot target process [32, 33, 34, 35, 36], where, during sputtering, the target temperature is elevated to produce fairly uniform compositions throughout a majority of the film. Many of the devices described in detail in this chapter are based on a hot target deposition process [1, 3, 35, 37, 38]. In the following section a brief review of pertinent material attributes that are important in the fabrication process of SMA MEMS devices is discussed.

13.3   Thin film properties

As stated previously, TiNi thin films possess several desirable attributes for the design of microactuators. In the vast majority of these devices either a shape memory effect or a superelastic effect is employed. The shape memory effect is simply the return of the structure's shape through heating the material above its austenite phase transformation temperature. The superelastic effect represents a non-linear large elastic deformation which is possible due to a stress-induced phase transformation from austenite to martensite. A number of researchers studied the influence of process parameters on shape memory and superelasticity properties of TiNi thin films during the last 20 years [39, 40, 41, 42, 43, 44, 45]. All of these studies point to the difficulties associated with compositional control and deposition parameters, as well as annealing temperatures/times. In the following paragraphs we briefly review several of the important properties of thin film TiNi and their relation to microscale devices. This review is followed by a description, with fabrication details, of a few microscale TiNi mechanical and biomedical devices.

One key issue in fabricating microscale devices is the residual stresses present in sputter-deposited SMA thin films attached onto a substrate. In general, the residual stress is absent in macroscopic devices. While the community has studied, and is well aware of, residual stresses, they have not been extensively studied in the context of fabrication processes. One report by Gill *et al.* [37] indicates that residual stresses strongly dictate the fabrication process for a particular device. In general, compressive stresses build up in thin film during deposition. Upon crystallization, the developed stress may relax and the film may develop tensile stresses on cooling. The residual stress developed in the film has a strong influence on the transformation temperatures and failure strength, as well as the fabrication process chosen for a particular device. To illustrate the complexity of this problem we cite a few studies on residual stresses. Shih *et al.* [30] have shown that the film deposited on Si wafer at Ar pressure of around 3.5 mtorr has a high tensile stress and the film exhibits extensive delamination at the TiNi/Si interface. Similar to Ar pressure, the substrate temperature also has a very strong influence on the residual stresses in the film. Shih *et al.* [30] also showed that TiNi film deposited at a substrate temperature of 230 °C has lower intrinsic stresses. Therefore, one can conclude that processing parameters strongly influence the residual stresses in a film, which also strongly influences the fabrication process for a microdevice.

For microdevices relying on shape memory actuation, a two-way shape memory effect (TWSME) is desirable. To achieve this in thin film structures both intrinsic and extrinsic methods have been used. The intrinsic method relies on thermo mechanical processes following deposition to modify the material microstructure by introducing precipitates and defects. These precipitates and defects influence the martensitic variant nucleation to produce a TWSME. The thermo-mechanical processes that lead to the TWSME are also called "training processes". Several TiNi film MEMS devices that rely on TWSME for actuation have been proposed. For example, Zhang *et al.* [46] developed temperature-controlled reversible surface protrusions using TiNi thin films as a result of indentation-induced TWSME where the protrusions appear upon heating and disappear upon cooling. Another approach for TWSME is to functionally grade the film through the thickness of the material. A description of this process is described in Section 13.4.1.1. A variety of MEMS devices also use an extrinsic approach to produce the TWSE. Extrinsic approaches rely on external stresses (i.e. residual stresses described in the previous paragraph) to reset the martensitic variants upon cooling. For example, Krulevitch *et al.* [26] demonstrated this concept in a micro gripper made of TiNiCu thin film deposited on a thin Si beam to use a bimorph-like effect to obtain the TWSME. The Si substrate produces residual stresses in the film which act to bias the structure and move twin boundaries in the martensite phase. While these residual stresses (i.e. pre-stress in the thin film) are some fundamental examples of TWSME, a general conclusion may be that researchers have found more seamless approaches with introduce TWSME in MEMS devices when compared to macroscopic SMA devices.

13.4   TiNi thin film devices

The unique attributes of TiNi SMA have made it possible to begin studying innovative microscale actuators for both the industrial and medical fields. While these devices are possible, considerable work remains in developing fabrication processes required for individual applications. The microdevices being studied are broadly classified into two groups in this chapter depending on their field of application, mechanical devices and biomedical devices. Under mechanical devices, SMAs are used as thermal actuators where they convert thermal energy into mechanical energy typically producing both large forces and displacements. Some devices have utilized a TWSME, with reversible motion achieved as the temperature either increases or decreases. Similarly, due to superior mechanical stability, corrosion resistance, biocompatibility, etc., macroscopic TiNi has become the "material of choice" for many biomedical devices. While many commercial devices are available for macroscopic TiNi, a vast amount of the literature has focused on studying microscale mechanical devices rather than microscale biomedical devices. This may be attributed to the lack of collaboration between medical researchers and basic microdevice developers or to intrinsic problems with fabricating the high quality thin film TiNi necessary for medical application. That is, medical applications require superior film composition and uniformity when compared with mechanical applications to prevent the real or perceived release of Ni into the body. In the following two sections, a brief review of some mechanical and biomedical micro devices being studied with TiNi thin films are given.

13.4.1   Mechanical devices

In recent years, with the development of MEMS, TiNi based thin films have received considerable attention due to their unique properties. Walker *et al.* [19] used TiNi film in what is considered the first report of a TiNi MEMS component in 1990. Later in the same year Kuribayashi *et al.* [47] developed microactuators using the two-way shape memory effect of TiNi thin film. As the demands for microactuators increased, research on using TiNi films in MEMS components increased. Several microactuators using thin film TiNi have been developed during the last decades. Among them micropumps [48], microgrippers [49, 50], micro-valves and micro-optics switches [51, 52, 53] represent some microdevices studied. Most of these devices use a TiNi membrane (e.g. diaphragm/microbubble) for actuation [54, 55, 56, 57].

One of the fundamental actuator concepts employing TiNi thin film is the bilayer cantilever beam as described by Fu *et al.* [5]. They fabricated a shape memory effect microbeam made of r.f. sputtered TiNi thin film onto a 15 μm thick silicon cantilever. A cosine model proposed by Liang [58] was used to simulate the shape memory behavior of TiNi thin film deposited on a silicon beam under thermal cycling. During the simulations, the TiNi/Si cantilever beams were heated up to 120 °C then cooled to room temperature to demonstrate the TWSME. Fundamentally, this concept could be used in a wide variety of actuator concepts including optical switches. Below, we describe this in more detail along with fabrication processes for a bubble actuator [38], a micropump [35], a microwrapper [37] and microvalves [1] using sputter-deposited TiNi thin film.

These specific devices were chosen to illustrate specific fabrication processes as well as to span the gap between simple devices and more complicated systems.

13.4.1.1   Bubble actuator

In the late 1990s an intrinsic two-way SME was observed in a functionally graded TiNi thin film. This discovery spurred on the concept of engineering a pseudo-monolithic membrane forming a bubble actuator [35] having two stable states, one geometry at high temperature and another at low temperature. The fabrication process for the bubble actuator represents a good example of a basic or simple fabrication process for a TiNi microdevice as well as illustrating some operational issues. In most fabrication processes the TiNi is built upon a silicon wafer. In this particular example, fabrication consists of masking, etching the silicon, etching the TiNi and hot shaping. This entire process contains all the basics of manufacturing a TiNi microstructure and serves as a good example to illustrate both fabrication and actuation. The only point here is that very few microscale devices rely on a separate hot shaping step.

The fabrication of the bubble actuator and most other microdevices begins with a silicon wafer. In this example [59] the backside of a (100) silicon wafer 0.5 mm thick is initially wet-oxidized and coated with silicon nitride using low-pressure chemical-vapor deposition (LPCVD). The silicon nitride on the backside of the silicon wafer is patterned to form a square opening (~12 mm × 12 mm) using reactive ion etching (RIE). The patterned wafer is etched in KOH at 80 °C, removing silicon until a silicon membrane of approximately 40 μm thickness is obtained on the top side. The silicon nitride on the front side of the wafer is then removed by RIE, followed by sputter-deposition of 5 μm thick TiNi film using a heated target. During deposition the target temperature is increased, altering the ratio of Ni/Ti through the thickness of the film producing a pseudo-monolithic TiNi structure. After deposition, a thick photoresist (AZ-4620, Clariant, ~5 μm thick) is spin coated on the top surface to mask the TiNi for etching. The TiNi film is wet etched using a mixture of hydro-fluoric acid (HF), nitric acid (HNO_3), and de-ionized (DI) water with a mix ratio of 1:1:20. This TiNi etchant represents a standard one used throughout the community. After etching the TiNi film, the remaining 40 μm thick Si from beneath the TiNi film is removed by RIE to obtain a pseudo-homogeneous 12 × 12 mm TiNi membrane supported along its square boundary by Si.

Following the fabrication process, the membrane structure is hot-shaped into the desired geometric configuration (see Fig. 13.1(a)). It should be pointed out that very few microdevices rely on a hot shaping step. To hot shape the structure, the membrane is sandwiched between two stainless steel jigs that form a bubble shape in the membrane. For this particular example, a dimple in the stainless steel jig produces a 3 mm diameter bubble in the thin film TiNi. Following the membrane clamping, the structure is annealed at 500 °C for 20 minutes in a vacuum better than 10^{-6} torr. Following this process, the membrane with Si as the support structure is removed from the steel jig. When this membrane is heated to 100 °C either by Joule heating or with a hot plate, the membrane forms an approximately

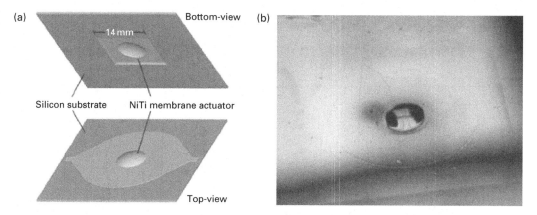

Figure 13.1   (a) Schematic diagram of bubble actuator. (b) Picture of a bubble [35]. (With permission from Elsevier.)

3 mm diameter bubble shape, as shown in Fig. 13.1(b). The researchers reported that the bubble shape vanishes when the material returns to room temperature [59]. When heated back to above 100 °C the membrane again regains its bubble shape, i.e. shows a TWSME.

The TWSME is attributed to a compositional gradation of Ni/Ti through the thickness of the film. It is believed that this intrinsic approach may be superior when compared to extrinsic bi morph designs relying on dissimilar materials layered Si and TiNi). The compositional gradation in this film was confirmed by performing RBS measurements on sputter-deposited 0.6 μm thick film. The researchers report that the Ni rich film adjacent to the Si wafer responds super-elastically, while the Ti rich film along the top surface provides the shape memory effect in the structure. By combining these two features, a two-way shape memory actuator was developed.

The mechanical performance of the hot shaped TiNi membrane was examined, including deflection height versus input power, and frequency response was characterized using a non-contact laser scanning system. The deflection height was measured at different input powers. Figure 13.2 provides a plot of membrane displacement as a function of time or the heating and cooling cycle. The particular results shown represent the maximum power applied, or in simpler terms the maximum deflection achieved. The negative displacement merely indicates that the membrane measurements were in the downward direction for actuation. For a 3 mm diameter bubble, the membrane reaches a maximum displacement of −2200 μm when heated. When cooled, the membrane returns to its initial flat shape at zero displacement. A 2200 μm displacement requires an average strain of approximately 2% based on geometric considerations [60] using the calculation shown below.

$$\varepsilon = \frac{\left\{\left[\left(\frac{c^2}{8h}+\frac{h}{2}\right)^2-\frac{c^2}{4}\right]^{0.5}+h\right\}\sin^2\left\{0.5\left|\dfrac{c}{\left(\left(\frac{c^2}{8h}+\frac{h}{2}\right)^2-\frac{c^2}{4}\right)^{0.5}+h}\right|\right\}-c}{c},$$

(13.1)

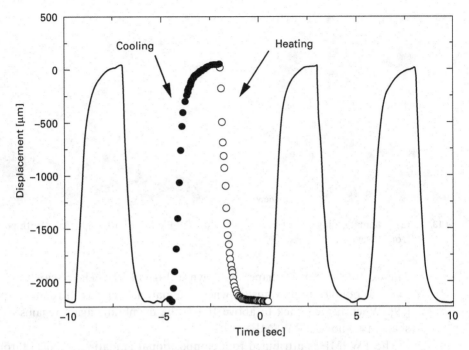

Figure 13.2   Deflection of a bubble actuator [35]. (With permission from Elsevier.)

where ε is the strain, h is the deflection height and c is the diameter of the dome. The researchers report that Joule heating is somewhat problematic due to heat conduction to the clamped edges (i.e. Si) as well as heat convection to the surrounding air. For input powers less than 70% of maximum, the dome shape does not form, indicating that the TiNi film is not reaching the austenite finish temperature. That is, the films is cooling as quickly as it is being heated and not allowed to reach the austenite finish temperature.

The deflection of the membrane and the actuation frequency can be optimized by providing proper cooling and Joule heating to the membrane. Cooling can be in the form of forced convection or placing the membrane in contact with a liquid. While beneficial for cooling, these scenarios require more power to heat the membrane. The study shows that de-ionized water surrounding the membrane provides sufficient mass to cool the membrane and reach at least 40 Hz cyclic response with a deflection of 280 μm [35]. However, the authors noted that actuation frequency is strongly dependent upon the percentage of heating time applied during a given cycle (i.e. quicker is better because it reduces heat to the surroundings) as well as movement of the surrounding liquid. Later the researchers found that the liquid directly adjacent to the bubble acted as a thermal insulating layer inhibiting higher frequency response. This can be remedied through appropriate design considerations such that the 40 Hz reported should not be considered an upper bound of frequency response.

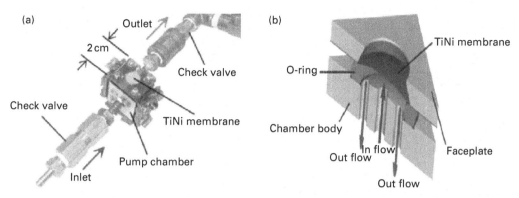

Figure 13.3   (a) Illustration of a TiNi pump. (b) Schematic diagram of the pump's cross-section [38]. (With permission from Elsevier.)

13.4.1.2   Micropump

A concept similar to the bubble actuator was used to develop a compact fluidic pump. The pump demonstrated that TiNi thin film combined with standard mechanical designs could produce a rather remarkable hydraulic based actuator. The small hydraulic actuator produces large flow rates at high pressures [38], i.e. large power density. An illustration of a pump and its schematic cross-sections are given in Figs. 13.3(a) and (b) respectively. The pump consists of a reservoir, two conventional passive check valves attached to the inlet and outlet ports, a cubic pump chamber (note several versions were studied), and four TiNi membranes on each cube face. The liquid entering the pump chamber from the pressurized reservoir through the inlet valve is divided among four cavities each containing a TiNi membrane. The pressurized fluid causes spherical deformations in the TiNi membrane. When a current is passed through the membrane, it is heated and the shape memory effect returns the deformed shape to the flat configuration and creates a net volume displacement in the pumping chamber. The fluid is subsequently pumped out of the outlet valve and the cycle is repeated. One key issue is the frequency (i.e. thermal cycling time) at which the TiNi membrane can be actuated which is related to both the amount of energy supplied to the TiNi membrane and the cooling time for the membrane.

In its base state, the TiNi membrane is under a bias pressure and thus takes the form of a spherical cap, as shown in Fig. 13.4 where the radius of the membrane is denoted by a, the central deflection by d, and the thickness by h. The thickness of the film used was 8 µm and the radius of the bubble was 6 mm. Joule heating was used to transform the membrane from a martensite to an austenite phase. The transformation caused the membrane to reduce the central deflection to d'. Note this actuation is in exactly the opposite direction to that described for the bubble actuator. The amount of liquid displaced by the membrane per stroke is the net volume change under the spherical cap when the

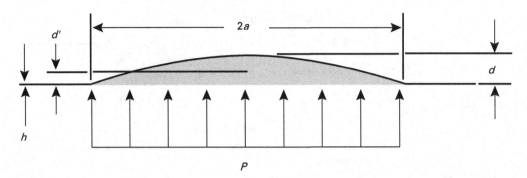

Figure 13.4   TiNi membrane under bias pressure P [38]. (With permission from Elsevier.)

central deflection changes from d to d'. While Joule heating is applied to the membrane, the temperature of the liquid immediately adjacent to the membrane increases. The heated liquid is forced out of the cavity through outflow portholes as the center deflection decreases. As the central deflection returns to d, cool liquid enters the cavity through an inflow porthole and impinges on the hot TiNi membrane.

The researchers quickly discovered that the optimal Joule heating sequence was to supply the highest current in the shortest amount of time. This process reduced undesirable heating of the fluid and the surrounding structure. While the highest current was desirable, the researchers also discovered that the electrical interconnections between the current source and the thin film represent a major failure location. Thus in actuation devices, the electrical interface represents an important design concern. For this particular pump, the researchers report that a thermal barrier layer of fluid existed next to the film. In order to flush this thermal barrier layer, the researchers designed a port system that allowed the incoming "cool" fluid to directly impinge on the thin film while allowing the "hot" fluid to exit through the outflow. This forced convection allowed the film to be operated at frequencies above 100 Hz with some results suggesting 300–500 Hz to be possible. For this particular configuration the optimum operating load was 98 N with a blocking force of 198 N. A peak velocity of 5.85 mm/s was obtained while operating the membrane at 100 Hz. Therefore, this device demonstrates that thin film structures that are produced through micro machining processes can be used in relatively macro-devices.

13.4.1.3   Microwrapper

The previous two examples did not focus on small-scale features on the order of tens of microns. In this section we describe a device that contains relatively small-scale features. One example is a MEMS microwrapper constructed from thin film TiNi. The fabrication process for this device is considerably more complex than described in the previous two examples. The TiNi microwrapper

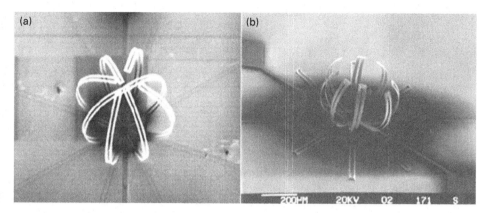

Figure 13.5 Microwrapper using (a) thin film TiNi and polyimide film and (b) graded TiNi thin film [37]. (With permission from Elsevier.)

was suggested for use for grabbing microsize objects for *in situ* analysis in a living organism or to be used in the removal of microscopic shards of glass during eye surgery. Micrograbbing devices using electrostatic and pneumatic actuation mechanisms have been previously reported [61] but typically either produce small forces (electrostatic) or require substantial infrastructure (pneumatic lines). TiNi actuation systems do not suffer from these problems and they also represent a material compatible with embedding into the human body (e.g. stents). The particular microwrapper described in this section uses micromachined arms to form a cage structure (i.e. grabbing) in the room temperature state and the micromachined arms return flat (i.e. release) when current (i.e. Joule heating) is applied.

Researchers have developed both bimorph MEMS machined microwrappers consisting of TiNi film with polyimide film and pseudo-homogeneous micro-wrappers using functionally graded TiNi film [37]. Figure 13.5 shows SEM micrographs of the two microwrappers developed. The following text provides a description of the fabrication process as well as lessons learned from these researchers. First, a silicon wafer is wet-oxidized and coated with a polysilicon sacrificial layer (200 nm thick). The polysilicon layer is patterned with xenon-difluoride (XeF_2) photolithographically. Following the sacrificial layer patterning, the TiNi film is sputter-deposited onto the substrate and crystallized. Following deposition, a thick photoresist (AZ-4620, Clariant, 5 μm thick) is used as an etch mask and the TiNi film is etched into the appropriate structures. As one can see from Fig. 13.5, the etched TiNi feature in the microwrapper is less than 20 microns. For the bimorph version, a polyimide (PI) layer is deposited on top of the TiNi film to induce curvature (i.e. bias spring). Finally, the polysilicon sacrificial layer is removed with XeF_2 and the arms of the microwrapper are released. Due to residual stress, the arms curl up and form a cage-like structure once released.

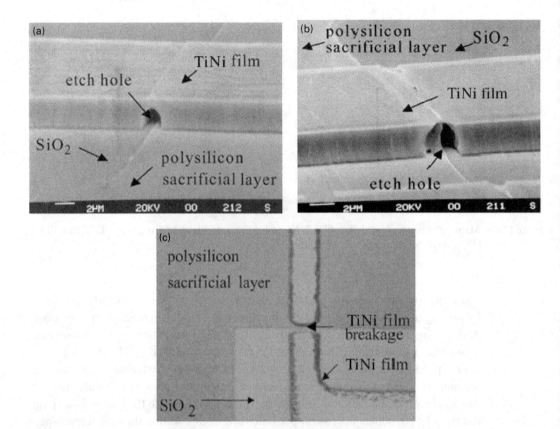

Figure 13.6   Pictures showing the TiNi film breakage at the polysilicon sacrificial layer step: (a) an etch hole grows between the TiNi film and the substrate; (b) almost broken TiNi film; (c) completely broken TiNi film [37]. (With permission from Elsevier.)

The researchers report several problems when patterning the TiNi film. The first problem is a non-uniform undercut of the TiNi film, as shown in Fig. 13.6. In some regions, the undercut is so severe the film is completely etched away. The researchers suggest that a better approach would be to use amorphous film rather than crystallized film during the etching. By etching the amorphous film and subsequently crystallizing it they reduced and almost eliminated the undercutting problem. In addition to the undercut problem, it is observed that the TiNi etches at an unusually fast rate near a sacrificial layer step (Fig. 13.6). Scanning electron microscopy (SEM) shows that a hole is produced during etching at the interface of the TiNi film and sacrificial layer step, producing an unzipping of the film. The researchers associate the unzipping of the film or breakage with the large stresses in the film at the step. It is believed that the large stresses near geometric anomalies produce unusually high etch rates and should thus be avoided during fabrication processes. The large unzipping near steps has been further confirmed in more recent unpublished studies.

Figure 13.7   TiNi film patterned by ion-milling (no film breakage is shown) [37]. (With permission from Elsevier.)

To resolve this problem, the researchers replaced the wet etching of TiNi with a dry etching technique using ion-milling to remove the TiNi thin film. Unlike chemical processes that etch isotropically, ion-milling is an anisotropic etching technique that can produce a vertical side-wall in a polycrystalline film. Therefore, fast etching near a step or gap does not occur. A thick photoresist (AZ-4620) was used as the etch mask and the TiNi film was ion-milled with the following conditions: (a) working gas was argon; (b) acceleration of the argon beam was at 150 V; (c) beam current was 8 mA; (d) etching time was 90 min and over-etching time was 30 min. Figure 13.7 shows the result of ion-milling of the TiNi film at the sacrificial layer step. Also, ion-milling produces a clean etch profile for crystallized film, however, recent studies suggest other fabrication approaches such as shadow masking could be used to solve this problem without the need for an ion-mill which is sometimes unavailable and expensive. However, reverting to a shadow mask technique requires a different fabrication process which also presents problems of itself.

13.4.1.4   Microvalves

One of the more common structures studied in the MEMS field is microvalves, with a few reports describing TiNi thin film based microvalves, [62, 63]. Microvalves are potentially useful in microfluidics, pumps, thermal switches and a wide range of other applications. The main purpose of microvalves is to open and allow fluid to flow or close and prevent fluid from flowing. Both shape memory microvalves (i.e. active valves) and superelastic microvalves (i.e. passive check valves) have been fabricated. Below we provide a description of two such valves [1, 3] containing either a tetragonal or a pentagonal flap.

These particular structures were chosen to illustrate different fabrication processes to build TiNi microdevices rather than focus on operational principles of the valve. The tetragonal flap structure was fabricated using wet-etching techniques while the pentagonal flap was constructed using a shadow masking

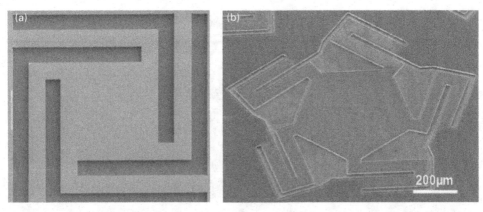

Figure 13.8   Micrographs of (a) tetragonal flap (b) pentagonal flap [1, 3] ([1] with permission from the Institute of Physics (IOP), [3] with permission from Elsevier).

technique, thus eliminating the need to wet etch TiNi. Photographs of both valves are provided in Fig. 13.8. Both valve sets use an array of microvalves in parallel. Each microvalve consists of an SMA TiNi flap connected with either four (tetragonal) or five (pentagonal) tethers anchored to the Si substrate, as shown in Fig. 13.8. A through channel is made below each flap to allow fluid flow. The flap status determines the closing and opening of the channel. When the flap is flat, the closed channel restricts the fluid flow. The working of the microvalve is based on the stress–strain behavior of the TiNi thin film [3] and both shape memory and superelastic valves have been fabricated and the fabrication process (with the exception of annealing) is identical for both. For the superelastic valve, at room temperature the microvalve is in the closed position. When the stress in the tether increases beyond the stress to induce phase transformation in the TiNi film due to the increase in pressure difference between the inlet and outlet side of the channel, the channel opens and allows the fluid to flow. When the load is reduced, the valve returns to the martensite phase and the valve closes. For the shape memory valve, at room temperature the microvalve is open and allows fluid to flow. When the temperature of the fluid exceeds the austenite transformation temperature the valve closes and restricts fluid flow, i.e. it is a thermal switch.

The tetragonal microvalve is fabricated as follows. Figure 13.9(a) shows the flow chart of the fabrication process. First deposit a 500 nm of wet thermal oxide on a 500 μm thick silicon substrate. The oxide layer on top of the substrate is removed and deposited with 15 nm of chromium (adhesion layer), followed by 600 nm of sacrificial copper layer. An AZ-4620 photoresist is then spin coated and patterned over the copper layer. The exposed copper is removed using a solution of $1(FeCl_3):5(H_2O)$ followed by removing chromium with CRE-473® chromium etchant. The substrate is cleaned and photoresist removed prior to depositing a TiNi layer. The TiNi is subsequently deposited on the front side of the substrate in a d.c. magnetron sputtering system. The backside of the substrate is spin coated

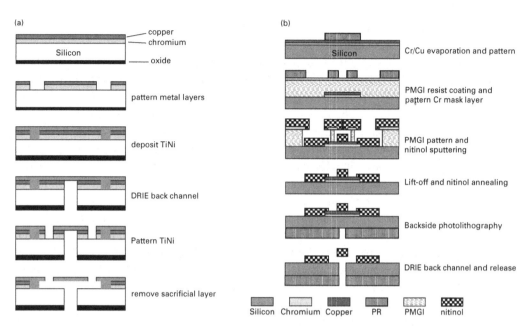

(a)

copper
chromium
oxide

Silicon

pattern metal layers

deposit TiNi

DRIE back channel

Pattern TiNi

remove sacrificial layer

(b)

Silicon   Cr/Cu evaporation and pattern

PMGI resist coating and
pattern Cr mask layer

PMGI pattern and
nitinol sputtering

Lift-off and nitinol annealing

Backside photolithography

DRIE back channel and release

Silicon   Chromium   Copper   PR   PMGI   nitinol

Figure 13.9   Fabrication process flow of TiNi MEMS valve: (a) tetragonal flap, (b) pentagonal flap [1,3] ([1] with permission to use from the Institute of Physics (IOP), [3] with permission from Elsevier).

and developed with NR5–8000 negative photoresist over an oxide layer. This provides a pattern for the channels through the silicon substrate. The oxide patterns are etched with RIE and the silicon is etched with DRIE with final channel diameters of 250 μm. Once the channels are etched, the TiNi on the front side is spin coated and developed with AZ-4620 photoresist to produce patterns for the microvalve's flaps and tethers. A solution of $HF(1):HNO_3(1):H_2O(20)$ is used to bulk etch the TiNi. The flaps are released from the silicon substrate by removing the copper sacrificial layer with HNO_3. HNO_3 does not attack TiNi because it forms a protective thin layer of oxide on the surface, and the TiNi is subsequently annealed at 500 °C for 30 minutes.

The researchers changed to a pentagonal flap design to achieve ortho-planar deflection without torsional stresses, thereby increasing the strength of the valve. Also the researchers used a different and reportedly superior fabrication process. It should be noted that other fabrication processes exist which may be superior to that described in this chapter. The pentagonal flaps are fabricated using a "bilayer lift-off" method to overcome fabrication issues such as undercutting at discontinuities and non-planar features associated with isotropic etching. The process takes advantage of the limited step coverage of metal deposition suggested by Nakamura *et al.* [64] who used chromium and polyimide bilayers to fabricate a TiNi loop actuator. However, Nakamura's fabrication method required reactive ion etching of a polymer which is inefficient and expensive compared to classical photolithography. Roch *et al.* [65] manufactured a TiNi actuator using two different resists, however, this technique is unsuitable for high temperature deposition processes (i.e. $> 250\,°C$) required for TiNi film [65].

Figure 13.9 shows the schematic diagram of process flow in both tetragonal and pentagonal microvalve fabrication. The modified bilayer lift-off method is developed by using a PMGI polymer layer as an underlayer while a chromium layer is used as a top masking layer. The initial steps in both tetragonal and pentagonal flap microvalve fabrication are identical until the deposition of TiNi. In this method, prior to TiNi deposition a thick SF-11 PMGI layer is coated using multiple spin processes and a 150 nm chromium layer is subsequently deposited. The PMGI is spin coated at 500 rpm (3.4 μm thick) and baked at $195\,°C$ for 2 minutes. Repeating the PMGI coating three or four times makes it possible to achieve a thick PMGI layer over 10 μm. An AZ5214 photoresist is used to pattern the chromium with the chromium etchant. The patterned chromium layer represents the masking layer of the PMGI layer which is patterned by deep UV exposure and XP 101A PMGI developer. After PMGI development, O_2 plasma etching is used to minimize the polymer residues. Following this, TiNi is deposited on the front side of the substrate. After deposition of the TiNi film, a lift-off process is followed to remove unnecessary TiNi film deposited on top of the chromium and PMGI bilayer. This lift-off process produces patterns for the MEMS valve flaps and tethers without etching of the TiNi film. The TiNi film is crystallized at $500\,°C$. After crystallization, the backside of the substrate is etched similarly to the tetragonal flap design. The final step removes the sacrificial layers and releases the TiNi pentagonal flap.

13.4.2   Biomedical devices

A large fraction of applications for macroscopic TiNi are in the medical field due to the biocompatibility of TiNi [16, 17, 66, 67, 68, 69, 70]. Typically, these bulk applications utilize superelasticity combined with some shape memory attributes to allow devices to be easily deployed within the body such as through a catheter. For these applications, the material's A_f is designed at or near body temperature. Devices falling into this category include stents, ASD closure devices, and vena cava filters to name a few [67, 69, 71]. TiNi is arguably superior to conventional materials such as surgical steel, because the TiNi device can be collapsed into a substantially smaller profile than surgical steel. For example, TiNi stents are collapsed into small diameter catheters (e.g. as small as 3 French or 1 mm, 1 French (F) being 1/3 mm) and guided along the vascular system to the appropriate location using imaging hardware. Once at the site, such as an occluded artery, the stent is forced out of the catheter with a push rod and exposed to the surrounding body temperature. The surrounding temperature causes the stent to fully deploy and open the occluded artery without requiring balloon expansion. Balloon expansion is routinely used to aid in deploying surgical steel stents that do not have the recovery strain associate with TiNi. While this and other vascular applications represent a significant portion for macroscopic TiNi, few papers are available in the open literature specifically addressing thin film TiNi medical devices. Some papers such as Strong *et al.* [72] report on TiNi devices for external applications such as measuring pH, glucose and urea [73, 74], but this is distinctly different from most of the commercial applications for macroscopic TiNi devices used in the vascular system. While there are a limited number of papers, we do point out that the number of patents on TiNi thin film devices for embeddable microdevices is growing rapidly. One explanation for the relatively small number of research publications on this topic may be due to the problems associated with fabricating consistent high quality thin film TiNi required for medical devices, as described in the previous sections.

13.4.2.1 Heart valve

One device recently proposed for thin film TiNi was a heart valve [70]. The TiNi film in the valve is used as the heart leaflets for a surgically placed heart valve replacing the mechanical or bio-prosthetic heart valves currently on the market. The arguments for using TiNi thin film in the heart valve leaflets are as follows. First the film may provide increased longevity compared to current bio-prosthetic valves (i.e. average 10 year lifespan). Second, the TiNi thin film valve may not require the anti-coagulation therapy necessary in mechanical valves. The researchers also suggest that TiNi film could potentially be used in a percutaneously placed heart valve. For percutaneously placed heart valves, the TiNi leaflets could be substantially thinner than the bio-prosthetic, allowing the heart valve to be collapsed into a substantially smaller catheter for delivery. This is extremely important in pediatric patients.

The surgically placed TiNi prosthetic heart valve (see Fig. 13.10) was designed and constructed by inserting an elliptically shaped piece of shape memory TiNi thin film into a Teflon scaffold made of two Teflon tubes with different diameters. The heart valve leaflets were manufactured similarly to the bubble actuator described in the preceding section, however, the Si was entirely removed, providing free-standing thin film TiNi. The TiNi film was also produced from a heated target during sputter deposition. Furthermore, the authors deposit the film onto SiO_2 to prevent reactions between the Si and the TiNi film. Following deposition and fabrication of a TiNi leaflet, the leaflet was wrapped around the anchoring bar, made of TiNi wire, and fixed in place by the insertion of the smaller tube into the inferior portion of the large tube. Figure 13.10 shows the closed and open positions of the heart valve.

Following heart valve fabrication, the researchers conducted both *in vivo* and *in vitro* tests on the surgical valve. The *in vitro* tests indicated that the heart valve had superior corrosion resistance compared to macroscopic TiNi [71] and that the operational characteristics were similar to that of existing heart valve solutions. These tests were performed in a pulsatile flow generation system mimicking the pressures and flow rates in a typical human. The preliminary *in vivo* and corrosion test results suggest that TiNi thin film is biocompatible when placed in the blood stream. *In vivo* results indicate that TiNi thin film is less susceptible to fibrin deposition, calcification, and neointimal formation than

Figure 13.10   Closed (left) and open (right) positions of a TiNi thin film heart valve [70]. (With permission from ASME.)

surgical steel, and preliminary biocompatibility tests show that the TiNi thin film does not corrode after one month in an artificial physiological solution. SEM images in Fig. 13.11 show only red blood cells on the surface of the film and there is no evidence of any embolus formation. In short term laboratory studies, a prototype thin film TiNi valve was shown to perform at the same level as a commercially available surgically implantable tissue valve. Although long-term durability tests and further biocompatibility studies are necessary to determine TiNi thin film's capabilities as a heart valve leaflet, it is certainly possible that this material may be very well suited for both surgical heart valves and transcatheter heart valves.

13.4.2.2   Covered stent

Although traditional stents consist of non-occlusive metal scaffolds, the treatment of many disease processes requires the use of a "covered stent". These stents are able to open vessels and provide a circumferentially occlusive boundary between the stent and the vessel. Thus, covered stents are ideal for re-establishing the integrity of aneurysmal vessels or for minimizing the risk of in-stent stenosis [75]. The potential applications of such covered stents include the treatment of coronary artery disease, aortic and central nervous system vascular aneurysms, carotid artery or pulmonary artery stenoses, and even treatment of ruptured vessels [76, 77]. In the palliation of congenital heart disease, specifically, the appropriate covered stent would be of tremendous value in stenting the ductus arteriosus, coarctation of the aorta, or in the stenting of pulmonary veins, an intervention often plagued by in-stent stenosis. Various materials have been used

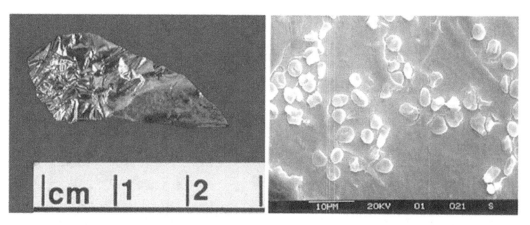

Figure 13.11   The piece of thin film TiNi shown on the left is free from tissue growth. An SEM picture of the same film (right) reveals only RBCs on the surface of the TiNi thin film [70]. (With permission from ASME.)

to cover stents, including silicone, polyurethane and polytetrafluoroethylene [78]. To date, the production of a highly flexible, durable and thrombus-resistant material has not been achieved for all applications. The researchers explored the use of a new thin film TiNi material for covering a stent. Once again the fabrication process for the TiNi film was based on a heated target concept.

Standard photolithography and etching techniques (see Section 13.4.1.1) were used to generate precise two-dimensional TiNi thin film shapes for covering the stents. The fabricated thin film shapes were mechanically removed from the wafer. Both self-expanding TiNi stents and balloon-inflatable stainless steel stents were mechanically covered with the thin film TiNi. Specifically, balloon-inflatable PG1910B and PG2910B stents (19 and 29 mm in length, respectively) from Cordis (Johnson and Johnson, Miami, FL), and prototype *pfm* (*pfm* AG, Hamburg, Germany) TiNi self-expanding stents (20 mm in length) were used for laboratory and animal testing. During the fabrication the researchers adopted a "wrap" design, a two-dimensional piece of TiNi thin film wrapped circumferentially around the commercial stent. To cover the self-expanding *pfm* stents, each stent was cooled with liquid nitrogen to $-195.95\,^{\circ}$C and compressed into 5 French sheaths. The thin film cover was then wrapped around this sheath (1 French = 3 mm). The 5 French sheath wrapped in TiNi film was subsequently placed into a larger 7 French sheath. Finally, the 5 French sheath was then pulled back, leaving the thin film cover in place while simultaneously exposing the stent. The stent became covered by self-deploying into the thin-film TiNi wrap. The final 7 French sheath was then attached to a modified *pfm* stent delivery system.

Both *in vivo* and *in vitro* testing of the TiNi thin film covered stents were conducted. The in vitro tests demonstrate the feasibility of successful stent deployment and the stent immobility under pulsatile flow of 1.5 l/min. Similarly, for in vivo testing of the covered stent a swine animal model was used. Four

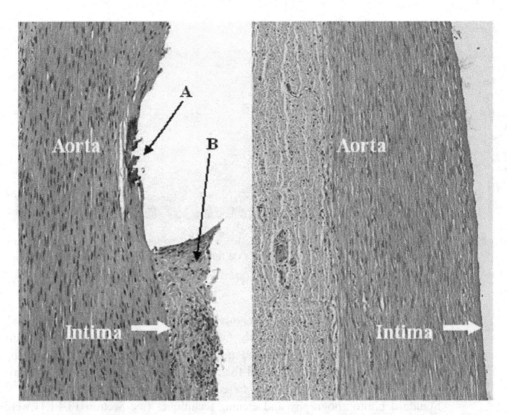

Figure 13.12 Left A: Damage on aorta wall from bare stent strut. Left B: Neointimal formation where bare stent contacted aorta wall. Right: Healthy aortic wall where thin film TiNi covered stent was placed [70]. (With permission from ASME.)

animals were used for this study. Thin film TiNi covered stents were percutaneously implanted into appropriately sized vessels of each swine. All arterial plantations were performed in the descending aorta. Venous implants were performed in either the superior vena cava or inferior vena cava and evaluated at different time intervals.

After animal euthanization, stent-containing vessels were harvested, as well as specimens from major organs (heart, lungs, liver, stomach, kidney, pancreas and spleen). Light microscopy was employed to analyze the covered and uncovered portions of the harvested stents (Fig. 13.12). Gross examination, trichrome staining and hematoxylin and eosin staining (H&E) were used to analyze the recovered organ specimens and stent implantation sites. No significant in-stent neointimal hyperplasia was found on thin film TiNi covered stents placed in the arterial circulation. Moderate-to-severe neointimal proliferation was seen in all venous stent implants after three weeks. Notably, several of the stent covers were found to have defects that were likely caused during deployment. In the arterial circulation, the thin film TiNi cover did prevent in-growth of neointima and supported growth of endothelial cells over the first month after implantation. In the venous system, there was angiographic evidence of less in-stent stenosis in the first three weeks after implantation (when compared with an uncovered stent). However, by four and six weeks all venous stents (covered and uncovered) had significant in-stent stenosis from neointimal growth. The vastly different hemodynamic environments encountered in the arterial versus the venous circulation likely contributed significantly to this differential result. The pulsatile arterial circulation, reaching peak flow velocities of approximately 1 m/s, is likely to have had a significant influence on the type and extent of neointimal proliferation when compared with the continuous low flow state present in the venous circulation. It is likely that the covered stents placed in the venous circulation prevented direct growth of the vessel wall through the stent but not onto and around the device. It may be possible to obtain improved results by covering both the outside and inside of the stent with thin film TiNi. Because thin film TiNi seems to promote a very thin layer of endothelialization, an inner covering may improve results by shielding the lumen of the vessel from exposure to the bulk TiNi or stainless steel present in the stent scaffolding. As thin film TiNi is much smoother and contains many less contaminants than bulk TiNi, this inner covering could allow for a more favorable biological response when compared with bulk TiNi.

13.5   Summary

This chapter has provided descriptions of several fundamental microdevices reported in the open literature. While the number of publications is increasing at a significant rate, the principal issue for thin film TiNi to become more accepted by the mechanical and biomedical community is process control and uniformity control of the thin film material. Fabrication of microdevices is fairly complex and difficult due to the etching properties of TiNi. Typically it is observed that the amorphous film etches qualitatively better than the crystalline forms of TiNi. Furthermore, the isotropic nature of wet etchants on polycrystalline thin film TiNi produces problems in micromachining. Two approaches to avoid this are lift-off methods and shadow masking methods, both of a similar context. While indeed other approaches may be available, each method tends to have a certain amount of complexity associated with the process so that it becomes device dependent. Furthermore, while many potential applications exist for thin film TiNi, based on macroscopic TiNi applications the biomedical device for vascular repair may represent the most fertile area of study in the next decade.

References
[1] M. Seong, K. P. Mohanchandra and G. P. Carman, Development of a 'bi-layer lift-off' method for high flow rate and high frequency nitinol MEMS valve fabrication. *J. Micromech. Microeng.*, **18** (2008) 075034.
[2] Y. Bellouard, Shape memory alloys for microsystems: a review from a material research perspective. *Mater. Sci. Eng. A*, **481** (2008) 582–589.

[3] D. D. Shin, K. P. Mohanchandra and G. P. Carman, Thin film NiTi microthermostat array. *Sensors and Actuators A: Physical*, **130–131** (2006) 37–41.

[4] B. Winzek, S. Schmitz, H. Rumpf, *et al.*, Recent developments in shape memory thin film technology. *Mater. Sci. Eng. A*, **378** (2004) 40–46.

[5] Y. Fu, W. Huang, H. Du, X. Huang, J. Tan and X. Gao, Characterization of TiNi shape-memory alloy thin films for MEMS applications. *Surface & Coatings Technology*, **145** (2001) 107–112.

[6] A. Olander, An electrochemical investigation of solid cadmium-gold alloys. *J. American Chemical Society*, **54** (1932) 3819–3833.

[7] W. J. Buehler, R. C. Wiley and J. V. Gilfrich, Effect of low-temperature phase changes on mechanical properties of alloys near composition TiNi. *J. Appl. Phys.*, **34** (1963) 1475–1477.

[8] F. E. Wang, W. J. Buehler and S. J. Pickart, Crystal structure and a unique martensitic transition of TiNi. *J. Appl. Phys.*, **36** (1965) 3232–3239.

[9] T. Tadaki, Y. Nakata, K. Shimizu and K. Otsuka, Crystal-structure, composition and morphology of a precipitate in an aged Ti-51 at. percent-Ni shape memory alloy. *Transactions of the Japan Institute of Metals*, **27**(1986) 731–740.

[10] K. Otsuka and X. B. Ren, Recent developments in the research of shape memory alloys. *Intermetallics*, **7**(1999) 511–528.

[11] Z. A. Andrzej, R. A. Bogdan and S. Miyazaki, Stress induced martensitic transformation kinetics of polycrystalline NiTi shape memory alloy. *Mater. Sci. Eng. A*, **378** (2004) 86–91.

[12] Y. C. Luo and D. Y. Li, New wear-resistant material: nano-TiN/TiC/TiNi composite. *J. Mater. Sci.*, **36** (2001) 4695–4702.

[13] L. M. Qian, X. D. Xiao, Q. P. Sun and T. X. Yu, Anomalous relationship between hardness and wear properties of a superelastic nickel-titanium alloy. *Appl. Phys. Lett.*, **84** (2004)1076–1078.

[14] D. E. Hodgson, Damping applications of shape-memory alloys. *Materials Science Forum*, **394–395** (2002) 69.

[15] E. Hornbogen, Thermo-mechanical fatigue of shape memory alloys. *J. Mater. Sci.*, **39**(2004) 385–399.

[16] J. Ryhanen, Biocompatibility of Nitinol. *Minimally Invasive Therapy & Allied Technologies*, **9** (2000) 99–105.

[17] S. A. Shabalovskaya, Surface, corrosion and biocompatibility aspects of nitinol as an implant material. *Bio-Medical Materials and Engineering*, **12** (2002) 69–109.

[18] K. Ikuta, Micro/miniature shape memory alloy actuator, *Proc. IEEE. Conference on Robotics and Automation* (1990.) 2156–2161.

[19] J. A. Walker, K. J. Gabriel and M. Mehregany, Thin-film processing of TiNi shape memory alloy. *Sensors and Actuators A: Physical*, **21** (1990) 243–246.

[20] J. D. Busch, A. D. Johnson, C. H. Lee and D. A. Stevenson, Shape-memory properties in Ni-Ti sputter-deposited film. *J. Appl. Phys.*, **68** (1990) 6224–6228.

[21] E. Makino, M. Uenoyama and T. Shibata, Flash evaporation of TiNi shape memory thin film for microactuators. *Sensors and Actuators A: Physical*, **71** (1998) 187–192.

[22] T. Sam and K. T. Davies, Ion beam sputter deposition of TiNi shape memory alloy thin films. *Proc. SPIE*, **3874** (1999) 8.

[23] H. Y. Noh, K. H. Lee, X. X. Cui and C. S. Choi, The composition and structure of TiNi thin film formed by electron beam evaporation. *Scripta Materialia*, **43** (2000) 847–852.

[24] V. Martynov, A. D. Johnson and V. Gupta, Effect of magnetic field configuration on the properties of thin films sputter-deposited from a dc magnetron TiNi target. *Journal De Physique IV*, **112** (2003) 845–848.

[25] Akihiro Ohta, Shekhar Bhansali, Isao Kishimoto and Akira Umeda, Development of TiNi shape memory alloy film deposited by sputtering from separate Ti and Ni targets, *Proc. SPIE*. **3512** (1998) 8.

[26] P. Krulevitch, P. B. Ramsey, D. M. Makowiecki, A. P. Lee, M. A. Northrup and A. D. Johnson, Mixed-sputter deposition of Ni-Ti-Cu shape memory films. *Thin Solid Films*, **274** (1996) 101–105.

[27] R. Hassdorf, J. Feydt, S. Thienhays, *et al.*, Microstructure, phase sequence, and superelasticity in highly oriented MBE-grown NiTiCu shape memory thin films, *Materials Science Forum*, **475**–479 (2005) 3827–3830.

[28] T. Lehnert, H. Grimmer, P. Boni, M. Horisberger and R Gotthardt, Characterization of shape-memory alloy thin films made up from sputter-deposited Ni/Ti multilayers. *Acta Materialia*, **48** (2000) 4065–4071.

[29] N. Yaakoubi, C. Serre, S. Martinez, Growth and characterization of shape memory alloy thin films for Si microactuator technologies, *J. Mater. Sci.: Materials in Electronics*, **12** (2001) 323–326.

[30] C. L. Shih, B. K. Lai, S. M. Philips and A. H. Heuer, A robust co-sputtering fabrication procedure for TiNi shape memory alloys for MEMS, *J. Microelectromechanical Systems*, **10** (2001) 69–79.

[31] A. Ishida, M. Sato, O Tabata and W Yoshikawa, Shape memory thin films formed with carrousel-type magnetron sputtering apparatus. *Smart Materials & Structures*, **14** (2005) S216–S222.

[32] K. K. Ho and G. P. Carman, Sputter deposition of NiTi thin film shape memory alloy using a heated target. *Thin Solid Films*, **370** (200) 18–29.

[33] K. K. Ho, K. P. Mohanchandra and G. P. Carman, Examination of the sputtering profile of NiTi under target heating conditions. *Thin Solid Films*, **413** (2002) 1–7.

[34] K. P. Mohanchandra, K. K. Ho and G. P. Carman, Electrical characterization of NiTi film on silicon substrate. *J. Intelligent Material Systems and Structures*, **15** (2004) 387–392.

[35] D. D. Shin, K. P. Mohanchandra and G. P. Carman, High frequency actuation of thin film NiTi. *Sensors and Actuators A – Physical*, **111** (2004) 166–171.

[36] K. P. Mohanchandra, K. K. Ho and G. P. Carman, Compositional uniformity in sputter-deposited NiTi shape memory alloy thin films. *Mater. Lett.*, **62** (2008) 3481–3483.

[37] J. J. Gill, D. T. Chang, L. A. Momoda and G. P. Carman, Manufacturing issues of thin film NiTi microwrapper. *Sensors and Actuators A – Physical*, **93** (2001)148–156.

[38] D. D. Shin, K. P. Mohanchandra and G. P. Carman, Development of hydraulic linear actuator using thin film SMA. *Sensors and Actuators A – Physical*, **119** (2005) 151–156.

[39] T. Yamazaki, T. Yoshizawa, H. Takada and F. Takeda, Dependence of composition distribution of NiTi sputtered films on Ar gas pressure. *Japanese Journal of Applied Physics Part*, **40** (2001) 6936–6940.

[40] J. M. Ting and P. Chen, Dependence of compositions and crystallization behaviors of dc-sputtered TiNi thin films on the deposition conditions. *Journal of Vacuum Science & Technology A*, **19** (2001) 2382–2387.

[41] Y. Nakata, T. Tadaka, H. Sakamoto, A Tanaka and K. Shimizu, Effect of heat treatments on morphology and transformation temperatures of sputtered Ti-Ni thin films. *Journal De Physique IV*, **5**(C8) (1995) 671–676.

[42] A. Ishida, M. Sate, A. Takei and S. Miyazaki, Effect of heat treatment on shape memory behavior of Ti-rich Ti-Ni thin films. *Mater. Trans. JIM*, **36** (1995) 1349–1355.

[43] M. Bendahan, P. Canet, J. L. Seguin and H. Carchano, Control composition study of sputtered Ni-Ti shape-memory alloy film. *Mater. Sci. Eng. B*, **34** (1995) 112–115.

[44] A. Ishida, M. Sate, A. Takei and S. Miyazaki, Effect of heat treatment on shape memory behavior of Ti-rich Ti-Ni thin Films, *Mater. Trans.*, **36** (1995)1349–1355

[45] S. Miyazaki and A. Ishida, Shape-memory characteristics of sputter-deposited Ti-Ni thin-films. *Mater. Trans. JIM*, **35** (1994) 14–19.

[46] Y. J. Zhang, Y. T. Cheng and D. S. Grummon, Shape memory surfaces. *Appl. Phys. Lett.*, **89** (2006) 041912.

[47] K. Kuribayashi, M. Yoshitake and S. Ogawa, Reversible SMA actuator for micron sized robot. *Proceedings of Micro Electro Mechanical Systems* (MEMS-90), (1990) 217–221.

[48] H. T. G. Vanlintel, F. C. M. Vandepol and S. Bouwstra, A piezoelectric micropump based on micromachining of silicon. *Sensors and Actuators*, **15** (1988) 153–167.

[49] M. Kohl, E. Just, W. Pfleging and S. Miyazaki, SMA microgripper with integrated antagonism. *Sensors and Actuators A – Physical*, **83** (2000) 208–213.

[50] W. M. Huang, J. P. Tan, X. Y. Gao and J. H. Yep, Design, testing, and simulation of NiTi shape-memory-alloy thin-film-based microgrippers. *Journal of Microlithography Microfabrication and Microsystems*, **2** (2003) 185–190.

[51] B. Sutapun, M. Tabib-Azar and M. A. Huff, Applications of shape memory alloys in optics. *Applied Optics*, **37** (1998) 6811–6815.

[52] M. Tabib-Azar, B. Sutapun and M. Huff, Applications of TiNi thin film shape memory alloys in micro-opto-electro-mechanical systems. *Sensors and Actuators A – Physical*, **77** (1999) 34–38.

[53] M. Kohl, D. Dittmann, E. Quandt and B. Winzek, Thin film shape memory microvalves with adjustable operation temperature. *Sensors and Actuators A – Physical*, **83** (2000) 214–219.

[54] H. Kahn, M. A. Huff and A. H. Heuer, The TiNi shape-memory alloy and its applications for MEMS. *J. Micromech. Microeng.*, **8** (1998) 213–221.

[55] W. L. Benard, H. Kahn, A. H. Heuer and M. A. Huff, Thin-film shape-memory alloy actuated micropumps. *J. Microelectromechanical Systems*, **7** (1998) 245–251.

[56] M. Kohl, D. Dittmann, E. Quandt, B. Winzek, S. Miyazaki and D. M. Allen, Shape memory microvalves based on thin films or rolled sheets. *Mater. Sci. Eng. A*, **275** (1999) 784–788.

[57] M. Kohl, K. D. Skrobanek and S. Miyazaki, Development of stress-optimised shape memory microvalves. *Sensors and Actuators A – Physical*, **72** (1999) 243–250.

[58] C. Liang and C. A. Rogers, One-dimensional thermomechanical constitutive relations for shape memory materials. *J. Intelligent Material Systems and Structures*, **1** (1990) 207–234.

[59] J. J. Gill and G. P. Carman, Thin film NiTi shape memory alloy microactuator with two-way effect, *Proc. Inter. Mech. Eng. Conf. and Exhibition, Orlando, FL* (2000) 243–246.

[60] J. J. Gill, K. K. Ho and G. P. Carman, Three-dimensional thin-film shape memory alloy microactuator with two-way effect. *J. Microelectromechanical Systems*, **11** (2002) 68–77.

[61] J. Ok, M. Chu and C. J. Kim, Pneumatically driven microcage for micro-objects in biological liquid. *Proceedings of the IEEE Annual Workshop of Microelectromechanical Systems, FL, USA* (1999) 459–463.

[62] G. Hahm, H. Kahn, S. M. Phillips and A. H. Heuer, Fully microfabricated silicon spring biased, shape memory actuated microvalve, *Solid-State Sensors and Actuator Workshop Hilton Head, SC, 4–8 June* (2000) 230.

[63] B. K. Lai, G. Hahm, L. You, *et al.*, The characterization of TiNi shape-memory actuated microvalves. *Materials Research Society*, **657** (2001) EE8.3.1–6.

[64] Y. Nakamura, S. Nakamura, L. Buchaillot and H. Fujita, A three-dimensional shape memory alloy loop actuator. *Proc. IEEE: Micro Electro Mechanical Systems*, MEMS **97** (1997) 262–266.

[65] I. Roch, P. Bidaud, D. Collard and L. Buchaillot, Fabrication and characterization of an SU-8 gripper actuated by a shape memory alloy thin film *J. Micromech. Microeng.*, **13** (2003) 330–336.

[66] G. Airoldi and G. Riva, Innovative materials: the NiTi alloys in orthodontics. *Bio-Medical Materials and Engineering*, **6** (1996) 299–305.

[67] T. W. Duerig, D. E. Tolomeo and M. Wholey, An overview of superelastic stent design. *Minimally Invasive Therapy & Allied Technologies*, **9** (2000) 235–246.

[68] A. R. Pelton, D. Stockel and T. W. Duerig, Medical uses of nitinol, *Shape Memory Materials. Proceedings of the International Symposium on Material Science Forum* **327–328** (2000) 63–70.

[69] D. Mantovani, Shape memory alloys: properties and biomedical applications. *JOM – Journal of the Minerals Metals & Materials Society*, **2** (2000) 36–44.

[70] L. L. Stepan, D. S. Levi and G. P. Carman, A thin film nitinol heart valve, *J. Biomech. Eng.*, **127** (2005) 915–918.

[71] L. L. Stepan, D. S. Levi, E. Gans, K. P. Mohanchandra, M. Ujiharai and G. P. Carman, Biocorrosion investigation of two shape memory nickel based alloys: Ni-Mn-Ga and thin film NiTi. *J. Biomedical Materials Research Part A*, **82A** (2007) 768–776.

[72] Strong, A. W. Wang and C. F. McConaghy, Hydrogel-actuated capacitive transducer for wireless biosensors. *Biomedical Microdevices*, (2002). 4(2): 97–103.

[73] A. Kikuchi, K. Suzuki, O. Okabayashi, *et al.*, Glucose-sensing electrode coated with polymer complex gel containing phenylboronic acid. *Analytical Chemistry*, **68** (1996) 823–828.

[74] T. Schalkhammer, C. Lobmaier, F. Pittner, A. Leitner, H. Brunner and F. R Aussenegg, The use of metal-island-coated Ph-sensitive swelling polymers for biosensor applications. *Sensors and Actuators B – Chemical*, **24** (1995) 166–172.

[75] I. Saatci, H. S. Cekirge, M. H. Ozturk, *et al.*, Treatment of internal carotid artery aneurysms with a covered stent: experience in 24 patients with mid-term follow-up results. *American Journal of Neuroradiology*, **25** (2004) 1742–1749.

[76] R. P. Cambria, Stenting for carotid-artery stenosis. *New England Journal of Medicine*, **351** (2004) 1565–1567.

[77] R. W. Hobson, Carotid artery stenting. *Surgical Clinics of North America*, **84** (2004) 1281–1294.

[78] A. Repici, M. Conio, C. De Angelis, *et al.*, Temporary placement of an expandable polyester silicone-covered stent for treatment of refractory benign esophageal strictures. *Gastrointestinal Endoscopy*, **60** (2004) 513–519.

14 Shape memory microvalves

M. Kohl

Abstract

Microvalves are a promising field of application for shape memory alloy (SMA) microactuators as they require a large force and stroke in a restricted space. The performance of SMA-actuated microvalves does not only depend on SMA material properties, but also requires a mechanically and thermally optimized design as well as a batch fabrication technology that is compatible with existing microsystems technologies. This chapter gives an overview of the different engineering aspects of SMA microvalves and describes the ongoing progress in related fields. Different valve types based on various design–material–technology combinations are highlighted. The examples also demonstrate the opportunities for emerging new applications.

14.1 Introduction

Shape memory alloys (SMAs) belong to the category of smart materials as they exhibit multi-functional properties, which can be used simultaneously to generate actuation, sensing and adaptive functions. The high energy density and favorable scaling behavior of their mechanical properties upon miniaturization make these materials particularly attractive for actuator applications in micro-dimensions.

In this realm, however, the close neighborhood of functional parts as well as fabrication constraints have a strong impact on physical properties and overall performance. In addition, the related research fields of *materials development*, *design engineering* and *microtechnology* are strongly interlinked and cannot be considered separately. Currently, each of these research fields faces a number of obstacles and constraints.

(1) Despite the tremendous success in the development of SMA thin films and foils, considerable limitations due to temperature range, hysteresis and fatigue exist.

Thin Film Shape Memory Alloys: Fundamentals and Device Applications, eds. Shuichi Miyazaki, Yong Qing Fu and Wei Min Huang. Published by Cambridge University Press. © Cambridge University Press 2009.

(2) SMA microactuators usually consist of a monolithic structure with quasi-two-dimensional shape. In order to make an optimum use of the shape memory effect, homogeneous stress distributions should be generated upon loading, which requires some kind of shape optimization. In general, SMA microactuators are strongly thermally coupled to their environment. Therefore, optimum design of the thermal interfaces is crucial for the microactuator performance to ensure that the right temperature profiles are reached within optimum time periods. The key requirement for mechanical and thermal design engineering are suitable SMA models and simulation tools, taking the interaction with neighboring functional parts into account.

(3) Integration of SMA microparts into a microsystem is still a challenge for the realization of SMA microactuators, since the combination of conventional microfabrication technologies and SMA materials leads to various fabrication process incompatibilities.

The following review addresses these issues for SMA microvalves. Usually, engineering of a device starts from the specific requirements of a given application in a top-down approach. However, for development of SMA microactuators, the right materials, design tools and technologies are not readily available at present. Therefore, the microvalves which will be presented here have been developed rather in a bottom-up approach by starting from the materials and engineering aspects. Still, their performance can be quite easily adapted to real applications. Several variants of microvalve will be discussed, highlighting the specific advantages and prospects of SMA microactuators.

14.2   Overview

Microvalves may be realized in two different ways, as sketched in Fig. 14.1. Most microvalves are of the seat-valve type, which requires an actuation force that counteracts the pressure force applied to the valve. Moreover, to allow for a useful flow, a valve stroke of at least several tens of micrometers is needed. Therefore, the major challenge for this type of valve is the integration of an actuator which provides sufficiently large force and stroke. SMA microactuators are ideal candidates due to their superior work density. For high pressure applications, the force requirements can be partly relaxed by special valve designs including pressure compensation.

For the case of gate valves, pressure force and actuation force are perpendicular to each other ("cross-flow") and therefore do not counteract each other. Gate valve design therefore removes a major actuator requirement. The challenge for this type of valve is therefore the integration of mechanically robust microactuators that deliver a large stroke, such as cantilevers of thick SMA films or foils. Depending on the valve condition in the absence of driving power, valves are classified as either normally open or closed, which demands different directions of

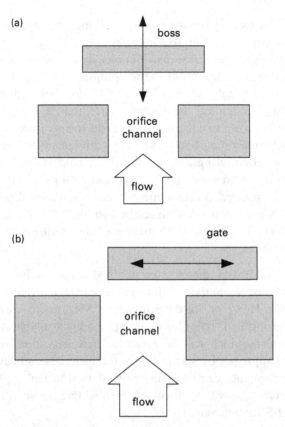

Figure 14.1. The two orifice/obstruction configurations common in microsystems: (a) the seat valve, (b) the cross-flow or gate valve.

actuation and reset mechanism. Generally, gate valves cannot maintain a leak-proof closed condition.

Since their introduction in 1979 [1], a large variety of MEMS-based microvalves have been developed using various actuation principles and mechanisms. For a recent review, see e.g. [2]. In particular, several different SMA microvalves have been presented up to now. In most cases, microvalve actuation is realized by a combination of bending and tensile motion of an SMA microdevice as sketched in Fig. 14.2. In general, bending actuation provides a large displacement at a low force, while tensile actuation allows for a large force and small displacement. However, the actual force and stroke strongly depend on the specific design of SMA microactuator, biasing mechanism and other microvalve components and, thus, can be adjusted in a wide range.

Bending actuation is achieved, e.g., by using monomorph or bi-/multimorph SMA beam cantilevers, see Figs. 14.2(a) and (b), respectively. In the case of a monomorph SMA beam, an additional reset mechanism is required to pre-strain the beam in the martensitic state. This can be realized by the applied pressure force or an external biasing spring. In principle, also the intrinsic two-way shape

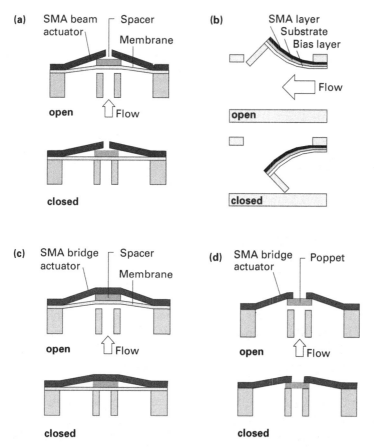

Figure 14.2   Schematic representations of different actuation and reset mechanisms for control of SMA microvalves: (a) bending actuation by monomorph SMA beams (seat microvalve) and (b) by trimorph SMA/Mo/FeCrNi beams (gate microvalve); (c) tensile actuation by an SMA microbridge with both ends being fixed on a substrate, and (d) with one of the ends being fixed on a movable poppet.

memory effect may be used by creating intrinsic stress states through a training process. The membrane is used to separate the fluid and actuation chamber. In the present case, it is also used to transfer the pressure force to the SMA microactuator. Figure 14.2(b) shows an example of a trimorph SMA beam consisting of the layer sequence Ti(Ni,Cu)/Mo/Fe(Cr,Ni). Due to the built-in reset mechanism, concave and convex curvatures upon heating and cooling are created, respectively [3]. By the martensitic transformation a strong change of bimorph stress is induced mainly in between the TiNiCu and Mo layers due to the difference in thermal expansion coefficients. The third layer of stainless steel is introduced in order to bias the initial state of curvature, thereby creating a considerable change of deflection angle.

Tensile actuation is achieved by using an SMA microbridge with both ends being clamped. Figures 14.2(c) and (d) show two examples. Both ends of the SMA microbridge are either fixed on the substrate (c) or one of the ends is fixed on a

Table 14.1 Different design–material–technology combinations for SMA microvalves and corresponding reports in the literature; no and nc denote normally-open and normally-closed operation, respectively

	Bending microactuators	Bridge microactuators
Si gate valve, no	(TiNi/SiO/Si thin foil) [6]	-
Si gate valve, nc	-	-
Si seat valve, no	-	-
Si seat valve, nc	-	TiNi thin film [9]
Polymer gate valve, no	-	-
Polymer gate valve, nc	-	-
Polymer seat valve, no	TiNi thin foil [7]	TiNi(Cu) thin film [10,11]
		TiNiPd thin film [11,12]
		Ni_2MnGa thin film [13]
Polymer seat valve, nc	TiNi thin foil [8]	TiNi thin foil [14]

movable poppet (d). A reset mechanism is required to pre-strain the microbridge in the martensitic state. Again, this can be realized, e.g., by the applied pressure force or an external biasing spring. Another very attractive solution is the combination of two SMA microbridges counteracting each other. Thus, an antagonistic actuation system is obtained showing a low biasing force in the martensitic state and a large switching force in the austenitic state, which results in a particularly large actuation stroke.

Due to thermal actuation, SMA microactuators require relatively large power, which reduces their application potential considerably. Therefore, monostable or bistable microvalve designs are of particular interest; they are capable of maintaining either one or two switching states without applying external power. Possible bistable mechanisms include clamped membranes [4] or ferromagnetic microstructures [5]. Due to the rather high complexity of these systems, no such SMA microvalves have become mature up to now.

Table 14.1 gives an overview of possible design–material–technology combinations for SMA microvalves. Two major technology approaches may be distinguished. Silicon-based technologies rely on bulk and surface micromachining technologies. Polymer technologies make use of injection moulding or hot embossing, which allows the parallel microfabrication of valve housings from a large variety of thermoplastics. Various design–material–technology combinations have been reported in the literature, as indicated in Table 14.1. In the following sections, selected examples will be presented in more detail.

14.3   Valve layout

Figure 14.3 shows a schematic layout of a polymer seat valve [13]. The main valve components are polymer housing, polyimide membrane to separate actuation and fluid chamber, heat sink and SMA microactuator and spacer to pre-strain the

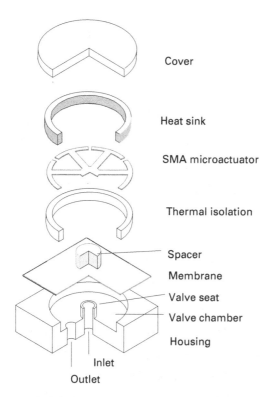

Cover

Heat sink

SMA microactuator

Thermal isolation

Spacer

Membrane

Valve seat

Valve chamber

Housing

Inlet

Outlet

Figure 14.3 Exploded view of a polymer seat valve for normally-open operation. The SMA actuator is designed as a circular array of SMA microbridges [13].

microactuator. In this example, the microactuator is designed as a circular array of SMA microbridges. The microvalve works as a normally-open valve making use of the pressure difference between inlet and outlet to deflect the membrane and thereby the center of the SMA microactuator. Upon heating the microactuator by an electrical current, the shape recovery force pushes the membrane against the valve seat to close the valve. Upon cooling, the microvalve is reset by the pressure difference. By this layout, fluid and actuation chamber remain thermally isolated well enough that gases and temperature-sensitive liquids can be controlled. The heat sink consists of a material with sufficient thermal conductivity to allow for efficient heat-transfer and, thus, fast response time.

In order to realize a polymer seat valve with normally-closed operation, the motion direction of the microactuator has to be reversed. Figures 14.4 and 14.5 present two schematic layouts of a normally-closed microvalve, which differ by their reset mechanism. In the simplest case, a passive microspring may be used as a reset element. Figure 14.4 shows, for instance, a mechanically coupled system of microspring and SMA microactuator, which are pre-deformed with respect to each other [14]. In the closed condition, the microactuator is in the martensitic state, allowing for a relatively large deformation by the microspring. A spacer and membrane are used to transfer the resulting net-force onto the valve orifice. The

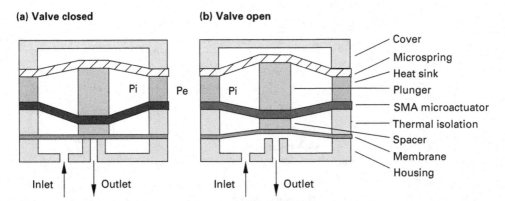

Figure 14.4   Schematic layout of a polymer seat valve for normally-closed operation with biasing spring; Pi and Pe denote internal and external pressure [14].

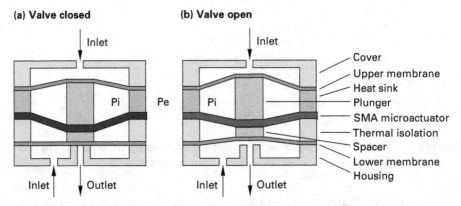

Figure14.5   Schematic layout of a polymer seat valve for normally-closed operation with pressure compensation mechanism [14].

deformation of the microactuator is partly recovered by direct electrical heating to the austenitic state. Thereby, the spacer and membrane are lifted from the valve orifice and the valve opens. The deformation of the microspring increases correspondingly. In this design, the maximum controllable pressure difference of the microvalve is determined by the force of the microspring. The maximum flow is determined by the actuation stroke. Similar design concepts have been implemented in Si microvalves [9].

An alternative design concept without a microspring is illustrated in Fig. 14.5. In this case, the reset mechanism makes use of the external pressure itself [14]. The pressure is applied to two mechanically coupled membranes, which are located above and below the SMA microactuator. In the closed condition, the area of the lower membrane is smaller than the area of the upper membrane due to the size of the orifice. Both membranes are exposed to the same inlet pressure, thus creating a net-force on the SMA microactuator and a deformation in the martensitic condition. Again, a spacer is used to transfer the net-force onto the valve orifice.

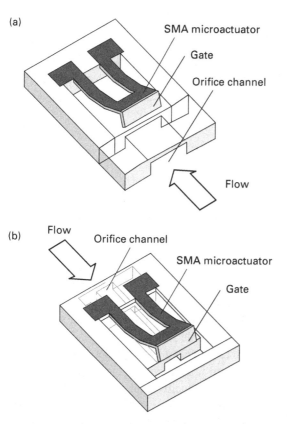

(a)

SMA microactuator

Gate

Orifice channel

Flow

(b) Flow

Orifice channel

SMA microactuator

Gate

Figure 14.6 Schematic layout of a front gate (a) and a back gate SMA microvalve (b) for normally-open operation.

Opening of the valve is caused by heating the microactuator with an electrical current. In the open condition, the areas of lower and upper membrane being exposed to the applied pressure are identical, which results in almost complete pressure compensation. By this design concept, the maximum controllable pressure difference of the microvalve is determined by the pressure-induced force mediated by the coupled membranes. The maximum flow is again determined by the stroke of the microactuator.

Gate microvalves can be designed in different layouts. Depending on the directions of actuation and flow, front gate, back gate and side gate microvalves may be distinguished [15]. Layouts of a front gate and a back gate SMA microvalve are sketched in Fig. 14.6. So far, prototypes of such valves have been fabricated in silicon technology using thermal bimorph microactuators for valve control [15], which can be easily adapted to SMA bimorph actuators. The channel cross-sectional area determines the pressure/flow characteristics and defines the pressure force on the closed gate. For the side gate valve, the pressure force is orthogonal to the beam and induces torsional forces along the beam. The back gate and front gate valve offer a better mechanical stability, since the flow is parallel to the beam and no torsional forces along the beam are created. Any friction due to sliding of the microstructures must be avoided. Therefore, a gap between gate and orifice has to be considered. Although microfabrication allows for minimizing this gap, a certain leakage remains in closed state. The back gate valve design has the largest footprint area of the three designs, but offers the best mechanical stability.

14.4   SMA materials

The basic materials for the development of SMA microvalves are in most cases SMA thin films and foils in the thickness range 10–100 µm. SMA thin films are usually fabricated by magnetron sputtering, which is described in detail in Chapter 3. When the sputtering process is performed close to room temperature, amorphous films are obtained. In this case, a subsequent thermo mechanical treatment is necessary to adjust the one-way or two-way shape memory effect. By removing the films from the substrate prior to heat treatment, possible delamination and interdiffusion effects can be avoided. For microactuator fabrication, it is therefore recommended that one sputters the films on a substrate with a sacrificial layer or on a chemically soluble substrate.

Compared to sputtered thin films, rolled SMA foils may offer some advantages. TiNi foils, for instance, have been fabricated by cold-rolling down to a thickness of a few micrometers, offering reproducible bulk material characteristics [16].

The application range of the microvalves strongly depends on the phase transformation temperatures and hysteresis width of the used SMA materials. Most SMA microvalves developed up to now make use of sputtered TiNi(Cu) thin films or cold-rolled TiNi foils, which only allow for operation close to room temperature. Several attempts have been made to increase the temperature range by introducing high-temperature SMA thin films, such as TiNiPd and TiNiHf films [11, 12, 17, 18]. Figure 14.7 presents differential scanning calorimetry measurements of TiNiPd thin films with different Pd content, showing that the transformation temperatures can be adjusted in a wide range. In binary NiPd thin films, transformation temperatures M_f/A_f of 498/570 K are achieved [17]. More details on TiNiPd thin films can be found in Chapter 4.

Figure 14.8 shows strain–temperature characteristics of TiNiPd thin films representing shape memory behavior for a variety of constant stresses. For increasing stress, the transformation temperatures shift to higher temperatures reflecting Clausius–Clapeyron behavior with a stress rate of about 8 MPa K^{-1}. The strain ε_M is associated with the martensitic transformation upon cooling and ε_A with the reverse transformation upon heating. The strain ε_p represents plastic strain due to slip deformation induced during the preceding transformations. The critical stress for onset of plastic deformation is about 100 MPa. The maximum transformation strain ε_A is determined to be about 3.2%.

The use of high-temperature SMA thin films also offers an improvement of the valve dynamics due to reduced cooling times in the presence of increased

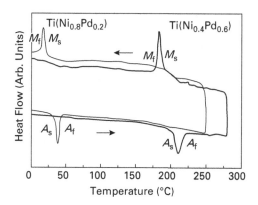

Figure 14.7 Differential scanning calorimetry measurements of sputtered TiNiPd thin films with different Pd content [17].

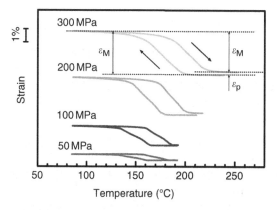

Figure 14.8 Strain–temperature characteristics of a sputtered $Ti_{51.0}Ni_{25.5}Pd_{23.5}$ thin film for various stress loads [12]; ε_M and ε_A denote the strain of martensitic and reverse transformation; ε_p is the plastic strain.

temperature gradients, see Section 14.5. This effect, however, is partly compensated by the increased hysteresis width in these materials. In addition, increased temperature gradients may result in increased regions of incomplete phase transformation, e.g. due to conductive cooling at interconnections, which constricts the actuation performance.

14.5 Modeling and simulation aspects

The material behavior of SMAs depends on stress and temperature and is intimately connected with the crystallographic material properties and the thermodynamics of the phase transformation. A variety of constitutive models have been developed, which may be classified into the categories of micromechanics-based

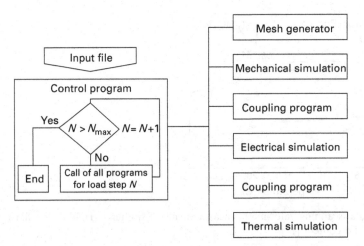

Figure 14.9   Procedure for coupled FEM simulation [32].

models [19, 20], microscopic thermodynamic models [21, 22] and macroscopic phenomenological models [23, 24, 25, 26, 27]. In recent years, a few models for SMA materials have been implemented in a finite element method (FEM) program [26, 28, 29, 30, 31]. For engineering purposes, FEM simulation based on a macroscopic phenomenological model already provides many insights, even though the predictive power remains limited.

Understanding the performance of electrically driven SMA actuators requires self-consistent simulation of mechanical, thermal and electrical fields in order to take electrothermomechanical coupling effects into account. For this purpose, an iterative simulation procedure has been developed, which links different FEM simulation routines by a superior control program as illustrated in Fig. 14.9 [32]. Between the simulation steps coupling programs are executed, which are responsible for the exchange of material data, geometry data and boundary conditions. The execution may be run through in an arbitrary sequence to obtain self-consistent results. For time-dependent calculations, an individual time is attributed to each sequence.

A two-phase macromodel may be used to describe the thermo-elastic properties of SMA materials displaying a single-stage martensitic or rhombohedral (R-) phase transformation. One class of macroscopic model, so-called Tanaka-type models, couple a phenomenological macro-scale constitutive law for the relation between stress, strain, temperature and martensitic phase fraction ξ with a kinetic law describing the evolution of ξ as a function of stress and temperature [23]. In this approach, both the martensitic transformation and the reorientation of martensite variants are described by a single internal variable ξ. More sophisticated models distinguish between the two processes by introducing two internal variables [25].

Figures 14.10(a) and (b) show, for instance, simulated distributions of R-phase fraction of a TiNi bridge microactuator of a polymer seat valve in stationary

(a)   (b)

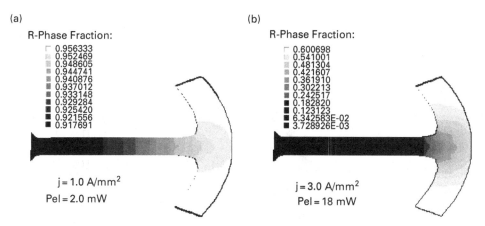

Figure 14.10 Simulation of distributions of R-phase fraction in a microbridge of TiNi in stationary equilibrium for a heating power of 2 mW (a) and 18 mW (b).

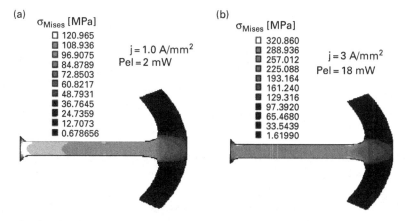

Figure 14.11 Simulated von Mises stress profiles corresponding to Figs. 14.10(a) and (b), respectively [31].

equilibrium. Details of the simulation procedure and parameters are given in [31]. Due to symmetry, only one segment of the microbridge array is considered. The austenite–R-phase transformation in the microbridge occurs in the range of heating power between 2 and 18 mW. At 18 mW, about 95% of the microbridge has transformed to austenite. In this power range, a small transition region between microbridges and conduction pads remains, showing a partial phase transformation.

Corresponding simulations of von Mises stress distributions are shown in Figs. 14.11(a) and (b). For a heating power of 2 mW, a maximum von Mises stress of about 120 MPa occurs close to the end of the SMA microbridge, which is strained by the spacer from below. In this case, most parts of the microbridge are in the R-phase condition. At 18 mW heating power, the phase transformation in

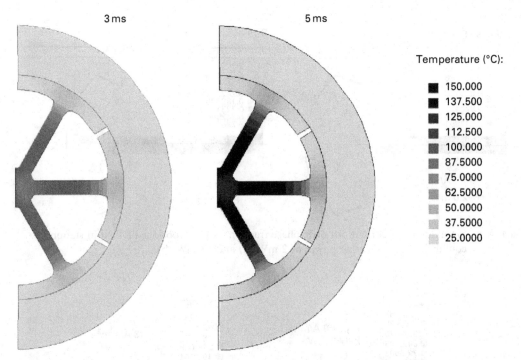

Figure 14.12   Simulated temperatures profiles of a Ni₂MnGa microactuator and a section of the heat sink for different heating times as indicated. The heating power is 760 mW [13].

the microbridge to austentite is almost complete, which is associated with an almost homogeneous stress distribution along the microbridge. This condition is very favorable for actuation, as it allows for an optimum use of the shape memory effect. Furthermore, fatigue due to stress peaks is avoided. The maximum von Mises stress reaches 320 MPa.

SMA microactuators are often driven by direct electrical heating, which causes a certain spatial distribution of heating power. In a first electrical simulation step, the distribution of the current density is determined by solving the Laplace equation for the electrical potential taking into account the Neumann boundary condition at the electrical connections [31]. Then the heating power is calculated from current density and electrical conductivity by volume integration. Subsequently, the temperature distribution can be determined by solving the heat transfer equation taking into account the sensible heat, the heat of phase transformation, the heating power generated by electrical current and losses due to heat conduction. At the actuator surface, the convective heat exchange with the environment has to be considered as well [31].

Figure 14.12 shows, for example, two simulated temperature profiles along the top surface of a Ni₂MnGa bridge microactuator and a section of the heat sink of a polymer seat valve for different heating times in stress-free conditions. The phase transformation temperatures of the microactuator M_f/A_f have been

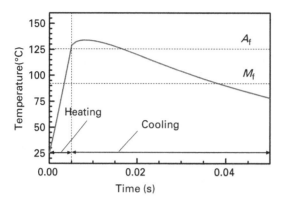

Figure 14.13 Heating and cooling characteristic in the center of the Ni_2MnGa microactuator corresponding to Fig. 14.12 [13].

92/125 °C. Details of the simulation procedure are reported in [13]. Due to symmetry, only one half of the system is taken into account. The corresponding heating and subsequent cooling characteristic in the centre of the microactuator are shown in Fig. 14.13. The cooling characteristic is obtained by switching the heating power off after 5 ms. The temperature profile along the surface of the microbridges is almost homogeneous except in the transition regions close to the bond pads of the SMA microactuator. After a heating time of about 5 ms, the actuator center reaches the A_f temperature. The transition from austenitic to martensitic region occurs at a distance of about 200 µm from the bond pads. This corresponds to about 85 % of the SMA microbridges having transformed to austenite. Thus, within 5 ms heating time, most parts of the microbridges contribute to the closing of the valve, while cooling effects have no significant influence.

The cooling performance is obtained by assuming an effective heat-transfer coefficient of 70 W m^{-2} K^{-1} for convective cooling and a thermal conductivity of 22 W m^{-1}K^{-1}. In this case, the actuator center and thus the whole microactuator are cooled below the M_f temperature within 33 ms after heating. TiNi microactuators show a similar performance. However, the heating power required for a heating time of 5 ms is about a factor of three smaller and the cooling time is roughly three times longer due to the smaller difference in phase transformation temperature and room temperature.

The simulated heat transfer times have been verified by time-resolved gas flow experiments [13]. Figure 14.14 shows a time-resolved gas flow characteristic of a TiNi microvalve at a pressure difference of 250 kPa. The valve is operated by applying periodic heating pulses. When no extra heating power is supplied after closing the valve, a minimum opening time of about 60 ms is observed, allowing a maximum operation frequency of 15 Hz. A similar characteristic is obtained for the Ni_2MnGa microvalve as shown in Fig. 14.15. Due to the higher phase transformation temperatures, the opening time is improved by a factor of about

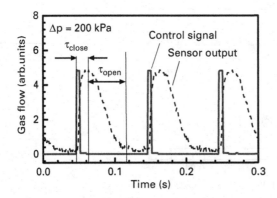

Figure 14.14 Time-resolved gas flow during closing and opening of a TiNi microvalve [13].

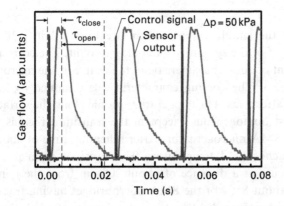

Figure 14.15 Time-resolved gas flow during closing and opening of a Ni₂MnGa microvalve [13].

three, which agrees very well with the simulated time constants for cooling. In this case, the maximum operation frequency is about 40 Hz.

14.6 Fabrication

For microvalve fabrication, a number of technologies have been developed, which can be classified into silicon- and polymer-based technologies. Silicon-based technologies stem from the electronics industry making use of monolithic integration processes in a batch fabrication manner. However, monolithic integration and thermo-mechanical treatment of SMA thin films are hardly compatible with each other. Therefore, novel transfer bonding technologies for integration of SMA microparts in a silicon or polymer microsystem have been developed recently [33, 34].

Transfer bonding relies on bonding and release technologies, which can be performed on the wafer scale. This procedure allows the performance of critical

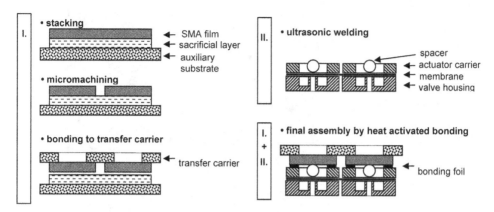

Figure 14.16 Process flow for transfer bonding movable SMA microparts from an auxiliary substrate to a final polymer substrate on the wafer level [34].

fabrication steps on separate optimized substrates, eventually using additional intermediate layers, or even avoiding any substrate, e.g., in the case of heat treatment of SMA films. When both the SMA microparts to be transferred and the receiving bonding site have similar size, full wafer-level integration can be cost-effective. In the case of transferring tiny devices, i.e. SMA micro-actuators, that are much smaller than the receiving bonding site, a lot of unused space and SMA material may be lost on the wafer carrying the devices to be transferred. In this case, a selective transfer technique appears to be an adequate solution [33].

A typical process flow of transfer bonding on a full wafer level is sketched in Fig. 14.16. After sputtering, the SMA thin film is released from the substrate to allow for heat treatment in a free-standing condition. In the subsequent stacking process, the thin film is integrated on an auxiliary wafer to perform the micromachining step by optical lithography and wet-chemical etching. Then, a micromachined carrier wafer is bonded onto the auxiliary wafer. After selective removal of the sacrificial layer, freely suspended SMA microparts are formed. In parallel, the final polymer substrate is micromachined, e.g., by hot embossing to form the valve housings. Actuator carrier and membrane are integrated using ultrasonic welding in this case. Finally, the composite of transfer carrier and micromachined SMA thin film is lifted off the auxiliary wafer and transfer bonded onto the final polymer substrate. This technology is particularly compatible to final polymer substrates of high temperature-sensitivity, since critical steps of thermal annealing and chemical micromachining are performed separately.

Intermediate fabrication steps of this process are documented in Fig. 14.17. In this case, a cold-rolled TiNi foil of 20 μm thickness is used as the active material. The transfer of the TiNi microactuators between etching substrate and actuator carrier is performed by a transfer carrier of a micromachined Ni foil. Final microvalves are obtained after transfer bonding by integration of electrical contacts and fluidic interfaces.

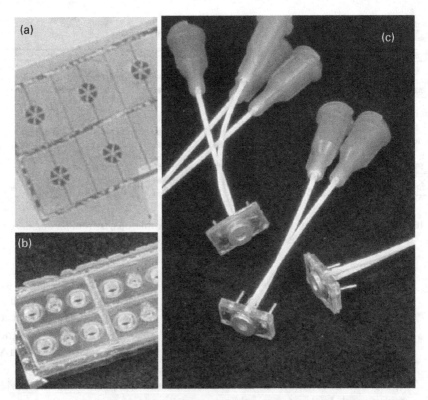

Figure 14.17 Intermediate fabrication steps for SMA polymer microvalves: (a) TiNi microactuators after chemical micromachining, (b) stack of valve housing and actuator carrier, (c) final SMA microvalves including electrical contacts and fluidic interfaces [34].

14.7   Performance characteristics

14.7.1   Polymer seat microvalves

Most SMA microvalves have been developed for on/off operation, which avoids control problems due to non-linearity and hysteresis. Figure 14.18 shows, for example, typical flow characteristics of a normally-open microvalve in a stationary condition using nitrogen gas [13]. The flow is controlled by a TiNi bridge microactuator. The electrical resistance characteristics of the bridge microactuator demonstrate the simultaneous progression of flow change and phase transformation in the material. The microactuator is fabricated from a TiNi foil of 20 μm thickness produced by cold rolling. The final heat treatment is performed at 723 K for 30 min to adjust the one-way shape memory behavior. Upon cooling, the material shows a two-stage phase transformation from austenite via R-phase to martensite. However, only the R-phase transformation occurs above room temperature. Thus, electrical actuation allows the selective use of the R-phase transformation, which has some advantages in terms of narrow hysteresis and negligible fatigue.

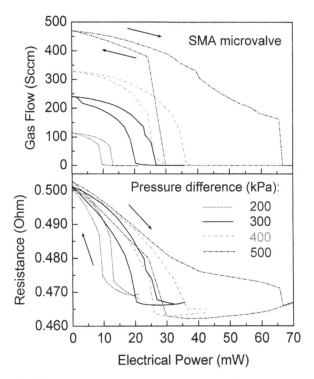

Figure 14.18 Gas flow and electrical resistance characteristics of a normally-open TiNi microvalve in a stationary condition [13].

At zero power, the maximum flow is observed depending on the pressure difference. For increasing heating power, the flow decreases until a critical heating power is reached, above which the valve is closed. The hysteresis widths only reflect the narrow hysteresis of the R-phase transformation at low pressure differences. For increasing pressure differences, increased cooling as well as stress-induced martensite formation affect the observed characteristics.

Figure 14.19 shows gas flow and electrical resistance characteristics of a normally-closed TiNi microvalve in a stationary condition [14]. In this case, the pressure compensation mechanism illustrated in Fig. 14.5 is used to reset the microvalve and to maintain the closed condition at zero power. In the investigated pressure range, the closed condition is maintained for all pressure differences. The resistance characteristics demonstrate again that the observed change of gas flow is caused by the R-phase transformation. The critical heating power required to open the valve is about 20 mW. However, full opening requires a considerably higher heating power, since conductive cooling of the microactuator hampers complete phase transformation close to the onset of the microbridges. From the observed change in flow rate of 330 standard ccm for 250 kPa, the corresponding actuation stroke is estimated to be about 20 μm.

Some attempts have been made to implement TiNiPd thin films in seat microvalves for high temperature applications. Figure 14.20 shows, for example, gas flow

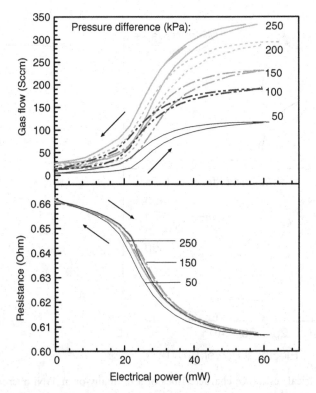

Figure 14.19   Gas flow and electrical resistance characteristics of a normally-closed TiNi microvalve in a stationary condition [14].

and electrical resistance characteristics of a normally-open TiNiPd microvalve in a stationary condition [12]. The flow characteristics of the microvalve have been determined at various ambient temperatures between room temperature and 120°C in a thermostat. The resistance characteristics indicate that the change of gas flow is caused by martensitic phase transformation. Due to the high transformation temperatures of the TiNiPd thin films, M_f/A_f of 112°C/171°C, complete actuation cycles are achieved for ambient temperatures up to 110°C. For higher temperatures, partial phase transformations give rise to reduced actuation strokes and flows. At 120°C, the critical heating power required to close the valve is about 80 mW. For decreasing temperature, the critical heating power increases, reaching about 260 mW at room temperature.

Figure 14.21 shows the corresponding time-resolved gas flow characteristics for a pressure difference of 30 kPa at various ambient temperatures between room temperature and 120°C. In this case, the microvalve is operated by applying periodic heating pulses of 1 ms duration. The heating power is adjusted to close the microvalve within about 10 ms. If no extra heating power is supplied after closing the valve, a minimum opening time of 22 ms is observed at 25°C. For increasing ambient temperature, the opening time increases to 75 ms at 120°C. Thus, the microvalves can be operated at maximum frequencies between 10 and 40 Hz depending on the ambient

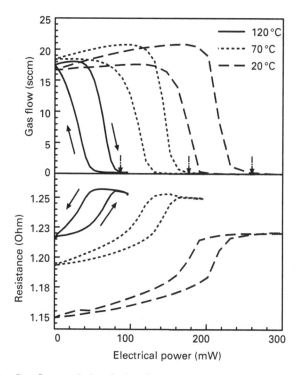

Figure 14.20 Gas flow and electrical resistance characteristics of a normally-open TiNiPd microvalve in a stationary condition [12].

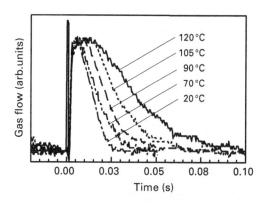

Figure 14.21 Time-resolved gas flow characteristics of the normally-open TiNiPd microvalve shown in Fig. 14.20 for a pressure difference of 30 kPa at various ambient temperatures [12].

temperature. Compared to TiNi microvalves of similar layout, an improvement in cooling time by a factor of about 3.5 is achieved at room temperature.

The performance tests have mainly been made with nitrogen gas. In fact, pneumatic applications like gas chromatography or IP converters for pneumatic control, which convert a current signal (current I) into a pneumatic signal (pressure P), are considered to be promising targets for the development of SMA microvalves. Nevertheless, control of liquids is also possible even though some design adaptations may be necessary [16]. Currently, emerging application fields like bioanalytics or fuel cells requiring precise liquid handling systems appear to be another target for SMA microvalves.

14.7.2   Microfluidic controller

Many applications in microfluidics demand flow regulation of microvalves instead of on/off operation. The driving signal of the SMA actuator may be controlled, e.g., by a sensor for SMA actuator deflection or for fluid flow. The use of an intrinsic sensor principle, such as the electrical resistance of the active SMA parts, appears to be very attractive, since this would take advantage of the multi-functional properties of the SMA material itself. However, this approach is rather limited due to the lack of understanding of the relationship between electrical and mechanical hysteresis in arbitrarily shaped microparts. Therefore, in general, an extrinsic deflection or flow sensor is used, which can be quite easily implemented, but requires additional space.

Figure 14.22 shows a prototype of a microfluidic controller consisting of a TiNi microvalve, a flow sensor and integrated fluidic channels connecting

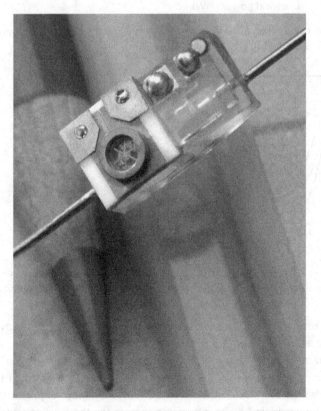

Figure 14.22 Prototype of a microfluidic controller consisting of a TiNi microvalve and a flow sensor [35].

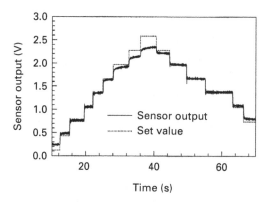

Figure 14.23 Time-resolved flow characteristic of the microfluidic controller shown in Fig. 14.22 operated in closed-loop mode for various set values. The pressure difference is 230 kPa [35].

both components [35]. The flow sensor makes use of the thermal anemometric principle. It consists of a freely suspended polyimide membrane with integrated heater in its neutral axis. The fluidic channels are designed to be as small as possible without limiting the maximum flow of the valve orifice. In combination with the integrated design of microvalve and flow sensor, small dead volumes are achieved.

For adjustment of arbitrary flow levels, the microfluidic controller has been operated in closed-loop mode. Figure 14.23 shows the response of the modular fluidic controller to an arbitrary sequence of set values. The proportional-integral- (PI-) control parameters have been optimized for maximum control velocity and minimum tolerance for increasing set values. For set values close to the maximum and minimum flow, a rather large tolerance is observed. Best performance is obtained in the range between 25 and 70% of maximum flow. In this range, a control tolerance of 1% and a time constant of about 70 ms are found.

14.8   Summary

Various SMA microvalves have been realized up to now, based on different design–material–technology combinations. In most cases, these microvalves make use of TiNi-based microactuators, which are fabricated either by magnetron sputtering or by cold rolling. Sophisticated design tools have been developed allowing the development of microvalves with optimized mechanical and electro-thermal behavior. Thus, small high-performance microvalves have been realized allowing the control of relatively large pressure differences and flows at moderate frequencies of the order of 10 Hz. By the use of TiNiPd and Ni_2MnGa thin film actuators with transformation temperatures of the order of 100 °C, the operation frequency range has been extended to about 40 Hz.

Despite these achievements, SMA microvalves and SMA microactuators in general have received only very limited acceptance in the industrial market. The reasons are partly due to the limitations of available SMA materials and the relatively high costs of SMA microtechnology. Recent progress in transfer bonding technologies allowing for batch fabrication of SMA microactuators and devices on the wafer scale will reduce the technological barriers. Further limitations exist due to restricted availability of suitable design tools. In order to comply with the needs of specific application fields, future research and development will have to concentrate further on systems solutions and corresponding technology combinations.

References

[1] S. C. Terry, J. H. Jerman and J. B. Angell, A gas chromatographic air analyzer fabricated on a silicon wafer, *IEEE Trans. Electron Devices* **26** (1979) 1880–1886.
[2] K. W. Oh and C. H. Ahn, A review of microvalves, *J. Micromech. Microeng.* **16** (2006) R13–R39.
[3] B. Winzek, S. Schmitz, H. Rumpf, T. Sterzl and E. Quandt, SMA-thin film composites providing traveling waves, *Proc. Int. Conf. on Smart Structures and Materials 2003: Active Materials: Behaviour and Mechanics*, SPIE Vol. **5053** (2003), pp. 110–118.
[4] C. Goll, W. Bacher, B. Bustgens, *et al.*, Microvalves with bistable buckled polymer diaphragms, *J. Micromech. Microeng.* **6** (1996) 77–79.
[5] M. Capanu, J. G. Boyd and P. J. Hesketh, Design, fabrication, and testing of a bistable electromagnetically actuated microvalve, *J. Microelectromechanical Systems* **9**, No. 2 (2000) 181–189.
[6] S. Braun, T. Grund, S. Ingvarsdottir, W. van der Wijngaart, M. Kohl and G. Stemme, Robust trimorph bulk SMA microactuators for batch manufacturing and integration, *Proc. Transducers 07*, Lyon, France, (2007), pp. 2191–2194.
[7] M. Kohl, K. D. Skrobanek and S. Miyazaki, Development of stress-optimised shape memory microvalves, *Sensors and Actuators* **A72** (1999), pp. 243–250.
[8] M. Kohl, J. Gobes and B. Krevet, Normally-closed shape memory microvalve, *Int. J. Appl. Electromagnetics and Mechanics* **11** (2000) 71–77.
[9] C. A. Ray, C. L. Sloan, A. D. Johnson, J. D. Busch and B. R. Petty, A silicon-based shape memory alloy microvalve, *Mat. Res. Soc. Symp. Proc.* Vol. **276** (1992), pp. 161–166.
[10] M. Kohl, I. Hurst and B. Krevet, Time response of shape memory microvalves, in *Proc. Actuator 2000, Bremen, Germany*, ed. H. Borgmann, (2000), pp. 212–215.
[11] M. Kohl, D. Dittmann, E. Quandt and B. Winzek, Thin film shape memory microvalves with adjustable operation temperature, *Sensors and Actuators* **A83** (2000) 214–219.
[12] Y. Liu, M. Kohl, K. Okutsu and S. Miyazaki, A TiNiPd thin film microvalve for high temperature applications, *Mater. Sci. Eng.* **378** (2004) 205–209.
[13] M. Kohl, Y. Liu, B. Krevet, S. Durr and M. Ohtsuka, SMA microactuators for microvalve applications, *J. Phys. IV France* **115** (2004) 333–342.
[14] M. Kohl, M. Popp and B. Krevet, Shape memory micromechanisms for microvalve applications, *Proc. Int. Conf. on Smart Structures and Materials*, San Diego, California, USA, ed. Dimitris C. Lagoudas, SPIE Vol. 5387, (2004), pp. 106–117.
[15] S. Braun, S. Haasl, S. Sadoon, A. S. Ridgeway, W. van der Wiijngaart and G. Stemme, Small footprint knife gate microvalves for large flow control, *Proc. Transducers '05*, Seoul, Korea (2005), pp. 329–332.
[16] M. Kohl, *Shape Memory Microactuators*, Springer series on Microtechnology and MEMS, Berlin and Heidelberg: Springer-Verlag, (2004).

[17] E. Quandt, C. Halene, H. Holleck, *et al.*, Sputter deposition of TiNi, TiNiPd and TiPd films displaying the two-way shape memory effect, *Sensors and Actuators* **A 53** (1996), 434–439.

[18] T. Sterzl, B. Winzek, J. Rumpf and E. Quandt, Bistable shape memory composites for switches, grippers and adjustable capacitors, *Proc. Actuator 2002*, Bremen, Germany, ed. H. Borgmann, (2002), pp. 91–94.

[19] E. Patoor, A. Eberhardt and M. Berveiller, Micromechanical modelling of the shape memory behavior, *Proceedings of the Symposium on Phase Transformation and Shape Memory Alloys, AMD*, Vol. 189, (1994) pp. 23–37.

[20] X. Gao, M. Huang and L. C. Brinson, A multivariant micromechanical model for SMAs, *Int. J. Plast.* **16** (2000) 1345–1390.

[21] M. Achenbach and I. Muller, Simulation of material behaviour of alloys with shape memory, *Arch. Mech.* **37**, 6 (1985) 573–585.

[22] Y. Huo and I. Mueller, Nonequilibrium thermodynamics of pseudoelasticity, *Continuum Mechanics and Thermodynamics* **5** (1993) 163–204.

[23] K. Tanaka, S. Kobayashi and Y. Sato, Thermomechanics of transformation pseudoelasticity and shape memory effect in alloys, *Int. J. Plast.* **2** (1986) 59–72.

[24] C. Liang and C. A. Rogers, One-dimensional thermomechanical constitutive relations for shape memory materials, *J. Intell. Mat. Syst. & Struct.*, **1** (1990) 207–234.

[25] L. C. Brinson, One-dimensional constitutive behaviour of shape memory alloys: thermomechanical derivation with non-constant material functions and redefined martensite internal variable, *J. Intell. Mat. Syst. & Struct.* **4**(2) (1993) 229–242.

[26] K. Ikuta and H. Shimizu, Two-dimensional mathematical model of shape memory alloy and intelligent SMA-CAD, *Proc. MEMS 93*, Fort Lauderdale, USA, IEEE Catalog No. 0–7803–0957–2/93 (1993) pp. 87–92.

[27] J. G. Boyd and D. C. Lagoudas, Thermomechanical response of shape memory composites, *J. Intell. Mat. Syst. & Struct.*, **5** (1994) 333–346.

[28] L. C. Brinson and R. Lammering, Finite element analysis of the behavior of shape memory alloys and their applications, *Int. J. Solids & Struct.*, **30** (1993) 3261–3280.

[29] F. Auricchio and R. L. Taylor, Shape memory alloy superelastic behaviour: 3D finite element simulations, *Proc. ICIM 96*, Lyon, France, SPIE Vol. 2779, (1996), pp. 487–492.

[30] S. Leclercq, C. Lexcellent and J. C. Gelin, A finite element calculation for the design of devices made of shape memory alloys, *Journal de Physique IV*, Colloque C1 **6**, (1996) 225–234.

[31] M. Kohl and B. Krevet, 3D Simulation of a shape memory microactuator, *Materials Transactions* **43**, No. 5 (2002) 1030–1036.

[32] B. Krevet and W. Kaboth, Coupling of FEM programs for simulation of complex systems, *Proc. MSM 98*, Santa Clara, USA, (1998), pp. 320–324.

[33] T. Grund, R. Guerre, M. Despont and M. Kohl, Transfer bonding technology for batch fabrication of SMA microactuators, *European Physics Journal – Special Topics*, **158** (2008) 237–242.

[34] T. Grund, T. Cuntz and M. Kohl, Batch fabrication of polymer microsystems with shape memory microactuators, *Proc. MEMS 08*, Tucson, Arizona, USA, (2008), pp. 423–426.

[35] M. Kohl, Y. Liu and D. Dittmann, A polymer-based microfluidic controller, *Proc. MEMS 04*, Maastricht, The Netherlands, IEEE Catalog Number: 04CH37517, (2004), pp. 288–291.

15 Superelastic thin films and applications for medical devices

Christiane Zamponi, Rodrigo Lima De Miranda and Eckhard Quandt

Abstract
Superelastic shape memory materials are of special interest in medical applications due to the large obtainable strains, the constant stress level and their biocompatibility. TiNi sputtered tubes have high potential for application as vascular implants, e.g. stents, whereas superelastic TiNi-polymer-composites could be used for novel applications in orthodontics and medical instrumentation as well as in certain areas of mechanical engineering. In orthodontic applications, lowering the forces which are applied on the teeth during archwire treatment is of special importance due to tooth root resorption, caused by the application of oversized forces. Furthermore, the use of superelastic materials or composites enables the application of constant forces independent of diminutive tooth movements during the therapy due to the superelastic plateau. Superelastic TiNi thin films have been fabricated by magnetron sputtering using extremely pure cast melted targets. Special heat treatments were performed for the adjustment of the superelastic properties and the transformation temperatures. A superelastic strain exceeding 6% at 36 °C was obtained.

15.1 Introduction

TiNi based shape memory effects are related to a reversible phase transformation between a high temperature phase (austenite, B2) and a low temperature phase (martensite, B19′) showing a different crystal structure. Phase transformation can be achieved by temperature (one-way or two-way effect) and/or stress (superelasticity). In Ni-rich TiNi alloys, an intermediate phase can appear after a heat treatment below 500 °C, this phase is called the R phase (rhombohedral) and the existence of this R phase supports the formation of stress induced martensite. Superelastic shape memory materials are of special interest in medical applications

Thin Film Shape Memory Alloys: Fundamentals and Device Applications, eds. Shuichi Miyazaki, Yong Qing Fu and Wei Min Huang. Published by Cambridge University Press. © Cambridge University Press 2009.

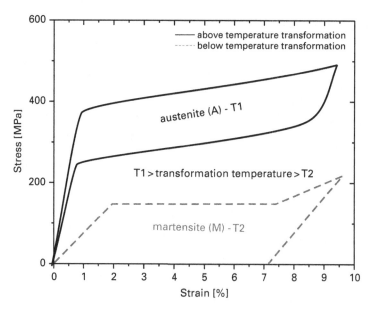

Figure 15.1 Typical stress–strain curve for the austenite and the martensite in TiNi above and below the transformation temperature.

due to the large obtainable strains (up to 8 %), the constant stress level and their biocompatibility. A typical stress–strain curve for the austenite and the martensite in TiNi is shown in the Fig. 15.1.

A commonly used method to determine transformation temperatures is the differential scanning calorimeter (DSC), which uses small and precise ovens to cool and heat the sample at a given rate. Figure 15.2 shows an example of a DSC curve. The upper line indicates the heat change in the specimen on cooling and the lower line that on heating. The phase transition in thin films is endothermic upon heating and exothermic upon cooling. Exothermic reactions are presented as downward pointing peaks and endothermic reactions as upward pointing peaks [1]. The transition from austenite to R phase and R phase to martensite transitions is exothermic, while the transformation from martensite to austenite is endothermic. The superelastic martensitic transformation in TiNi alloys is characterized by a closed loop hysteresis in the stress–strain diagram showing an almost flat loading and unloading plateau after an elastic deformation of approximately 1% strain. This results from a stress induced phase transformation from the austenite (B2) into the martensite phase (B19′). Upon reaching the temperature-dependent transformation stress the material is transformed into the martensite phase, which allows an elongation of the sample length at almost constant stress due to the reorientation of the martensitic variants. Upon removing the external stress the superelastic shape memory material returns with a stress hysteresis into the austenite phase and the elongated specimen returns to its previous length [2]. Thin film techniques are applied for the fabrication of stents for neurovascular blood vessels [3], as well as in

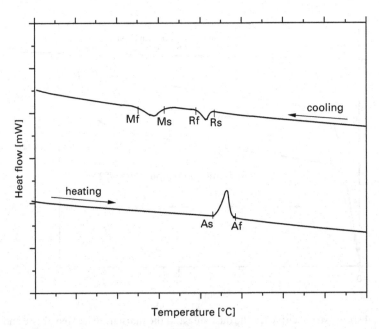

Figure 15.2 Typical DSC curve. The diagram shows the transformation temperatures of the austenite, R phase and martensite phase. The start of the peak indicates the beginning of the transformation (start s). At the end of the peak the whole sample is transformed (finish f).

the production of membrane based micropumps [4]. Further applications focus on the fabrication of superelastic orthodontic archwires [5], medical instrumentation and in certain areas of mechanical engineering.

15.2   Superelasticity in thin films

TiNi thin films were deposited at 450 °C in order to obtain a crystalline state on silicon substrates using a magnetron sputtering cluster machine. The base vacuum was about 1×10^{-7} mbar. Due to the different sputtering rates for nickel and titanium the film composition deviates from the composition of the target. This deviation of the stoichiometry was determined by energy dispersive X-ray microanalysis (EDX). The film thickness was determined by a surface profiler. Freestanding films were obtained by mechanical release of the TiNi films from the Si substrates. Tensile testing was performed in a universal tensile test machine with the sample being located within a temperature chamber (Fig. 15.3).The superelastic plateau during loading and unloading occurred at different stresses (σ_{trans}) depending on the sample temperature (T) [6], as presented in Fig. 15.4. Freestanding TiNi films (with thickness of 21 μm) showed a typical martensitic behavior at −10 °C (Fig.15.4 (a)). A strain of $\varepsilon = 5\%$ was achieved after application of 220 MPa stress which led to a permanent plastic deformation of 3.5%

Figure 15.3 Universal tensile test machine (Messphysik UTM Beta EDC 100N).

strain. Upon heating, the plastic deformation recovered and superelastic behavior appeared. Figures 15.4 (b), (c) and (d) show typical superelastic curves repeatedly exceeding an elastic strain of 6.5% at body temperature as well as at 57 °C, whereas no stress induced martensitic transformation is achieved when the temperature exceeds A_D (austenite deformation). The stress–strain diagram of the pure austenitic state at 120 °C is presented in Fig. 15.4(d), which shows a typical result of an ultimate tensile strength of 960 MPa. The same experiment carried out below A_D is shown in Fig. 15.4(e), indicating superelastic strain and an ultimate tensile strength of 1180 MPa at a maximum strain of 11.5%. The temperature dependence for reaching the stress-induced martensitic transformation (start of plateau) is summarized in Fig. 15.5. It can be found that the Clausius–Clapeyron equation

$$\frac{d\sigma_{trans}}{dT} = \rho \frac{\Delta H}{\varepsilon T} \qquad (15.1)$$

with

ΔH: transformation enthalpy,
ε: imposed strain,
σ: uniaxial stress,
T: temperature,
ρ: density,

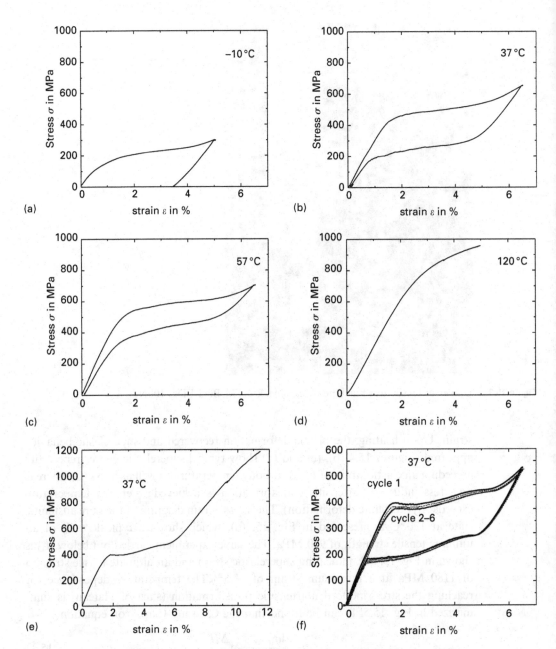

Figure 15.4   Stress–strain diagrams for tensile testing experiments of sputtered thin films at different temperatures (a)–(d). At 37 °C, the ultimate tensile strength in the superelastic state was determined (e) and a cycling dependence of the first six cycles was observed (f).

is fulfilled, leading to a linear relationship between plateau stress for the tensile test in the austenite state above A_D (120 °C) and at room temperature. The coefficient $d\sigma/dT$ is 7.0 ± 0.1 MPa/K, which is in the lower range of typical values known from the literature for bulk material (4–20 MPa/K), [7, 8].

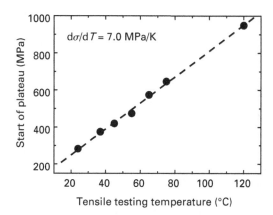

Figure 15.5   The onset of the superelastic plateau versus the tensile testing temperature. The value of the slope determines the coefficient ds/dT of the Clausius–Clapeyron equation.

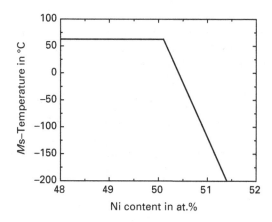

Figure 15.6   Martensite start temperature Ms depending on the Ni content [9].

15.3   Fabrication of planar superelastic thin films

In order to obtain superelastic TiNi films it is most important to control the composition of the sputtered film (Fig. 15.6) [9]. For superelastic properties at 37 °C – for medical implant applications – a Ni-rich stoichiometry of Ti49.2Ni50.8at% is required. The martensitic transformation temperature for this composition is approximately −42 °C as shown in the DSC related graphic (Fig. 15.7). In the case of magnetron sputtering – the most prominent deposition technique – the stoichiometry of the film is determined by the target stoichiometry and quality, by the geometry of the deposition process, by the vacuum conditions and by the purity of the sputtering gas supply. In general, it is most important to control the

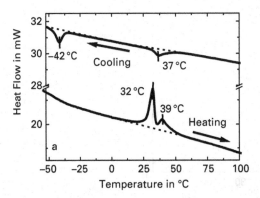

Figure 15.7   DSC analysis of freestanding Ni-rich TiNi films fabricated by magnetron sputtering. The diagram shows the transformation of TiNi from martensite to the R phase and to austenite during heating. The cooling curve shows the transformation from austenite to R phase and to martensite.

oxygen and carbon impurities since they result in a significant change of the phase transformation temperatures [10] leading to a lower ductility and an increased hysteresis upon loading/unloading [11]. Comparing the composition in the film with respect to the target composition, sputtering of TiNi is generally associated with a characteristic loss of titanium. This loss occurs mainly due to the difference of the sputter angular yields of the related materials. In different approaches the titanium deficiency is compensated for either by placing additional titanium on top of the alloyed target [12] or preferentially by using Ti-rich (54 at% Ti) alloy targets [13].

Thornton [14] introduced a structure zone model (SZM), which describes the microstructure of magnetron sputtering films based on the relation between substrate temperature and the melting point of the sputtered material (T_s/T_m) and also the variation of the inert sputtering gas pressure. This model can be used to obtain the parameters for the sputtering process of TiNi.

Freestanding thin films are obtained by sacrificial layer methods using chemical etching technology. Superelastic properties are observed in these freestanding crystalline TiNi films. These films can be obtained using two crystallization processes (*in situ* and *ex situ*). For the *in situ* process, the substrate is heated to the crystallisation temperature of nickel-titanium (450°C) during the sputtering process [15]. For the *ex situ* process, TiNi is sputtered at RT and afterwards crystallized by means of a rapid thermal annealing (RTA) system. In order to avoid oxygen contamination the thin films have to be crystallized in a high vacuum environment. The halogen lamp driven heating chamber enables typical heating ramps of 50 K/s in a vacuum environment of about 10^{-6}–10^{-7} mbar. Typical heat treatments for sputtered amorphous nickel-rich films consist of two steps. The first one takes place at a higher temperature to crystallize the film while the second one, at a lower temperature, is chosen to induce the formation of Ti_3Ni_4 precipitates in order

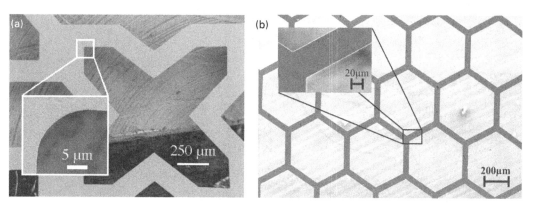

Figure 15.8　Scanning electron micrograph of TiNi net-shaped structures: (a) 5 µm wet etching and (b) 2 µm dry etching.

to adjust the transformation temperatures of the martensite to austenite phase transformation.

15.4　Patterning of planar films using lithography and etching

Patterning of planar films involves mainly two technologies: lithography and etching. Lithography is based on the transfer of a desired pattern to a photoresist. This photoresist can be coated on the substrate through the use of spin, dip and spray coating. The selection of a suitable resist will be directly related within the desired lithography resolution, and the related coating technique. For an enhanced resolution, electron, beam lithography (EBL) might be used and can deliver feature sizes smaller than 100 nm [16], using high energy electrons.

There are two different methods to structure a planar film, wet-chemical and dry etching. A wet-chemical process (isotropic) is mainly defined by dissolving a solid material in a liquid solution. For TiNi etching a hydrofluoric acid (HF) based solution is used at room temperature. In the case of dry etching (anisotropic) a solid material is transformed into a gas phase. Dry etching is mainly subdivided in to sputter etching, ion beam etching and plasma chemical etching. The sputter etching process uses the mechanical momentum of fast ions or neutral particles to knock out atoms or clusters from the solid into the gas phase. Ion beam etching follows the same physical fundamental of sputter etching differing only by the separation between ion generation and etching. Plasma chemical etching uses reactive or thermal particles in order to etch the desired layer.

Comparing both methods regarding resolution, etching rate and cost, it is easy to conclude that dry etching is a stable process with higher resolution due to its anisotropic behavior. The main disadvantage of this process is the low etching rate and higher system costs. Figure 15.8 shows the comparison of TiNi net-shaped structures prepared by photolithography and two different etching

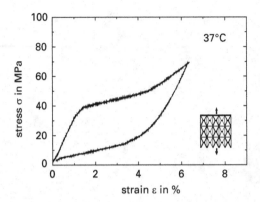

Figure 15.9   Tensile testing experiments on photo-etched structure at 37 °C, showing superelastic properties.

processes. Figure 15.8(a) corresponds to wet etching and Fig. 15.8(b) to the dry etching process [17].

A tensile test is performed in order to investigate the influence of the etching process on the properties of nickel-titanium, as shown in Fig. 15.9. The super-elastic properties of the structured films can still be demonstrated and the behavior of the films on tensile tests is described by the Clausius–Clapeyron relationship (Eq. 15.1). These net-shaped structures seem to be potential candidates for medical applications, e.g. embolic filters.

15.5   Fabrication of superelastic thin film tubes

In orthodontic applications, lowering the forces which are applied on the teeth during archwire treatment is of special importance due to tooth root resorption, caused by the application of oversized forces. Furthermore, the use of superelastic materials or composites enables the application of constant forces independent of diminutive tooth movements during the therapy due to the super-elastic plateau [5].

In a bending experiment a bulk material core region, which is called the neutral fiber, is not under any forces. Using a tubular material the stiffness can be decreased and thus the necessary forces to bend a wire or tube are reduced.

The sputtered TiNi tubes are filled with a polyamide wire in order to prevent buckling of the thin walled devices (Fig. 15.10) [5]. The Young's modulus of the soft core material is negligible compared to the TiNi Young's modulus ($E_{aus} = 40$ GPa). Bending experiments confirm the influence of the core material in the film device (a TiNi rectangular tube of dimension 400 µm × 560 µm). The composite generates a very low torque of about 1.0 N mm on the loading branch and 0.3 N mm on the unloading branch, while a solid wire of the same dimensions would generate a bending moment which is about seven times higher, as

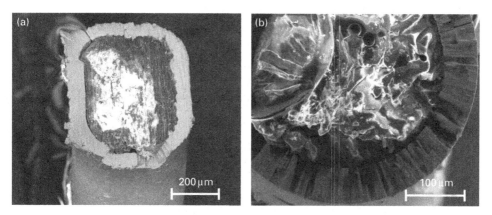

Figure 15.10 SEM images of sputtered TiNi tubes filled with polyamide: (a) rectangular cross-section with magnification 286 ×, (b) round cross-section with magnification 770 ×.

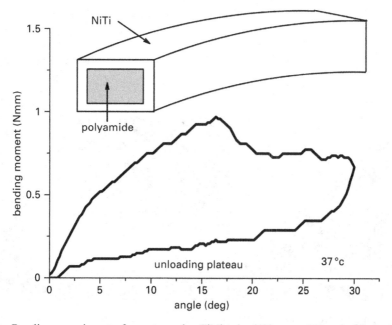

Figure 15.11 Bending experiment of a rectangular TiNi tube (400 µm × 560 µm), filled with polyamide.

shown in Fig. 15.11 [5].Two different methods for the production of three dimensional devices are possible. Either the substrate is static during the sputtering process, which then requires a three-dimensional arrangement of the sputtering cathodes [18], or the substrate moves during the sputtering process using planar cathode geometries. For the latter, the use of a rotational device is necessary, which allows the continuous rotation of substrates during the

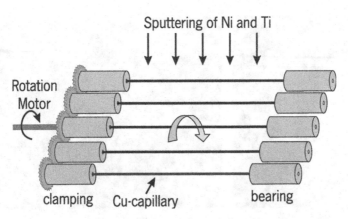

Figure 15.12   Device for continuous rotation during sputter deposition of TiNi on copper capillaries.

Figure 15.13   Scanning electron microscope image of a sputtered tube after etching the copper substrate ($0.016' \times 0.022'$). The wall thickness of the remaining TiNi is $25\,\mu m$.

sputtering process. A schematic set-up is shown in Fig. 15.12. Copper capillaries with round and rectangular cross-sections are used as substrates, which act as sacrificial materials. Figures 15.13 and 15.14 show scanning electron microscope images of sputtered TiNi tubes after etching the sacrificial copper substrates. Nitric acid (30%) was used as a solution to dissolve the copper sacrificial substrate. Due to the low concentration of this solution and the short etching time, the TiNi film was not affected.

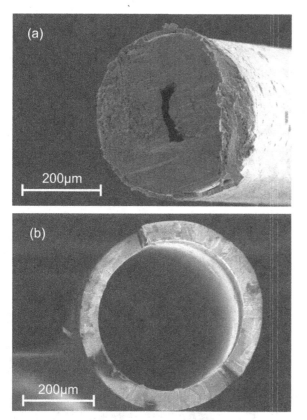

Figure 15.14 SEM images of sputtered TiNi tubes of dimensions 400 µm (outer diameter) before (a) and after (b) etching of the copper core.

15.6   Patterning of non-planar films using lithography and etching

The most prominent application example of superelastic TiNi is the endovascular stent which is made of tubes fabricated by deep-hole drilling from bulk material followed by subsequent wire drawing and annealing processes to reach the final wall thickness. Patterning is achieved by precision laser cutting technology followed by different polishing procedures. For neurovascular applications a wall thickness smaller than approximately 50 µm is required. The described bulk fabrication is envisioned to be replaced by thin film technology using sputtering and sacrificial layer techniques as well as photolithography for patterning of the stents. The patterning can be obtained through the use of a rotary stage motor, as describe by S. Jacobsen [19, 20], which is placed inside the scanning electron microscope chamber configured with a controlled beam direction (electro-beam lithography). The substrate follows the same procedure described for planar patterning and afterwards the substrate is etched using a wet-chemical method.

Another method used to pattern non-planar films uses planar photolithography as a starting principle, this means a Cr mask and two linear motors are used to

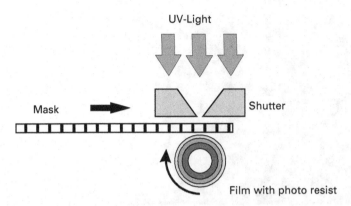

Figure 15.15   Schematic of rotational UV-lithography process

Figure 15.16   (a) Optical micrograph of a non-planar substrate after the photolithography process. (b) Scanning electron micrograph after the etching process.

bring the desired geometry to the substrate, as shown in the Fig. 15.15. The final resolution of the non-planar process can be compared to the planar one, although it complex is more. Figure 15.16 shows a non-planar substrate (a) after the photolithography process and (b) after the etching process.

15.7   Summary

Superelastic thin films can be fabricated achieving almost the same mechanical properties as their bulk counterparts. Superelastic strain up to 6% at 37 °C has been observed. The deposition and patterning technique allows the realization of smaller devices with higher precision compared to conventional fabrication routes. Feature sizes with a maximum diameter of 10 cm and a thickness up to 10 μm have been achieved using a special dry and wet etching technique. For medical application the usage of this technology will open several possibilities for the fabrication of endovascular/neurovascular devices, due to the miniaturization potential for other medical areas, e.g. as filters or valves. It can be expected that this technology will also be used for the fabrication of other devices. Up to now demonstrations of orthodontic archwires and stents have been realized.

References

[1] J. D. Busch and A. D. Johnson, Shape-memory properties in Ni-Ti sputter-deposited film, *J. Appl. Phys.* **68** 12 (1990) 6224–6228.

[2] T. W. Duerig, A. R. Pelton and D. Stockel, An overview of nitinol medical applications, *Mater. Sci. Eng.* **A273–275** (1999) 149–160.

[3] V. Gupta, V. Martynov and A. D. Johnson, Recent developments in SMA thin film based micro-actuators for biomedical and fiber optics applications, *Actuator* (2002) 355–358.

[4] M. Kohl, D. Dittmann, E. Quandt, B. Winzek, S. Miyazaki and D. M. Allen, Shape memory microvalves based on thin films or rolled sheets, *Mater. Sci. Eng.* **A273–275** (1999) 784–788.

[5] H. Rumpf, C. Zamponi, C. Bourauel, D. Drescher and E. Quandt, Superelastic thin film NiTi-polymer-composites and sputtered thin-walled tubes, *Mater. Res. Soc. Symp. Proc.* **855E** (2005) W1.5.1–W1.5.6.

[6] A. Ishida, A. Takei, M. Sato and S. Miyazaki, Stress–strain curves of sputtered thin films of Ti-Ni, *Thin Solid Films* **281–282** (1996) 337–339.

[7] K. Nomura, S. Miyazaki and A. Ishida, Effect of plastic strain on shape memory characteristics in sputter-deposited Ti-Ni thin films, *Journal de Physique IV*, **C8, 5** (1995) 695–700.

[8] J. L. Proft and T. W. Duerig, The mechanical aspects of constrained recovery, in *Engineering Aspects of Shape Memory Alloys*, ed. T. W. Duerig, Butterworth-Heinemann (1990) p. 115.

[9] W. Tang, Thermodynamic study of the low-temperature phase B19′ and the martensitic transformation in near-equiatomic Ti-Ni shape memory alloys, *Metal. Mater. Trans. A* **28A** (1997) 537–544.

[10] B. Winzek and E. Quandt, Shape-memory Ti-Ni-X-films (X = Cu, Pd) under constraint, *Z. Metallkd.* **90** (1999) 796–802.

[11] C. Zamponi, H. Rumpf, B. Wehner and E. Quandt, Superelastic NiTi thin films, *Proc. Actuator* (2004) 86–87.

[12] S. Miyazaki and K. Nomura, Development of perfect shape memory effect in sputter-deposited Ti-Ni thin films, *Proc. IEEE Micro Electro Mechanical Systems (MEMS94)*, Oiso, Japan (1994) 176–181.

[13] A. Gyobu, Y. Kawamura, H. Horikawa and T. Saburi, Martensitic transformations in sputter-deposited shape memory Ti-Ni films, *Mater. Trans. JIM* **37** (1996) 697–702.

[14] J. A. Thornton, High rate thick film growth, *Ann. Rev. Mater. Sci.* **7** (1977) 239–260.

[15] H. Rumpf, T. Walther, C. Zamponi and E. Quandt, High ultimate tensile stress in nano-grained superelastic NiTi thin films, *Mater. Sci. Eng. A*, 415, **1–2** (2006) 304–308.

[16] M. J. Madou, *Fundamentals of Microfabrication: the Science of Miniaturization*, second edn., Boca Raton, FL: CRC Press LLC, 2002.

[17] C. Zamponi, H. Rumpf, C. Schmutz and E. Quandt, Structuring of sputtered superelastic NiTi thin films by photolithography and etching, *Mater. Sci. Eng. A* **481–482** (2008) 623–625.

[18] D. Marton, C. T. Boyle, R. W. Wiseman and C. E. Banas, High strength vacuum deposited nitinol alloy films and method of making same, US Patent 7335426.

[19] S. Snow and S. C. Jacobsen, Microfabrication processes on cylindrical substrates – Part I: Material deposition and removal, *Microelec. Eng.* **83** (2006) 2534–2542.

[20] S. Snow and S. C. Jacobsen, Microfabrication processes on cylindrical substrates – Part II: Lithography and connections, *Microelec. Eng.* **84** (2007) 11–20.

16 Fabrication and characterization of sputter-deposited TiNi superelastic microtubes

Pio John Buenconsejo, Hee-Young Kim and Shuichi Miyazaki

Abstract

A novel method of fabricating TiNi superelastic microtubes with a dimension of less than $100\,\mu m$ is presented in this chapter. The method was carried out by sputter-deposition of TiNi on a Cu-wire substrate, and after deposition the Cu wire was removed by etching to produce a tube hole. The shape-memory/superelastic behavior and fracture strength of the microtubes were characterized. The factors affecting the properties and a method to produce high-strength superelastic TiNi microtubes are discussed.

16.1 Introduction

Superelastic microtubes are attractive materials for application in the medical industry due to the increasing demands for less-invasive surgical devices, such as stents, catheters, microneedles, etc. Among many alloy systems that exhibits superelastic behavior, TiNi shape memory alloy is considered the most prominent due to its superior properties, such as large and stable superelastic strain [1, 2], excellent mechanical properties and relatively good biocompatibility [3]. Commercially available TiNi microtubes are usually processed by a tube-drawing method [4, 5, 6, 7, 8]. The available size of tubes that can be fabricated by tube drawing is limited to larger than $200\,\mu m$ in outer-diameter and a wall thickness of $50\,\mu m$. A superelastic microtube with dimensions less than $200\,\mu m$ has the potential to further miniaturize microdevices and thereby increasing their applicability.

A novel method to fabricate TiNi microtubes with dimensions less than $100\,\mu m$ was suggested by utilizing a sputter-deposition method [9, 10]. The contents of this

Thin Film Shape Memory Alloys: Fundamentals and Device Applications, eds. Shuichi Miyazaki, Yong Qing Fu and Wei Min Huang. Published by Cambridge University Press. © Cambridge University Press 2009.

chapter are based on Ref. [9]. In this method a Cu-wire (diameter 50 μm) is used as a substrate and it was completely covered with TiNi by sputter-deposition. After deposition the Cu-wire was completely removed by etching to produce the tube hole. Therefore the inner diameter of a microtube depends on the diameter of the Cu-wire, and the wall thickness is controlled by the deposition time. The method of fabricating a superelastic TiNi microtube and the factors affecting its properties are discussed.

16.2   Fabrication and characterization method

16.2.1   Sputter deposition system

Figure 16.1 shows a schematic of the deposition system used to fabricate TiNi microtubes. A rotating jig is placed inside a sputter-deposition chamber and a Cu-wire is fixed such that its length is parallel to the surface of the alloy target. The rotation of the wire is controlled by a motor. The composition of the alloy target is Ti-50at%Ni and pure Ti-chips (5 mm by 5 mm) are placed on the surface to control the composition. The composition of the deposited TiNi was determined by electron probe microanalysis to be Ti-52at%Ni. The sputtering power of 500 W and an argon pressure of 0.6 Pa were fixed.

Two deposition methods are described in this chapter: (1) a two-step deposition method and (2) a rotating wire method. In the two-step deposition method the Cu-wire was not rotated during deposition. The first step deposition was carried out on one side of the Cu-wire for two hours, then it was rotated 180° and the second step deposition was performed. After the deposition process the Cu-wire was completely covered with TiNi. In the rotating wire method the Cu-wire was rotated at constant rotation speeds during the deposition process. After deposition, the Cu-wire was completely dissolved in a nitric acid solution, thus

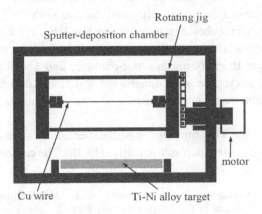

Figure 16.1   A schematic of the rotating jig placed inside the sputter-deposition chamber.

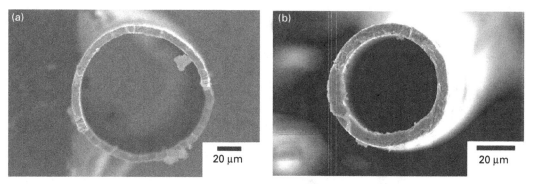

Figure 16.2   TiNi microtubes deposited on (a) a 100 μm diameter Cu-wire and (b) a 50 μm diameter Cu-wire.

producing a tube hole. The microtubes were sealed in an Ar-filled quartz tube and then crystallized by heat-treatment at 873 K for 3.6 ks followed by water-quenching without breaking the quartz tube.

Microtubes with different sizes were successfully fabricated using this method. For example, Fig. 16.2(a) shows a cross-section of a microtube deposited on a 100 μm diameter Cu-wire, and Fig. 16.2(b) shows a microtube deposited on a 50 μm diameter Cu-wire. For both microtubes it is clearly seen that the Cu-wire was completely removed after etching. The inner diameter is consistent with the diameter of the Cu-wire used. The wall thickness can be easily varied by changing the deposition time. Therefore the dimension of microtubes that can be fabricated by this method can be easily controlled. Depending on the available sizes of Cu-wire, microtubes with smaller dimensions can be fabricated.

16.2.2   Characterization of shape memory behaviour

Preparation of the tensile test specimen (8 mm in length) was carefully done by placing a Cu-wire at both ends of the tube. This is to prevent unnecessary deformation when both ends of the tube were fixed with chucks of a tensile test machine. A schematic of the microtube fixed with the chucks is shown in Fig. 16.3. The gauge length of the microtube was 5 mm. The lower chuck holder was fixed while the upper chuck holder was connected to an extensometer or load controller. This set-up was placed inside a thermally insulated chamber, with the capability of thermal cycling.

Shape memory behavior was characterized by tensile tests under various constant stresses during the thermal cycling (heating and cooling rate of 10 K per minute). This test was done by subsequently increasing the load after each thermal cycle until the specimen fractured, so as to obtain the fracture stress. Fracture surface morphologies were observed using scanning electron microscopy (SEM). Stress–strain curves at constant temperatures were also obtained by tensile testing.

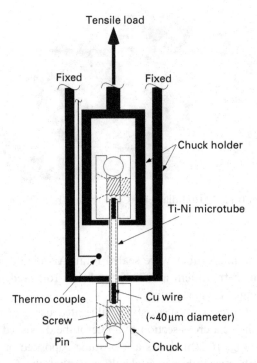

Figure 16.3   Schematic of a tensile test specimen placed in the chuck holders of a tensile test machine.

16.3   TiNi microtube fabricated by a two-step deposition method

16.3.1   Cross-sectional microstructure

Figure 16.4(a) shows a fracture surface of the TiNi microtube fabricated by the two-step deposition method observed by SEM. The deposition directions, which are normal to the surface of the alloy target, are indicated by arrows. It is seen that the microtube wall thickness is not uniform, ranging from 3 μm to 6 μm. The surfaces that were directly facing the surface of the target were thicker and the thickness gradually decreased to the thinnest region, where the surface was parallel to the deposition direction. It is also noted that the fracture surface is rather smooth in the thickest region, as shown in Fig. 16.4(b), which is an enlarged image of section A, while columnar grains were formed in the thinnest region, as shown in Fig. 16.4(c), which is an enlarged image of section B. Two layers of columnar grains with different directions are clearly distinguished in Fig. 16.4(c). A schematic of the observed microstructure in the thinnest region is shown in Fig. 16.4(d). It is considered that the inner layer of columnar grains was formed during the first step deposition, and the outer layer was formed on top of the previous layer during the second step deposition.

The formation of columnar grains and non-uniformity of film thickness were strongly dependent on the position of the surface normal relative to the deposition

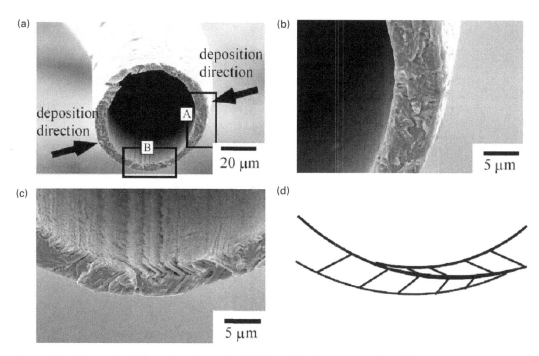

(a)

deposition
direction

deposition
direction

A

B

20 μm

(b)

5 μm

(c)

5 μm

(d)

Figure 16.4   Fracture surface of a microtube fabricated by the two-step deposition method.

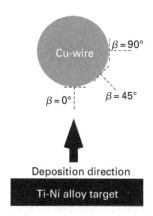

Cu-wire

$\beta = 90°$

$\beta = 0°$ $\beta = 45°$

Deposition direction

Ti-Ni alloy target

Figure 16.5   A cross-section of a Cu-wire positioned relative to the alloy target.

direction. As shown in Fig. 16.5, the surface normal of the curved section of the Cu-wire was inclined relative to the surface normal of the sputter target at an oblique angle defined here as deposition angle β. This shows that the microstructure of the microtube is strongly dependent on the geometry of the substrate. A detailed discussion on the effect of the β angle on the microstructure of thin films is given in Section 16.4.

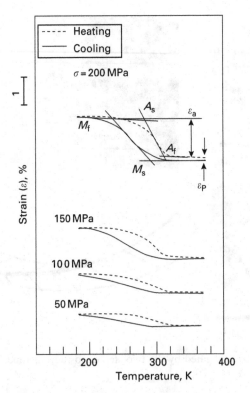

Figure 16.6   Strain–temperature curves for a TiNi microtube fabricated by a two-step deposition method and crystallized at 873 K for 3.6 ks.

16.3.2   Shape memory behavior and fracture strength

Figure 16.6 shows the strain–temperature curves for a TiNi microtube fabricated by the two-step deposition method and crystallized at 873 K for 3.6 ks. The strain–temperature curves were obtained during thermal cycling between 180 K and 390 K under various constant stresses. The tensile test was performed initially at a constant stress of 50 MPa and the applied stress was increased by 50 MPa after each thermal cycle. The cooling cycles are indicated by solid lines and the heating cycles are indicated by dashed lines. As shown in the figure, M_s, M_f, A_s and A_f denote the start and finish temperatures for the martensitic transformation and its reverse transformation, respectively. The recovery strain and plastic strain are denoted by ε_a and ε_p, respectively. Figure 16.6 confirms that the crystallized TiNi microtube exhibits the shape memory effect: the specimen elongates on cooling due to the martensitic transformation and the strain recovers on heating due to the reverse transformation. The M_s at zero stress was obtained by extrapolating the linear relationship between M_s and the applied stress. The M_s of the TiNi microtube fabricated by the two-step sputter-deposition method was 250 K. It is noted that the difference between the M_s and M_f is wide, indicating a broad transformation range. It is supposed that compositional variation and thickness variation due to different β angles have caused the broad transformation range in the microtube.

The microtube could be tensile tested up to 200 MPa without failure, but when the stress was increased to 250 MPa it fractured during thermal cycling. The low fracture stress was due to the columnar grains. Thus, in order to improve the fracture stress of the microtube fabricated by the two-step sputter-deposition it is necessary to eliminate the columnar grains.

16.4   Effect of deposition angle β

16.4.1   Film thickness

As discussed in Section 16.3.1, the microstructure of a sputter deposited microtube was strongly affected by the β angle. In order to systematically investigate the effect of β on the microstructure, TiNi thin film was deposited on a flat substrate made of SiO$_2$/Si (1 cm by 1 cm). The position of the substrate was inclined at various β angles relative to the surface of the sputter-target, as shown in Fig. 16.7. Similar deposition parameters mentioned in Section 16.2.1, i.e. power and Ar pressure, were fixed while depositing the TiNi.

The surface area of the substrate was defined by its dimension and denoted as A_{subs}. Another surface area was defined as an effective area (A_{eff}), which is the area projected on the surface of the sputter target. Geometrically the two areas are related to the deposition angle (β) as follows:

$$\cos \beta = \frac{A_{eff}}{A_{subs}}. \qquad (16.1)$$

If the amount of sputtered atoms coming from the target and deposited on the substrate area (A_{subs}) is confined by the effective area (A_{eff}) then the deposition

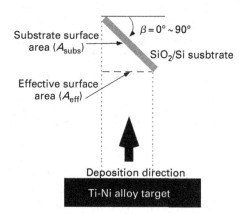

Figure 16.7   A schematic of an SiO$_2$/Si substrate inclined at various β angles relative to the sputter target.

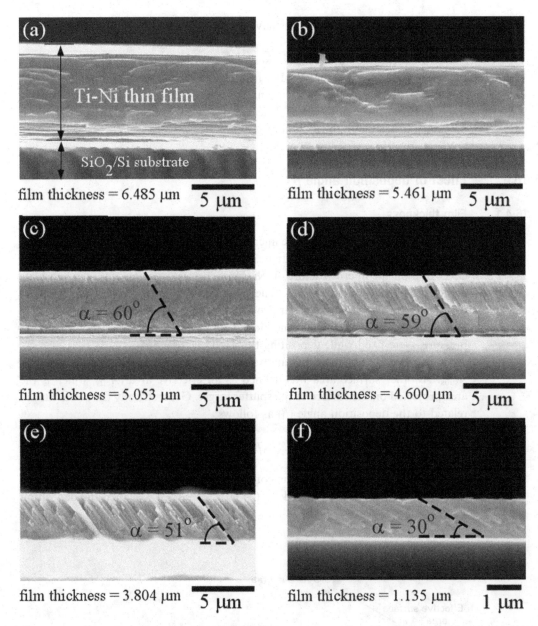

Figure 16.8 Cross-sections of thin films deposited for 3.6 ks at various β angles: (a) 0°, (b) 15°, (c) 30°, (d) 45°, (e) 60°, and (f) 90°. Angle α indicates the angle of inclination between the columnar grains and the substrate surface.

rate of thin film is influenced by the β angle. The film deposited at $\beta = 0°$ gives the highest deposition rate and with increasing β angle the deposition rate decreases.

Figure 16.8 shows a series of SEM images for TiNi films deposited for 3.6 ks at various β angles. The thickness measured from the cross-section for each specimen is also shown. The thickness of the film deposited at 0° was about 6.5 μm and the

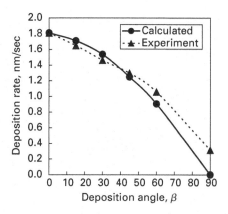

Figure 16.9 Deposition rate versus the β angle.

thickness of the film deposited at 90° was about 1.1 μm. Since the deposition times for all the films were the same, then this clearly shows that deposition rate decreases with increasing β angle.

Figure 16.9 shows a plot of deposition rate as a function of the β angle. The deposition rate at $\beta = 0°$ was considered as the reference or standard, and this value was multiplied by Eq.(16.1) to obtain the calculated deposition rate at various β angles. The experimental data were obtained from the measured film thickness in the SEM image. In general both plots revealed a similar trend, where deposition rate decreases with increasing β angle. However, the deposition rates between experimental and calculated data were found to deviate from each other with increasing β angle. The calculated deposition rate at $\beta = 90°$ is zero, but the experimental deposition rate was 0.2 nm/s. Suppose that the incident atoms travel a direct path to the substrate, then the number of incident atoms per area is confined by the effective area (A_{eff}), and they must be equal to the number of sputtered atoms within this area of the alloy target. In an actual sputter deposition process the sputtered atoms do not travel a direct path from the target to the substrate due to collisions with other atoms or Ar ions. Consequently incident atoms coming from outside the effective area (A_{eff}) are deposited.

16.4.2   Surface roughness

Figure 16.10 shows a series of STM images of the surface of TiNi deposited for 3.6 ks at various β angles. A scan size of 1 μm by 1 μm was obtained for each film. The variation of height of the film was measured by taking several line profile scans for each STM image. For example, a line profile scan (taken along the dashed line in the STM image) is shown below each image. It is easy to confirm that the surface of the film deposited at $\beta = 0°$ was smooth, and with increasing β angle the film became more rough. A more quantitative method to determine roughness is shown in Fig. 16.11. The surface roughness (R) can be measured from a line profile scan by measuring the variation of height relative to a median

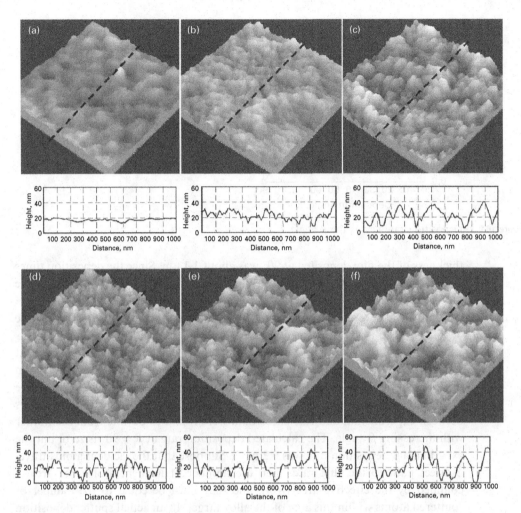

Figure 16.10   STM images of the surface of TiNi deposited for 3.6 ks at various β angles: (a) $0°$, (b)$15°$, (c) $30°$, (d) $45°$, (e) $60°$, and (f) $90°$.

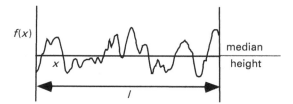

Figure 16.11 A schematic of a line profile scan to measure the surface roughness.

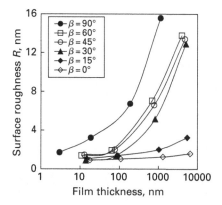

Figure 16.12 Surface roughness of films deposited at various β angles plotted with respect to film thickness.

height as a function of position x. This can be expressed in an equation as:

$$R = \sqrt{\frac{1}{l} \int_0^l f(x)^2 \mathrm{d}x} = \sqrt{\frac{1}{l} \sum_0^l f(x)^2}, \qquad (16.2)$$

where R is the roughness in nm, l is the length of the line scan, x is the distance and $f(x)$ is the height relative to a median height.

The measured surface roughness as a function of film thickness for films deposited at various β angles is shown in Fig. 16.12. A surface roughness of less than 3 nm was observed for the entire specimen when the thickness was about 10 nm. Significant difference in surface roughness of films deposited at various β angles was observed when the film thickness increased. The surface roughness of films did not significantly increase with increasing thickness for films deposited at $\beta = 0°$ and $\beta = 15°$, thus it can be considered that the surface of films deposited at low β angle always remains smooth during film growth. On the other hand the surface roughness of films deposited at higher β angles increases with increasing film thickness. Moreover, for the same film thickness the roughness of the film increases with increasing β angle.

(a)   (b)

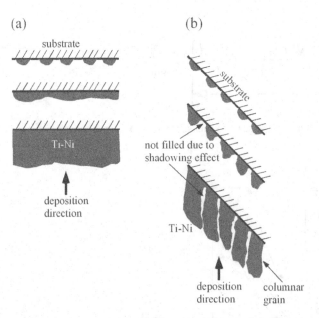

Figure 16.13   Schematic of film growth when the substrate is oriented (a) at $\beta = 0°$ and (b) at high β angle.

16.4.3   Columnar grain formation

Looking back again at Fig. 16.8, the photomicrographs revealed the presence of columnar grains for films deposited at β angles of 30° and above. On the other hand films deposited below $\beta = 30°$ did not show the presence of any columnar grains. The columnar grains formed at $\beta = 30°$ were very fine, and then they became more defined with increasing β angle. A schematic of film growth when the substrate is oriented at $\beta = 0°$ and at high β angle is shown in Fig. 16.13.

The early stages of the deposition process are essentially the same for both cases, where the deposited atoms would tend to accumulate forming islands. These islands are the nucleation sites for film growth. As shown in Fig. 16.12, at a film thickness of about 10 nm the surface roughness for the films deposited at various β angles was less than 3 nm and almost similar for all the films. This can be interpreted as the formation of islands. If the β angle is 0° then during continued deposition of atoms the gaps between islands are easily filled. Thus the surface will remain smooth throughout the deposition process. Similarly, for films deposited at low β angles ($\beta = 0°$ and 15°) the surface roughness remained low with increasing thickness. On the other hand for the film deposited at high β angle, there is a preferential deposition of atoms on the islands, due to a geometrical shadowing effect [11, 12, 13, 14, 15, 16]. The increase of surface roughness with continued deposition of atoms will promote further the geometrical shadowing effect and consequently voids or cavities will be created. The cavities will eventually grow into columnar grains.

16.5   Fabrication of high-strength superelastic TiNi microtubes

16.5.1   Effect of rotation speed on the microstructure

The presence of columnar grains and non-uniform wall thickness of the microtube fabricated by the two-step deposition method were found to be detrimental to its mechanical properties. These were caused by the influence of the β angle which is inherent to the geometry of the Cu-wire substrate. So in order to remove the influence of the β angle it is suggested that the Cu-wire is rotated during deposition.

Figure 16.14 shows the fracture surfaces observed by SEM for TiNi microtubes fabricated by deposition on a rotating wire at 0.6 rpm (Figs. 16.14(a) and (d)), 15 rpm (Figs. 16.14(b) and (e)) and 30 rpm (Figs. 16.14(c) and (f)), followed by crystallization at 873 K for 3.6 ks. The wall thickness for all the microtubes was uniform, with a thickness of about 6 μm. It was established in Section 16.4.1 that the deposition rate decreases with increasing β angle. The effect of rotating the Cu-wire was to create a similar deposition condition and thus a uniform deposition rate around the substrate surface.

Examination of the enlarged images of microtube cross-sections (Figs. 16.14(d), (e), (f)) revealed that the morphologies were different with respect to the rotation speed. For example, in the microtube deposited at 0.6 rpm, columnar grains were observed, while the microtubes deposited at 15 rpm and 30 rpm did not contain any columnar grains. It is suggested that the deposition time on the oblique surface area (high β angle) is long enough to form the columnar grains when the rotation speed is slow (0.6 rpm). The deposition time on the oblique surface area in one cycle decreases with increasing rotation speed, causing the columnar grains not to be formed for the microtubes deposited at 15 and 30 rpm.

16.5.2   Effect of rotation speed on shape memory behavior and fracture strength

Figure 16.15 shows the strain–temperature curves of TiNi microtubes fabricated by the rotating wire method at 0.6, 15 and 30 rpm. The strain–temperature curves were obtained during thermal cycling between 160 and 390 K, under various constant stresses. The martensitic transformation temperatures (M_s, M_f, A_s and A_f), shape recovery strain (ε_a) and plastic strain (ε_p) were obtained as shown in the figure. The tensile elongation during cooling and its recovery during heating confirm that the TiNi microtubes exhibit the shape memory effect. The M_s and A_f extrapolated at zero applied stress for each specimen are plotted as a function of rotation speed in Fig. 16.16(a). The M_s and A_f slightly varied from 160 to 170 K and from 250 to 260 K, respectively, indicating that M_s and A_f are not very sensitive to the rotation speed. Similarly the maximum recoverable strain (ε_a^{max}) is plotted as a function of rotation speed in Fig. 16.16(b). It is seen that the ε_a^{max} of microtubes exhibits a similar value of about 3%. These results suggest that the martensitic transformation temperatures and the maximum recoverable strain of

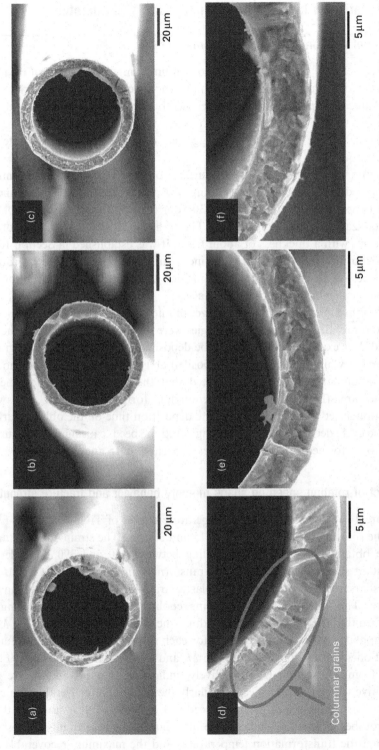

Figure 16.14 Fracture surfaces of TiNi microtubes fabricated by rotating wire method at 0.6 rpm (a), (d), at 15 rpm (b), (e) and at 30 rpm (c), (f).

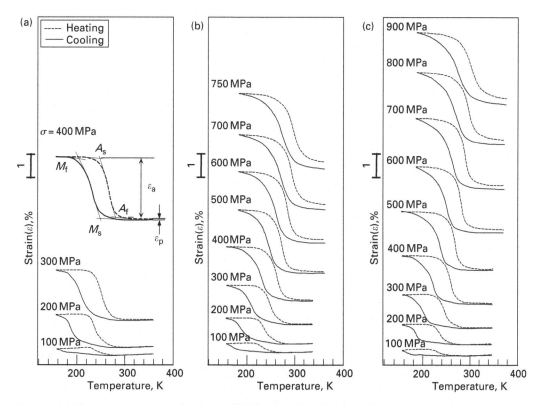

Figure 16.15 Strain–temperature diagrams of TiNi microtubes fabricated by the rotating wire method at (a) 0.6 rpm, (b) 15 rpm and (c) 30 rpm.

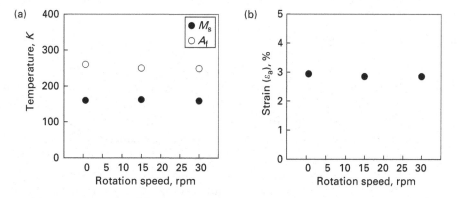

Figure 16.16 (a) M_s and A_f, and (b) ε_a^{max} versus the rotation speed.

the TiNi microtubes deposited by the rotating wire at various rotation speeds are not significantly affected by the rotation speed, because the sputtering condition, including the β angle, is similar for every part of the tube.

The TiNi microtube sputter-deposited at 0.6 rpm could be tensile tested up to 400 MPa without fracture but it failed during thermal cycling under the next

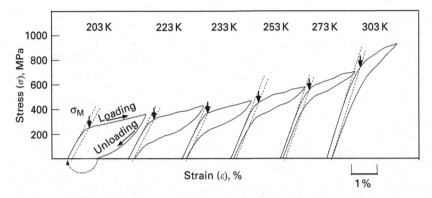

Figure 16.17 Stress–strain curves measured at various test temperatures for a TiNi microtube deposited on a rotating wire at 15 rpm.

constant stress of 450 MPa. For the microtubes sputter-deposited at 15 and 30 rpm, they failed at 800 and 950 MPa, respectively, indicating that the fracture stress increased with increasing rotation speed. From the microstructure analysis in Fig. 16.14, it is suggested that the low fracture stress for the microtube deposited at 0.6 rpm was due to the abnormally coarsened columnar grains. On the other hand, the microtubes deposited at higher rotation speeds of 15 and 30 rpm exhibited no columnar grains, which explains the remarkable increase in the fracture stress. In addition, ε_p for all the specimens was significantly small, reaching only about 0.3% prior to fracture. This suggests that high strength superelastic TiNi microtubes can be obtained by sputter-deposition at high rotation speeds of 15 rpm and above.

16.5.3   Superelasticity

In order to evaluate the superelastic behaviour of TiNi microtubes, the stress–strain curves were obtained by tensile testing (up to 3% strain) at different test temperatures. Figure 16.17 shows a series of stress–strain curves obtained at different test temperatures for TiNi microtubes fabricated by sputter-deposition on a rotating wire at 15 rpm and crystallized at 873 K for 3.6 ks. The microtube was loaded up to a tensile strain of 3% and then unloaded until the tensile stress reached zero. On loading, the microtubes underwent yielding and the apparent yield stress (σ_M) indicated by an arrow was obtained by drawing a straight line coinciding with the elastic portion of the stress–strain curves with offset to 0.2% strain. The microtube was tensile tested at different test temperatures from 203 to 303 K and, since the M_s is 170 K, the stable phase for each test temperature is the B2-parent phase. Therefore the σ_M observed indicates the critical stress for inducing the martensitic transformation. The σ_M increased with increasing test temperatures since the stress for inducing the martensitic transformation increases with increasing temperature. In this study

σ_M reached 740 MPa when the test temperature was 303 K. Partial superelasticity was observed at 203 K, since the test temperature was below A_f (250 K) and complete strain recovery was achieved by heating the microtube above A_f after unloading. The superelastic curves were obtained at temperatures between 223 K and 303 K.

16.6   Summary and remarks

TiNi microtubes with a dimension of 50 μm and a wall thickness of about 6 μm were successfully fabricated by the sputter-deposition method. The results are summarized as follows

(1) TiNi microtubes fabricated by the two-step sputter-deposition method were characterized by low fracture stress of 250 MPa due to the presence of columnar grains and a non-uniform wall thickness varying between 3 μm and 6 μm. The columnar grains were formed by a geometric shadowing effect due to the inclination of the substrate surface normal (curve surface of the Cu-wire) relative to the deposition direction. The inclination is defined as the deposition angle (β), and at high β angle columnar grains are easily formed. Also, with increasing β angle the deposition rate decreases, thus the film thickness on the surface oriented at a high β angle were thinner than that on the surface oriented at a low β angle.

(2) TiNi microtubes fabricated by the rotating wire method were characterized by a uniform wall thickness of 6 μm and the fracture stress increased with increasing rotation speed. The microtube fabricated at 0.6 rpm revealed the presence of columnar grains and had a low fracture stress of 450 MPa, whereas the microtubes fabricated at higher rotation speeds of 15 rpm and 30 rpm exhibited high fracture stress above 800 MPa, since no columnar grains were formed. The martensitic transformation temperatures and the recovery strains were not significantly affected with the rotation speed. The M_s and A_f slightly varied from 160 K to 170 K and 250 K to 260 K, respectively. Similarly ε_a^{max} was about 3.0% for all the specimens. The stress–strain curves revealed superelastic behavior at test temperatures between 223 K and 303 K.

As a final remark, we suggest that the TiNi microtubes fabricated in this study, which exhibited superelasticity at room temperature with a high σ_M and a high fracture stress, are suitable for practical applications as biomedical devices such as microstents and microcatheters.

Acknowledgement

The present authors would like to express their gratitude to Mr. K. Ito for his contribution to this chapter.

References

[1] S. Miyazaki. Thermal and stress cycling effects and fatigue properties of Ni-Ti alloys, in *Engineering Aspects of Shape Memory Alloys*, ed. T. W. Duerig, *et al.*, London: Butterworth-Heinemann (1990) 394–413.

[2] S. Miyazaki, Y. Ohmi, K. Otsuka and Y. Suzuki. Characteristics of deformation and transformation pscudelasticity in Ti-Ni alloys, *Journal de Physique*, **43** Suppl. 12 (1982) C4–255–260.

[3] T. Duerig, A. Pelton and D. Stockel. An overview of nitinol medical applications, *Mater. Sci. Eng.* **A273–275** (1999) 149–160.

[4] K. Yoshida, M. Watanabe and H. Ishikawa. Drawing of Ni-Ti shape-memory-alloy fine tubes used in medical tests, *J. Mater. Proc. Tech.* **118** (2001) 251–255.

[5] H. Horikawa, T. Ueki and K. Shiroyama. Superelastic performance of Ni-Ti thin tubes, *Proceedings of SMST-94*, Pacific Grove, USA (1994) 347–352.

[6] K. Yoshida and H. Furuya. Mandrel drawing and plug drawing of shape-memory-alloy fine tubes used in catheters and stents, *J. Mater. Proc. Tech.* **153–154** (2004) 145–150.

[7] Z. Q. Li and Q. P. Sun. The initiation and growth of macroscopic band in nano-grained NiTi microtube under tension, *Int. J. Plasticity* **18** (2002) 1481–1498.

[8] K. Muller. Extrusion of nickel-titanium alloys nitinol to hollow shapes, *J. Mater. Proc. Tech.* **111** (2001) 122–126.

[9] P. J. S. Buenconsejo, K. Ito, H. Y. Kim and S. Miyazaki. High-strength superelastic Ti-Ni microtubes fabricated by sputter deposition, *Acta Mater.* **56** (2008) 2063–2072.

[10] S. Miyazaki, M. Tomozawa, P. J. Buenconsejo, K. Okutsu, H. Cho and H. Y. Kim. Ti-Ni based SMA thin film microactuators and related science and technology, *Proceedings of SMST-2006*, Pacific Grove, USA. Materials Park, USA: ASM International (2006) 339–356.

[11] J. Thornton. Influence of apparatus geometry and deposition conditions on the structure and topography of thick sputtered coatings, *J. Vac. Sci. Technol.* **11**(4) (1974) 666–674.

[12] J. W. Patten. The influences of surface topography and angle of adatom incidence on growth structure in sputtered chromium, *Thin Solid Films* **63** (1979) 121–129.

[13] A. G. Dirks and H. J. Leamy. Columnar microstructure in vapor-deposited thin films, *Thin Solid Films* **47** (1977) 219–233.

[14] L. Dong, R. W. Smith and D. J. Srolovitz. A two-dimensional molecular dynamics simulation of thin film growth by oblique deposition, *J. Appl. Phys.* **80** (10) (1996) 5682–5690.

[15] F. Ying, R. W. Smith and D. J. Srolovitz. The mechanism of texture formation during film growth: the roles of preferential sputtering and shadowing, *Appl. Phys. Lett.* **69** (20) (1996) 3007–3009.

[16] R. P. U. Karunasiri, R. Bruinsma and J. Rudnick. Thin-film growth and the shadow instability, *Phys. Rev. Lett.* **62** (1989) 788–791.

17 Thin film shape memory microcage for biological applications

Y. Q. Fu, J. K. Luo, S. E. Ong and S. Zhang

Abstract

This chapter focuses on the fabrication and characterization of a microcage for biopsy applications. A microcage based on a free-standing film could be opened/closed through substrate heating with a maximum temperature of 90 °C, or Joule heating with a power less than 5 mW and a maximum response frequency of 300 Hz. A TiNi/diamond-like-carbon (DLC) microcage has been designed, analyzed, fabricated and characterized. The bimorph structure is composed of a top layer of TiNi film and a bottom layer of highly compressively stressed DLC for upward bending once it is released from the substrate. The fingers of the microcage quickly close through the shape memory effect once the temperature reaches the austenite transformation point to execute the gripping action. Opening of the microcage is realized by either decreasing the temperature to make use of the martensitic transformation or further increasing the temperature to use the bimorph thermal effect. The biocompatibility of both the TiNi and DLC films has been investigated using a cell-culture method.

17.1 Introduction

The wireless capsule endoscope (WCE) is a new diagnostic tool for searching for the cause of obscure gastrointestinal bleeding. A WCE contains video imaging, self-illumination, image transmission modules and a battery [1, 2]. The indwelling camera takes images and uses wireless radio transmission to send the images to a receiving recorder device that the patient wears around the waist. However, there are two drawbacks for the current WCE: (1) lack of ability for biopsy; and (2) difficulty in identifying the precise location of the pathology. Without tissue diagnosis, it is often difficult to differentiate inflammatory lesions from tumour infiltration. A biopsy is the removal of tissues or cells from the body and

Thin Film Shape Memory Alloys: Fundamentals and Device Applications, eds. Shuichi Miyazaki, Yong Qing Fu and Wei Min Huang. Published by Cambridge University Press. © Cambridge University Press 2009.

determination of any abnormality, such as cancer, infection, inflammation or swelling. Microgrippers are very important microtools for the applications in biopsy, tissue sampling, cell manipulation, nerve repair and minimally invasive surgery [3, 4, 5, 6, 7]. The requirements for the microgrippers for such applications include easy operation, low actuation temperature, low operation voltage and low power consumption. A microgripper actuated by high temperature or high voltage will damage or even kill living cells or tissues.

To manipulate the tissue, cells or other biological objects, there are different types of microactuation mechanisms. An optical mechanism is mainly for cell manipulation application. A highly collimated light source, such as a laser, can exert a focused radiation pressure that is substantial enough to manipulate large particles. Because the optical micromanipulation requires no physical contact, cells or tissues can be manipulated within enclosed glass chambers under sterile conditions. Optical scissors or optical tweezers are two examples of optical micromanipulation [8, 9]. However, photodamage incurred during manipulation could cause cell injury and death [10]. Microgrippers based on electrostatic, piezoelectric, electrophoresis or dielectrophoretic manipulation have been widely studied [11, 12, 13]. For example, dielectrophoresis can be used to trap, move, separate or concentrate cells/particles based on the interaction between a polarized cell and a non-uniform electric field around it [14]. However, interaction with the sample electrically is needed for electrokinetic micromanipulation, which may bring a negative influence on the cell or tissue. Deriving the gripping ability of a microgripper by way of thermal expansion is one of the most common technologies for micromanipulation. The principle is based on the fact that thermal expansion (upon heating) and contraction (upon cooling) of a material could provide actuation functions. However, the problems with this type of actuation include a relatively small displacement and a high working temperature. Some compliant mechanisms are normally designed and applied to amplify the displacement in thermal actuators [15, 16].

Compared to the above-mentioned microactuation mechanisms, thin film shape memory alloy is promising because of its high power density (up to 10 J/cm^3), low actuation temperature (less than 80 °C), the ability for large recovery stress and strain during actuation, as well as biocompatibility [17, 18]. The popular gripper design is an out-of-plane bending mode, mostly with two integrated TiNi/Si cantilevers (or other substrate, such as SU-8 or polyimide, etc.) with opposite actuation directions (see Fig. 17.1(a)) [19, 20, 21, 22, 23]. Figure 17.1(b) shows the patterned TiNi electrodes on silicon cantilevers [24]. When the electrodes are Joule heated, the cantilever bends significantly due to the shape memory effects of TiNi films, thus generating a gripping force. Another gripper design is the in-plane mode, in which the deformation of two arms (using freestanding TiNi films or TiNi/Si beams) is within a plane through a compliant structure design [25]. However, the force generated from the horizontal deformation of the TiNi films is not large enough to grasp large objects. The other problem in this design is how to prevent out-of-plane bending and beam deformation caused by intrinsic film

(a)

(b)

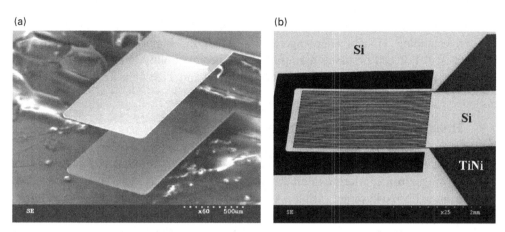

Figure 17.1 TiNi/Si microgripper with cantilever structure with out-of-plane bending mode: (a) micogripper; (b) the patterned TiNi electrodes on silicon cantilevers [19,24]. (Reproduced with permission from Elsevier, UK.)

stress. Most of the microgrippers, such as microtweezers, capture micro-objects through applying a force directly to the object, which may cause damage to the object. A microcage with multiple fingers of a bimorph structure fixed on a multi-degree-of-freedom robotic arm can be used to capture, transport and manipulate biocells for dissection and injection. The microcage captures the micro-object by confining or trapping it without applying a large force directly on it, thus avoiding the potential damage to the tissues or cells [26].

Diamond-like carbon (DLC) is a promising material for MEMS applications [27, 28]. DLC is biocompatible and mechanically and tribologically strong, thus is an excellent coating material for implantable medical devices. It has a large Young's modulus, excellent thermal conductivity and a low coefficient of thermal expansion (CTE), all of which are useful in developing bimorph thermal actuators with large displacements. In [29], a normally closed multi-finger microcage has been fabricated based on electroplated nickel and DLC bimorph structure with dimensions down to biocell sizes of ~40 µm. A highly compressively stressed (a few GPa) DLC expands and lifts the fingers upwards once they are released from the substrate, forming a closed microcage. The device can be opened by millisecond pulsed current at a power of a few tens of milliwatts. However, the operation temperature of the microcages is too high (300 to 400 °C) for practical biological applications. Theoretical analysis revealed that the operation temperature of the bimorph microcage is inversely proportional to the difference in the CTEs of the two-layer materials [29]. In order to reduce the operation temperature, it is better to select two materials with a large difference in CTE, for example, DLC and polymer. Based on this analysis, a microcage with an SU-8/Al/DLC trilayer structure has been fabricated [30], and experimental results confirmed that the temperature for operation can be lowered to nearly 100 °C. Further reduction in the operation temperature of the microcage can be realized by using the shape memory effect [31, 32, 33].

In this chapter, the design, analysis, fabrication and characterization of a freestanding TiNiCu microcage and a TiNi/DLC bimorph microcage are discussed. For cell or tissue manipulation, biocompatibility is an important issue, thus *in situ* cell-culture behaviors of both DLC and TiNi films are also discussed.

17.2   Freestanding TiNiCu microcage

Films of Ti50Ni47Cu3 alloy were deposited on 100 mm diameter Si(100) wafers by magnetron sputtering in an argon gas environment at a pressure of 0.8 mtorr from a Ti55Ni45 target (using a 400 W DC electric field) and a 99.99 % pure Cu target (using a 2 W electric field). The distance between the substrate and the target was 100 mm. Film thickness was 3.5 μm. After the deposition, the films were annealed at a temperature of 550 °C in a vacuum at 1×10^{-7} torr. Microactuators were fabricated by photolithographically patterning 4.8 μm thick layers of AZ4562 photoresist on top of the TiNiCu films. HF:HNO$_3$:H$_2$O (1:1:20) solution was employed to etch the TiNiCu films and form the microactuator patterns. With the remaining photoresist as the mask, the silicon substrate beneath the TiNiCu/ SiN patterns was isotropically etched by SF$_6$ plasma using an rf reactive ion etch (rf-RIE) system with an SF$_6$ plasma (power of 100 W, 80 sccm, 100 mtorr) until the free standing TiNiCu structures were released [33].

Figure 17.2 (a) to (c) show that the freestanding TiNiCu films possess an intrinsic two-way shape memory effect with large displacements: the film curls up at room temperature, but becomes flat when heated and curls up once more when cooled back to room temperature [34]. The freestanding film was obtained by peeling off the film from the Si substrate after the deposition/annealing process. For bulk TiNi alloy, a one-way shape memory effect is normally observed – the material returns to its original shape during heating after deformation in the martensitic phase, but not in cooling. The origin of the two-way shape memory effect observed in the TiNiCu films can be attributed to the difference in sputtering yields of titanium and nickel, which produces a compositional gradient through the film thickness [34, 35]. The film layer near the substrate is normally nickel-rich, and no shape memory effect is observed, but the material may possess superelasticity. As the Ti/Ni content changes through the film thickness, the material properties change from being superelastic to having a shape memory. A stress gradient is generated due to the changing microstructure and composition as a function of thickness. The bottom layer of material is under a compressive stress relative to the higher layers and so the film layer extends dramatically upon release from the substrate, causing the free standing structures to bend upward. When heated, the film layer returns to a flat position due to the shape memory effect.

The observed intrinsic two-way shape memory effect is of great practical importance, in particular for the design of microactuators [36]. Microactuators made from these free standing TiNi films can provide large deflections from simple structures and MEMS processes.

Figure 17.2   Two-way shape memory effect of free standing TiNiCu films at (a) 20 °C; (b) 65 °C and (c) 90 °C [34]. (Reproduced with permission from Elsevier, UK.)

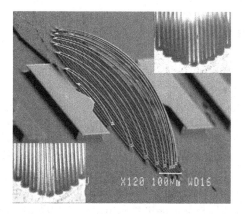

Figure 17.3   A downwards bending microcage structure; inset upper: microcage in bending-up position at 20 °C; inset below: microcage in flat position at 80 °C [33]. (Reproduced with permission from Institute of Physics, UK.)

The actuation performance of the released free standing microcages was evaluated by either heating the structure on a hot plate up to a maximum temperature of 100 °C or by passing a current through the patterns and resistively heating the film in air. In the latter case, the current was provided by a Keithley 224 A voltage/current power supply generating a square-wave voltage signal and this allowed actuation as a function of frequency to be analysed [33]. The displacement of the microgrippers was recorded on a video camera attached to a metallurgical microscope.

Figure 17.3 shows a fabricated microcage consisting of thin TiNiCu beams patterned to form a dome upon release from the substrate [33]. The stress gradient in the TiNiCu thin film causes the dome to have a height of between 180 and 220 μm at room temperature. When Joule heated, the dome microcage becomes almost flat, returning to its original state upon cooling, forming a confining or trapping action.

A modified microcage design consisting of thin TiNiCu beams patterned into microfingers is shown in Fig. 17.4 [33]. After release from the Si substrate, the

Figure 17.4   A TiNiCu bending-up microcage structure and examples of capturing of (a) an ant, (b) an aphid [33]. (Reproduced with permission from Institute of Physics, UK.)

fingers of the microcage curl up to form a cage structure due to the gradient stress and can be used to confine an object at room temperature. The microfingers uncurl when heated above 55 °C and become flat at a temperature above 80 °C, as shown in the series of Figs. 17.5 (a) to (e). Upon cooling, the structure returns to its original curled shape. Figure 17.6 plots the experimentally measured changes in the horizontal and vertical displacement of the microcage fingertips as a function of substrate temperature when heated with a hot plate [33]. Horizontal and vertical displacements of hundreds of micrometers are achievable at temperatures less than 100 °C. It should be noted that negative values of vertical displacement indicate that the cage fingers bend downwards upon actuation. The fabricated microcage has been thermally cycled between 20 °C and 100 °C for more than 100 cycles, and optical microscopy observation did not reveal apparent degradation.

The microcage can be actuated by passing a current through the TiNi microfingers. Figure 17.7(a) shows the horizontal displacement of the fingers achieved by varying the applied power [33]. Significant displacements are observed for input powers between 1.5 and 5 mW as the shape memory alloy heats up through the transition temperature. Above 5 mW, the increase rate of tip displacement gradually decreases up to a power of 7 mW, above which the displacement slightly decreases. Excessive heating through the application of very high powers (>20 mW) resulted in visible changes in the color of the TiNiCu – an indication of overheating and surface oxidation. The microcage can be used to capture microscale objects, with examples shown in Fig. 17.4, in which an ant and an aphid were captured [33].

Figure 17.7(b) shows the measured tip displacement produced by passing a current through the metallic layers as a function of the voltage amplitude and frequency of the square wave signal applied [33]. For frequencies below ~300 Hz, the tip displacement increases with applied voltage. However, above ~100 Hz, the displacement decreases with increasing frequency for all applied voltages, indicating that about 10 ms is required to cool the microstructure due to the thermal

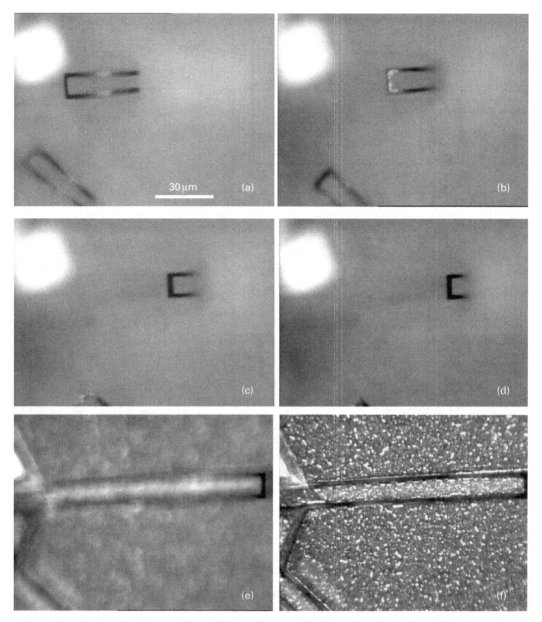

Figure 17.5   Optical microscopy images of microcage heated using a hot plate: (a) 20 °C; (b) 55 °C; (c) 65 °C; (d) 70 °C; (e) 75 °C; and (f) 85 °C [33]. (Reproduced with permission from Institute of Physics, UK.)

capacity of the system. This places a maximum operating frequency of about 300 Hz for this microcage design. The realization of high frequency micro-actuators utilizing TiNi-based thin films has been a challenge. The frequency reported in thin film TiNi based devices, such as microactuators and micropumps

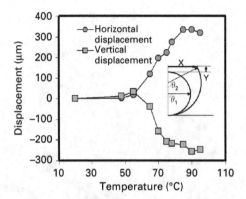

Figure 17.6 Horizontal and vertical displacement of the TiNiCu fingers in the microcage as a function of substrate temperature where negative displacements denote a downwards displacement [33]. (Reproduced with permission from Institute of Physics, UK.)

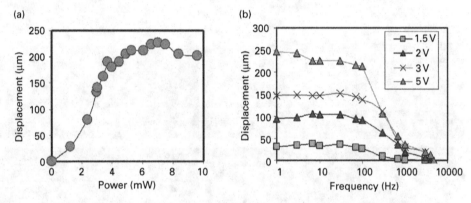

Figure 17.7 (a) The horizontal displacement of the fingers of the microcage as a function of input power applied. (b) The horizontal displacement of the fingers of the microcage as a function of voltage and frequency of the applied actuation square wave [33]. (Reproduced with permission from Institute of Physics, UK.)

[37, 38, 39] generally have maximum operating frequencies in the range 1 to 100 Hz. It should also be noted that the frequency is strongly dependent on the medium in which the device is placed, as well as the mass and dimension of the structure. Actuation in liquid, or thinner beams, and smaller structure dimensions may permit higher operating frequencies to be achieved due to easier thermal dissipation into the liquid.

The microcage structure has been Joule tested up to 18 000 cycles (with a fixed frequency of 20 Hz). With a low applied power less than 5 mW, upon increasing thermal cycles up to a few thousand cycles, there are slight decreases in the horizontal displacement, as well as the original positions, but it does not change much afterwards even up to 18 000 cycles [33]. The initial changes can be attributed to the training process, a common phenomenon for shape memory alloys.

For the TiNi based films, at the initial actuation stage, the repeated phase changes will alter the microstructure and hysteresis of the transformation and in turn lead to changes in transformation temperatures, transformation stresses and strains [17]. The recovery stress of TiNi films was found to decrease dramatically in the first tens of cycles, and became stable after hundreds of cycles [24]. However, at a power above 5 mW, the maximum displacement of the cantilever tip gradually decreases. There is a gradual shift in the room temperature position of the microfingers, and the tips of the microfingers do not return to their original position after 18 000 cycles. The reason is attributed to the fatigue and degradation problem, which may be attributed to the defects in the film. With the power increased to above 8 mW, there is a significant shift of finger tip, especially at the beginning of working cycles. This is attributed to the thermal degradation due to the excessive heating effect. It is apparent that the current or power applied to these microcages has a dramatic effect on the stability or fatigue properties. This issue should be addressed before application of free standing TiNiCu micro-actuators.

17.3   TiNi/DLC microcage fabrication

17.3.1   Design considerations

A schematic bimorph microfinger TiNi/DLC structure is shown in Fig. 17.8. The DLC film has a large compressive stress, and the TiNi film at room temperature (i.e., in the martensite state) normally shows small tensile stress. When the DLC is used as the bottom layer and TiNi as the top layer of the bimorph structure, significant up-bending of the fingers of the microcage could occur when it is released from the Si substrate. The radius of curvature, R, of the bimorph TiNi/DLC structure (shown in Figs 17.8(a) and (b)) after release from the Si substrate can be controlled by changing the thickness ratio of the two layers, or by changing the stress state in both the TiNi and DLC layers [29, 40, 41]:

$$\frac{1}{R} = \frac{6 \cdot \varepsilon_{eq}(1+m)^2}{d[3(1+m)^2 + (1+mn)(m^2 + (mn)^{-1})]} = \varepsilon_{eq} \cdot M, \qquad (17.1)$$

where M is a dimension-related parameter and ε_{eq} is the equivalent strain. Because the as-deposited TiNi film has a very low tensile stress, it is assumed that strain change of the bimorph structure is dominated by the DLC layer. Assuming that stress is uniaxial (since the bimorph is slender), the equivalent strain ε_{eq} in the DLC layer can be simply calculated using the as-deposited stress σ_d and the Young's modulus E_d of the DLC layer, i.e., $\varepsilon_{eq} = \sigma_d/E_d$. In Eq. (17.1), $d = d_1 + d_2$, in which d_1 and d_2 are the thicknesses of the DLC and TiNi layers respectively, n ($= E_1/E_2$) and m ($= d_1/d_2$) are the ratios of the Young's modulus and the layer thickness of the two layers. M is a constant for the fixed materials and structural configuration. Equation (17.1) implies that once the materials and

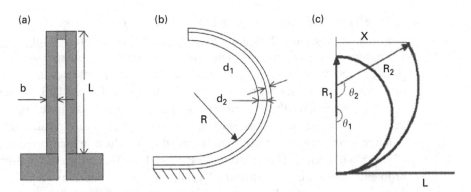

Figure17.8   Schematic drawing of a bimorph TiNi/DLC microfinger structure: (a) top view; (b) cross-section view after bending up; (c) illustration of bending angle and displacement [41]. (Reproduced with permission from Institute of Physics, UK.)

the layer thickness are fixed, the radius of curvature can be adjusted by varying the stress and strain of both the top TiNi layer and bottom DLC layer, which can be realized using different deposition methods and/or different process conditions, as well-documented in Ref. [29].

During thermal cycling, the opening/closing performance of the microcage is mainly determined by (1) thermal effect: temperature change and the difference in the CTEs of the two materials, and (2) the shape memory effect. First, the curvature changes due to a pure thermal effect are analyzed. If the temperature of the bimorph finger increases from T_1 to T_2, due to the difference of CTE between the TiNi and DLC, the thermal expansion mismatch leads to the TiNi layer expanding more than the DLC layer, thus opening the microcage and changing the radius of curvature from R_1 to R_2. The thermal strain, $\varepsilon_{\mathrm{th}}$, generated through resistive Joule heating of the bimorph layer is expressed as $\varepsilon_{\mathrm{th}} = (a_{\mathrm{d}} - a_{\mathrm{TiNi}})\,\Delta T$, where a_{d} and a_{TiNi} are the CTEs of the DLC and TiNi, respectively. The radius of curvature R, bending angle θ (in degrees) and the length of the finger, L, have the following relationship (see Fig. 17.8(c)) [29]:

$$\theta = \frac{180L}{\pi R}. \tag{17.2}$$

Combining Eqs. (17.1) and (17.2), the angle change from θ_1 (which refers to the initial degree of bending at room temperature) to θ_2 (a new position) due to the pure thermal effect from a temperature T_1 to a temperature T_2 ($\Delta T = T_2 - T_1$) can be expressed by [29]:

$$\Delta\theta = \theta_2 - \theta_1 = \frac{180L}{\pi}\left(\frac{1}{R_2} - \frac{1}{R_1}\right) = \frac{180L}{\pi}\Delta\varepsilon_{th}M$$
$$= \frac{180L}{\pi}\Delta a\Delta TM, \tag{17.3}$$

where M is a dimension parameter as given in Eq. (17.1). According to the design, the finger lengths are 50, 100 and 150 μm. The thickness of the DLC layer is

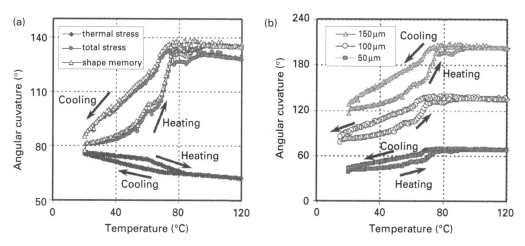

Figure 17.9 Comparison of curvature angular changes (a) due to total stress evolution, thermal effect and shape memory effect for the TiNi/DLC microfinger with beam length of 100 microns; (b) beam lengths of 50, 100 and 150 microns [41]. (Reproduced with permission from Institute of Physics, UK.)

100 nm, and the thickness of the TiNi layer is 800 nm. The stress of the as-deposited DLC film was determined to be 5 GPa. In calculation, the elastic modulus and thermal expansion coefficient of the TiNi film is a variable as a function of temperature, based on the rule of mixtures depending on the percentage of martensite/austenite at a given temperature. The CTE of the DLC is 1×10^{-6} K^{-1}. The elastic modulus of martensite and austenite of the TiNi film are chosen as 30 GPa and 75 GPa and the CTEs of martensite/austenite of the TiNi films are 11 and 9×10^{-6} K^{-1}, respectively. The Young's modulus of DLC is ~600 GPa. The calculated results are shown in Fig. 17.9 [41]. The bending angle decreases slightly with the increase of temperature during heating, indicating the microfingers slightly open owing to the bimorph thermal effect. A hysteresis is observed during the thermal cycle due to differences in forward and martensitic transformations upon heating/cooling (i.e., different contents of martensite and austenite at a certain temperature during thermal cycling).

To calculate the curvature change of the bimorph structure due to the shape memory effect, the stress evolution of the TiNi film on the Si substrate as a function of temperature was measured (see Fig. 17.10) [41]. The stress vs. temperature plot shows a closed hysteresis loop after heating and cooling: a sharp increase in tensile stress is seen in response to the phase transformation from martensite to austenite upon heating; upon cooling, the stress relaxes significantly due to the martensitic transformation. Since the thermal stress also contributes to the stress evolution of the TiNi/Si structure, the stress evolution due to pure shape memory effect can be obtained by deducting the thermal stress from the total stress evolution. The obtained stress change induced by the shape memory effect can be substituted into Eq. (17.1) to estimate the angle changes of the finger due to

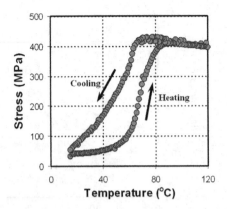

Figure 17.10 Stress evolution as a function of temperature for TiNi film on silicon substrate [41]. (Reproduced with permission from Institute of Physics, UK.)

the shape memory effect.

$$\Delta\theta_{SME} = \frac{180L}{\pi}\Delta\varepsilon_{SME}M$$
$$= \frac{180L}{\pi}\left(\frac{\Delta\sigma_{total}}{E_{TiNi}/(1-v)} - \Delta a \cdot \Delta T\right)M. \qquad (17.4)$$

The CTE of Si is 2.6×10^{-6} K^{-1}. Figure 17.9(a) shows the estimated curvature angular changes of the TiNi/DLC microfinger due to the shape memory effect as well as total stress (for a microfinger with a length of 100 microns). The curvature angular changes significantly (by about 60°) with temperature increased from room temperature to 80 °C. During heating, the bending angle increases with temperature indicating that the fingers of the microcage close, which is opposite to the opening caused by the thermal effect (Figure 17.9(a)). Clearly, the shape memory effect dominates the microfinger deformations within this temperature range, and a hysteresis for the angular change as a function of temperature can be observed. Figure 17.9(b) shows the effects of the finger beam length on the curvature angle changes during a thermal cycle. The longer the finger beam is, the larger the initial bending curvature angle after release from the Si substrate, and the more significant the changes in curvature angular during heating/cooling [41].

From the results shown in Figs. 17.9(a) and (b), it is predicted that there will be two possible curvature changes during heating/cooling. At initial heating, there will be insignificant opening of the microcage fingers due to the thermal effect. When heated above the austenite start transformation temperature, the microfingers close significantly due to the shape memory effect, which executes the action of capturing the micro-object. Opening of the microcage can be realized by either decreasing the temperature (due to the martensitic transformation) or by further increasing the temperature (thermal bimorph effect, but it may need a very high temperature, thus it is not desirable for practical application).

A rough estimation according to Eq. (17.1) indicates that if the thickness ratio of TiNi/DLC is 1, there will be a maximum bending effect. However, the thickness of TiNi should be larger than a few hundred nanometers, below which the shape memory effect will be too weak for an efficient actuation [42]. When the film is too thin, surface oxide and film/substrate interfacial diffusion layers exert a dominant constraining effect that renders high residual stress and low recovery capabilities [42,43]. The surface oxide and inter-diffusion layer restricts the phase transformation and alters the chemical stoichiometry of the remaining TiNi film, which effectively reduces the volume of the material available for phase transformation. It was reported that a maximum recovery stress and actuation speed can be realized with a TiNi film thickness at about 1 micron [29]. On the other hand, there is also a limit to the DLC thickness. When the thickness of DLC is above 100 nm, the DLC layer may peel off from the Si substrate due to intrinsic stress. This has severely restricted the usage of DLC of a few hundred nanometers thickness.

17.3.2   Fabrication and characterization

Microcages of five, six and seven fingers were designed. The width of the fingers and the gap between the beams were 4 μm. The fingers were connected to each other with bond pads. The central part of the microcage was large enough so that it remained attached to the substrate after the fingers were released from the substrate. A DLC film of 100 nm was deposited on an Si substrate using a filtered cathodic vacuum arc (FCVA) method with a graphite source. The compressive stress of the film was 5 GPa as determined via curvature measurement. A TiNi film 800 nm thick was deposited on top of the DLC layer by magnetron sputtering in an argon gas environment at a pressure of 0.8 mtorr from a Ti50Ni50 target (using a 400 W rf power) and a 99.99 % pure Ti target (using a 70 W DC power). Post-annealing of the TiNi/DLC bimorph layer was performed at 480 °C for 30 min in a high vacuum condition for crystallization.

The DLC/TiNi microcage was fabricated by photolithographical patterning of a 4.8 μm thick AZ4562 photoresist on top of the TiNi films. An HF:HNO$_3$:H$_2$O (1:1:20) solution was employed to etch the TiNi films to form the microcage patterns. The exposed DLC underlayer was etched off in oxygen plasma at a flow rate of 80 sccm and a power of 100 W. A deep reactive ion etching machine was used to isotropically etch the silicon substrate with SF$_6$ plasma at a flow rate of 70 sccm and pressure of 72 mtorr. The coil and platen powers were set at 50 W and 100 W, respectively. A time controlled etching was performed to release the fingers leaving the middle parts of the microcage attached to the Si-substrate. The actuation performance of the released microgrippers was evaluated using a Peltier device (with a temperature range 5 °C to 100 °C). The displacements of the TiNi pattern were measured using a video camera from the top view.

An SEM morphology of the fabricated TiNi/DLC microcages on a 10-cm silicon wafer is shown in Fig. 17.11(a) [41]. The microcages have different finger

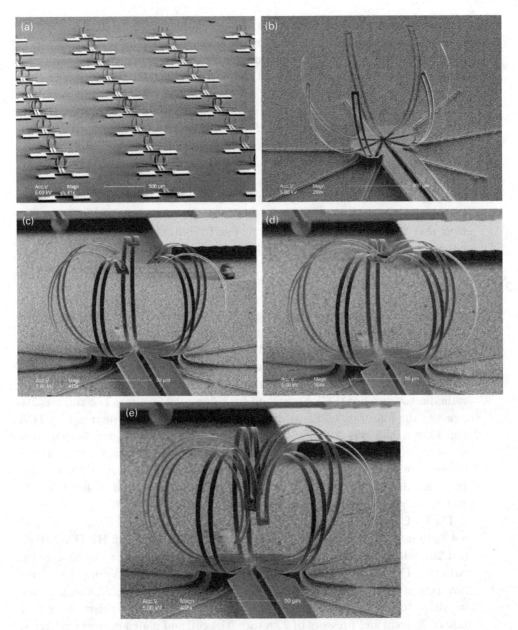

Figure 17.11 TiNi bending-up microcage structure (a) on a 10-cm wafer; with different finger lengths (b) 100 microns; (c) 120 microns; (d) 180 microns; (e) 200 microns [41]. (Reproduced with permission from Institute of Physics, UK.)

numbers and beam lengths. After released from the Si substrate, the microcages show significant curling up of the microfingers, depending on the beam lengths. The fabricated seven-finger microcages with different beam lengths are shown in Figs. 17.11 (b) to (e). As designed, with increase of beam length, the microfinger

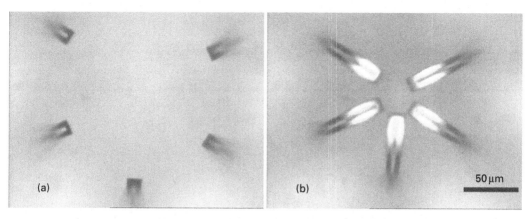

Figure 17.12 Optical microscopy images showing the closing of a five-finger microcage during heating (during cooling the process is reversed, beam length of 100 microns): (a) 20 °C (b) 100 °C [41]. (Reproduced with permission from Institute of Physics, UK.)

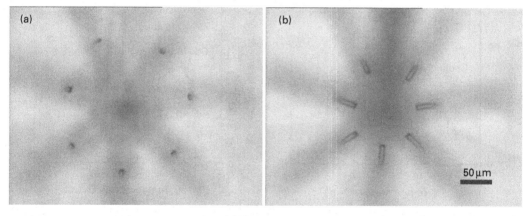

Figure 17.13 Optical microscopy images showing the opening/closing of a microcage during heating (during cooling the process is reversed; beam length: 150 microns). (a) 20 °C (b) 100 °C [41]. (Reproduced with permission from Institute of Physics, UK.)

patterns of the microcages change from fully open to under-closed to over-closed (see Figs. 17.11(b) to (e)) [41].

Figure 17.12 shows the top-view optical images of the deformation behavior of a 5-finger microcage upon heating [41]. Actuation of the microcage is mainly determined by the shape memory effect. With temperature increased above the martensitic start transformation temperature (about 50 to 60 °C), martensite (loose structure) changes to austenite (a dense structure), causing the closing of microcages, and capturing an object (see Figs. 17.12(a) and b). Further increase in temperature above 100 °C causes the slight opening of the microfingers due to thermal effects. Upon cooling, the microcage fingers quickly open due to the martensitic transformation. Figures 17.13(a) and (b) show the top-view optical

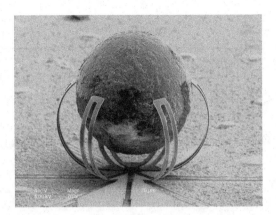

Figure 17.14 SEM image of a microcage capturing a micropolymer ball [41]. (Reproduced with permission from Institute of Physics, UK.)

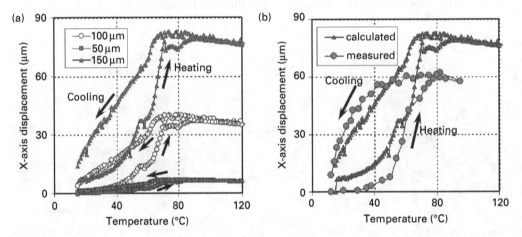

Figure 17.15 Horizontal displacement as a function of temperature for microfinger (a) with different finger lengths; and (b) experimental results and calculated results with a finger length of 150 microns [41]. (Reproduced with permission from Institute of Physics, UK.)

images of a 7-finger microcage, also showing a dramatic closing/opening of the microcage during heating/cooling. As one example of a real application, Fig. 17.14 shows an SEM image of the microcage capturing a micro-polymer ball with a diameter of 50 microns [41].

Based on the calculation results of curvature angular change for the beams with different lengths shown in Fig. 17.9(b), the microfinger displacement in the horizontal direction can be estimated from the simplified relationship: $\sin(180 - \theta) = x/R = (x\theta)/L$, and the results are shown in Fig. 17.15(a) [41]. Beam length has a dramatic influence on the opening displacement in the x-direction. Figure 17.15 (b) plots the experimentally measured changes in the horizontal displacement of a microcage fingertip (beam length of 150 microns) as a function of substrate temperature. Horizontal displacements of 50 to 60 micrometers can be achieved at temperatures less than 80 °C. The theoretical calculated results are comparable with the results, as shown in Fig. 17.15(b). However, there is a discrepancy between the estimated and the measured results. Several reasons are possible for this discrepancy. (1) In the theoretical calculation, the stress–temperature results measured from a film on a 10-cm Si wafer were used. However, the real stress–temperature relationship could be different from that on a TiNi /DLC bimorph structure. (2) In measurement, the temperature of the fingers of the microcage should be lower than those measured on the surface of the peltier device. (3) During the fabrication of DLC/ TiNi microcages, a post-annealing crystallization process was used which could have relaxed some of the DLC stress. Also the plasma releasing of the microfinger from the silicon substrate could deteriorate the shape memory effect of the TiNi film due to the plasma damage on the surface of the material. These effects should be considered when designing microcages for practical usages.

17.4 Biological study of the TiNi film

When TiNi based films are applied in biomedical fields, they must be capable of fulfilling functional requirements relating not only to mechanical reliability but also to chemical reliability (in vivo degradation, decomposition, dissolution and corrosion) and biological reliability (toxicity, antigenicity, etc.). Although TiNi has been recognized as a good material for biological applications, it is still not clear if the release of a small amount of nickel and copper could cause allergy and inflammation of human organs [44].

Surface chemistry of the Ti50Ni50 film was analyzed using a Kratos AXIS spectrometer with monochromatic Al Kα (1486.71 eV) X-ray radiation. The survey spectra in the range 0–1100 eV was recorded for each sample, followed by high-resolution spectra over different element peaks, from which the detailed compositions were calculated. The XPS survey spectrum of the as-deposited TiNi film (Fig. 17.16) shows that the dominant signals are from Ti, C and O, while Ni content is relatively low [45]. Elemental surface composition analysis indicates that oxygen content is about 36.9 at%, Ti 27.3 at%, carbon 26.6 at%, Ni 9.2 at%. As soon as the TiNi film is exposed to the ambient, oxygen and carbon are quickly adsorbed on the surface because titanium has a high affinity for oxygen and carbon (chemisorb barrier energy or enthalpy for Ti and Ni to react with oxygen at the surface are 241 KJ/mol and 956 KJ/mol respectively [45]). The high nickel content in the TiNi films often causes concern about its suitability for medical use. The presence of a TiO_2 oxide layer on the TiNi film is beneficial to its corrosion resistance and biocompatibility. The XPS survey spectrum on the TiNi film exposed to ambient atmosphere for three months is shown in Fig. 17.16 (the middle spectrum). The dominant signals are from C, O and Ti, and the Ni peak becomes very weak (1.64 at%). This indicates that the outermost layer of the TiNi film is a titanium oxide and carbon contamination layer. Titanium and nickel increase significantly in the first 20 nm. A Ti-rich layer is observed in the first

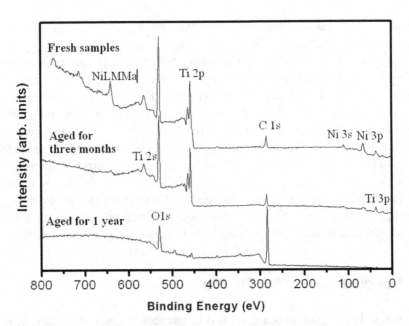

Figure 17.16   XPS survey spectrum for TiNi film aged for different periods [45]. (Reproduced with permission from Elsevier, UK.)

20 nm, while a slightly Ni-rich layer is observed beneath this layer. In Fig. 17.16 the bottom spectrum is the survey spectrum of the film exposed to the air for one year. The C 1s peak is very strong, and the Ni peaks disappear. High resolution spectral analysis reveals that carbon content is as high as 65.57%, oxygen content is about 26.37%, Ti content 8.06%, and that of the Ni is below the detection limit of XPS. Figure 17.17 shows the elemental depth profiling of the surface layer for a sample aged for one year. The oxygen diffusion layer is about 60 nm thick, and the carbon layer 30 nm. Titanium and nickel increase significantly in the first 25 nm.

These findings show that sputtered TiNi thin films are easily contaminated with carbon and oxygen in air. With exposure to the atmosphere, carbon and oxygen increase drastically at the surface, and at the same time diffuse deep into the film. The increase is very fast at the beginning but slows down with time after a long exposure. This is beneficial for the compatibility of the TiNi films.

For the biological evaluation of the devices, cell culture study was performed on both the TiNi and DLC films. The cell line used for the culturing was COS7 (African green monkey kidney fibroblast). The autoclave-sterilized samples (10 mm by 10 mm) were placed in a 24-well plate for fibroblast seeding at a set density of 1.4×10^4 cells/ml. The cells were incubated in Dulbecco's modified eagle's medium supplemented by 10% fetal calf serum and 1% penicillin for a total period of six days. The incubation was carried out at $37\,^{\circ}C/5\%$ CO_2 in air in a humidified incubator. The cells were trysinized to detach them from the sample surface. After being extracted from the culture plate, the cells and medium mixture

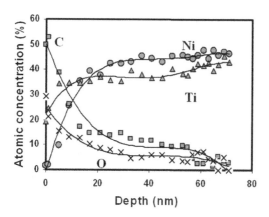

Figure 17.17 XPS depth analysis of the TiNi film exposed in ambient for one year [45]. (Reproduced with permission from Elsevier, UK.)

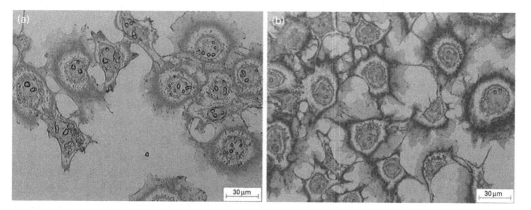

Figure 17.18 Typical morphology of cells attached on surfaces of (a) TiNi and (b) DLC [41]. (Reproduced with permission from Institute of Physics, UK.)

was centrifuged in order for them to be separated. The supernatant was removed and 1 ml of fresh medium was added and mixed well. 20 µl of cell suspension was then mixed with 180 µl of trypan blue, and injected into the chambers of a hemacytometer for counting. Three replicated samples were processed for each coating. For cell morphology observation, the fibroblasts attached on the coatings were fixed with 2.5% gluteraldehyde for 30 min, and then dehydrated.

Figure 17.18 shows the fibroblast cell attachment on both the TiNi and DLC film surfaces after two-day incubation; well-attached and proliferated cells can be observed, indicating fast proliferation and differentiation of the cells [41]. Analysis of variance (ANOVA) was used to determine if the growth is actually a true proliferation. This was conducted for the counts at different durations up to six days for the TiNi and DLC films. Results in Fig. 17.19 showed that the cells are truly proliferating on both surfaces, and the confidence level is much better than 99.99% [41]. There are no significant differences between the counts on

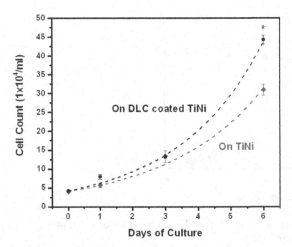

Figure 17.19 Cell growth curve on TiNi and DLC coated TiNi surfaces. * denotes significant difference from TiNi [41]. (Reproduced with permission from Institute of Physics, UK.)

both surfaces within the three-day culture, but the cell count on DLC coated TiNi on the sixth day is significantly higher than that on TiNi with a confidence level of 99%. Although the growth rate of the cells is higher on DLC, the exponential growth and the good attachment and spreading of the cells on both surfaces show that both materials are compatible to the cell line and do not exhibit any cytotoxicity. High nickel content in the TiNi films often causes suspicion of its suitability for medical use. However, as soon as the TiNi film is exposed to the ambient, oxygen and carbon are quickly adsorbed on the surface because titanium has a high affinity for oxygen and carbon. The presence of a TiO_2 oxide layer (about 50 nm) on the TiNi film is beneficial to its corrosion resistance and biocompatibility [45].

17.5   Summary

Based on the good mechanical properties and two-way shape memory effect of freestanding TiNiCu films, a freestanding TiNiCu microcage has been fabricated. With either substrate heating or Joule heating, the microcages can be opened and closed, enabling them to trap or hold micro-objects. Results showed that the microcage could be actuated with a maximum temperature of 90 °C, and a low power less than 10 mW with a maximum frequency of 300 Hz. Stability of movement of the freestanding structure depends much on the applied power. During plasma etching, the beam bending could be different due to the existence of an interlayer. A TiNi/diamond-like-carbon (DLC) microcage for biological application has also been designed, fabricated and characterized. A compressively stressed DLC film with TiNi pattern on top expands and lifts the fingers upwards once they are released from the substrate, and the microcage can be closed through the shape memory effect of the top TiNi film with a temperature below 80 °C. On further heating above 100 °C, the gradual opening of the microcage can be obtained due to the thermal bimorph effect. Cells proliferate and spread well on both TiNi and DLC films indicating no sign of cytotoxicity towards the cell-line.

Acknowledgement

Dr. A. Flewitt and Prof. Bill Milne from the University of Cambridge, UK are acknowledged.

References

[1] G. Iddan, G. Meron, A. Glukhovsky and P. Swain, Wireless capsule endoscopy, *Nature*, **405** (2000) 417.

[2] J. D. Waye, Small-bowel endoscopy, *Endoscopy*, **35** (2003) 15.

[3] C. S. Pan and W. Y. Hsu, An electro-thermally and laterally driven polysilicon microactuator, *J. Micromech. Microeng.* **7** (1997) 7–13.

[4] J. Cecil, D. Powell and D. Vasquez, Assembly and manipulation of micro devices – a state of the art survey, *Robotics and Computer-Integrated Manufacturing*, **23** (2007) 580–588.

[5] B. E. Volland, H. Heerlein and I. W. Rangelow, Electrostatically driven microgripper, *Microelectr. Engng.*, **61–62** (2002) 1015–1023.

[6] R. Wierzbicki, K. Houston, H. Heerlein, *et al.* Design and fabrication of an electrostatically driven microgripper for blood vessel manipulation, *Microelectr. Engng.*, **83** (2006) 1651–1654.

[7] J. K. Luo, A. J. Flewitt, S. M. Spearing, N. A. Fleck and W. I. Milne, Three types of planar structure microspring electro-thermal actuators with insulating beam constraints, *J. Micromech. Microeng.*, **15** (2005) 1294.

[8] C. Reichle, T. Schnelle, T. Muller, T. Leya and G. Fuhr. A new microsystem for automated electrorotation measurements using laser tweezers, *Biochim. Biophys. Acta*, **1459** (2000) 218–229.

[9] M. Ericsson, D. Hanstorp, P. Hagberg, J. Enger and T. Nystrom, Sorting out bacterial viability with optical tweezers, *J. Bacteriol.*, **182** (2000) 5551–5555.

[10] K. C. Neuman, E. H. Chadd, G. F. Liou, K. Bergman and S. M. Block. Characterization of photodamage to *Escherichia coli* in optical traps, *Biophys. J.* **77** (1999) 2856–2863.

[11] B. E. Volland, Electrostatically driven microgripper, *Microelectr. Eng.*, **61** (2002) 1015.

[12] C. J. Kim, A. P. Pisano and R. S. Muller, Silicon-processed overhanging microgripper, *J. Microelectromech. Syst.*, **1** (1992) 31–36.

[13] O. Millet, P. Bernardoni, S. Regnier, *et al.*, Electrostatic actuated micro gripper using an amplification mechanism, *Sens. Actuat. A*, **114** (2004) 371–378.

[14] P. R. C. Gascoyne and J. Vykoukal, Particle separation by dielectrophoresis, *Electrophoresis*, **23** (2002) 1973–1983.

[15] N. Chronis and L. P. Lee, Electrothermally activated SU-8 microgripper for single cell manipulation in solution, *J. MEMS*, **14** (2005) 857–863.

[16] J. K. Luo, A. J. Flewitt, S. M. Spearing, N. A. Fleck and W. I. Milne, Comparison of microtweezers based on three lateral thermal actuator configurations, *J. Micromech. Microeng.*, **15** (2005) 1294–1302.

[17] S. Miyazaki and A. Ishida, Martensitic transformation and shape memory behavior in sputter-deposited TiNi-base thin films, *Mater. Sci. Eng. A*, **273–275** (1999) 106–133.

[18] H. Kahn, M. A. Huff and A. H. Heuer, The TiNi shape-memory alloy and its applications for MEMS, *J. Micromech. Microeng.*, **8** (1998) 213–221.

[19] Y. Q. Fu, W. M. Huang, H. J. Du, X. Huang, J. P. Tan and X. Y. Gao, Characterization of TiNi shape-memory alloy thin films for MEMS applications, *Surf. Coat. Technol.* **145** (2001) 107–112.

[20] P. Krulevitch, A. P. Lee, P. B. Ramsey, J. C. Trevino, J. Hamilton and M. A. Northrup, Thin film shape memory alloy microactuators, *J. MEMS*, **5** (1996) 270–282.

[21] V. Seidemann, S. Butefisch and S. Buttgenbach, Fabrication and investigation of in-plane compliant SU8 structures for MEMS and their application to micro valves and micro grippers, *Sens. Actuator. A*, **97–98** (2002) 457–461.

[22] S. Takeuchi and I. Shimoyama, A three-dimensional shape memory alloy microelectrode with clipping structure for insect neural recording, *J. MEMS*, **9** (2000) 24–31.

[23] A. P. Lee, D. R. Ciarlo, P. A. Krulevitch, S. Lehew, J. Trevino and M. A. Northrup, A practical microgripper by fine alignment, eutectic bonding and SMA actuation, *Sens. Actuators A*, **54** (1996) 755–759.

[24] Y. Q. Fu, H. J. Du, W. M. Huang, S. Zhang and M. Hu, TiNi-based thin films in MEMS applications: a review, *Sens. Actuators A*, **112** (2004) 395–408.

[25] R. X. Wang, Y. Zohar and M. Wong, Residual stress-loaded titanium-nickel shape-memory alloy thin-film micro-actuators, *J. Micromech. Microeng.* **12** (2002) 323–327.

[26] J. J. Gill, D. T. Chang, L. A. Momoda, G. P. Carman, Manufacturing issues of thin film NiTi microwrapper, *Sens. Actuator, A*, **93** (2001) 148–156.

[27] G. Dearnaley J. H. Arps, Biomedical applications of diamond-like carbon (DLC) coatings: a review, *Surf. Coat. Technol.*, **200** (2005) 2518–2524.

[28] J. Robertson, Diamond like amorphous carbon, *Mat. Sci. Eng. R*, **37** (2002) 129.

[29] J. K. Luo, J. H. He, Y. Q. Fu, A. J. Flewitt, N. A. Fleck W. I. Milne, MEMS based digital variable capacitors with a high-k dielectric insulator, *J. Micromech. Microeng.* **15** (2005) 1406–1413.

[30] J. K. Luo, R. Huang, J. H. He, *et al.*, Modelling and fabrication of low operation temperature microcages with a polymer/metal/DLC trilayer structure, *Sens. Actuators A: Physical*, **132** (2006) 346–353.

[31] I. Roch, P. Bidaud, D. Collard and L. Buchaillot, Fabrication and characterization of an SU-8 gripper actuated by a shape memory alloy thin film, *J. Micromech. Microengng.*, **13** (2003) 330–336.

[32] M. Kohl, B. Krevet and E. Just, SMA microgripper system, *Sens. Actuators A*, **97–98** (2002) 646–652.

[33] Y. Q. Fu, J. K. Luo, A. J. Flewitt *et al.*, Micro-actuators of free-standing TiNiCu Films, *Smart. Mater. Struct.*, **16** (2007) 2651–2657.

[34] Y. Q. Fu and H. J. Du, RF magnetron sputtered TiNiCu shape memory alloy thin film, *Mater. Sci. Eng. A*, **342** (2003) 236–244.

[35] E. Quandt, C. Halene, H. Holleck, *et al.*, Sputter deposition of NiTi, NiTiPd and TiPd films displaying the two-way shape-memory effect, *Sens. Actuat. A*, **53** (1996), 434–439.

[36] A. Gyobu, Y. Kawamura, H. Horikawa and T. Saburi. Two-way shape memory effect of sputter-deposited Ti-rich Ti-Ni alloy films, *Mater. Sci. Engng. A*, **273–275** (1999) 749–753.

[37] J. L. Seguin, M. Bendahan, A. Isalgue, V. Esteve-Cano, H. Carchano and V. Torra, Low temperature crystallised Ti-rich NiTi shape memory alloy films for micro-actuators, *Sens. Actuators*, **74** (1999) 65–69.

[38] W. L. Benard, H. Kahn, A. H. Heuer and M. A. Huff, Thin-film shape-memory alloy actuated micropumps, *J. MEMS*, **7** (1998) 245–251.

[39] E. Makino, T. Mitsuya and T. Shibata, Micromachining of NiTi shape memory thin film for fabrication of micropump, *Sens. Actuators*, **79** (2000), 128–135.

[40] Y. C. Tsui and T. W. Clyne, An analytical model for predicting residual stresses in progressively deposited coatings. 1. Planar geometry, *Thin Solid Films*, **306** (1997) 23.

[41] Y. Q. Fu, J. K. Luo, S. E. Ong, S. Zhang, A. J. Flewitt and W. I. Milne, TiNi/DLC shape memory microcage for biological applications, *J. Micromech. Microeng.*, **18** (2008) 035026.

[42] Y. Q. Fu, S. Zhang, M. J. Wu, *et al.*, On the lower thickness boundary of sputtered TiNi films for shape memory application, *Thin Solid Films*, **515** (2006) 80–86.

[43] A. Ishida and M. Sato, Thickness effect on shape memory behavior of Ti-50.0at.%Ni thin film, *Acta Mater.*, **51** (2003) 5571–5578.

[44] M. Es-Souni and H. F. Brandies, On the transformation behaviour, mechanical properties and biocompatibility of two TiNi-based shape memory alloys: NiTi42 and NiTi42Cu7, *Biomaterials*, **22** (2001) 2153.

[45] Y. Q. Fu, H. J. Du, S. Zhang and W. M. Huang, XPS characterization of surface and interfacial structure of sputtered TiNi films on Si substrate, *Mater. Sci. Engng. A*, **403** (2005) 25–31.

18 Shape memory thin film composite microactuators

Eckhard Quandt

Abstract

Shape memory thin film composites consisting of at least one shape memory thin film component are of special interest as microactuators since they provide two-way shape memory behavior without any training of the shape memory material. Furthermore, they allow the realization of novel concepts like bistable or phase-coupled shape memory actuators. The potential of shape memory thin film composite microactuators is discussed in view of possible applications.

18.1 Introduction

The availability of suitable actuators is of particular importance for the advanced development of novel micro- or nanosystems. An attractive approach is based on smart materials that directly transduce electrical into mechanical energy, e.g., materials exhibiting a shape-memory effect, piezoelectricity or magnetostriction. Actuator mechanisms on these smart materials allow easy down-scaling to the micrometer or even nanometer range and can be realized by a cost-effective fabrication technique compatible with micro- or nanosystem technology using thin-film technologies. As shape memory alloys combine high output forces with large motions and can be controlled by Joule heat in a microelectronic-compatible way, they are of increasing interest for microactuator applications as long as only low frequencies or slow response times are required. Due to smaller grain sizes compared with those of bulk materials, magnetron sputtered shape memory thin films can be fabricated between approximately 1 and $50\,\mu m$ in thickness with properties not affected by the film thickness.

In this chapter the mechanism, fabrication and features of shape memory thin film composite microactuators are discussed in view of special designs (bistable, phase-coupled actuators) and applications.

Thin Film Shape Memory Alloys: Fundamentals and Device Applications, eds. Shuichi Miyazaki, Yong Qing Fu and Wei Min Huang. Published by Cambridge University Press. © Cambridge University Press 2009.

18.2   Mechanism of shape memory composites

Of special interest for applications in micro- and nanosystems are shape memory materials which show a two-way behavior. This can be achieved by combining the one-way shape memory with the bimetallic effect [1]. In this case one of the two layers of a bimetallic strip is a shape memory material and the other one can be the substrate material. Thermally induced stresses generated by different thermal expansion coefficients of the two materials can be resolved by the martensitic transformation of the shape memory component within a narrow temperature range of 10–30 °C. According to the expansion coefficients of the substrate materials, tensile stresses can be induced in shape memory films on Si and Mo substrates, while compressive stresses are induced in films on stainless steel sheets. The work output is defined by the elastic energy of the shape memory film that can be reduced by the martensitic transformation. In the case of Ti(Ni,Cu)/Mo composites, the work output is about ten times larger than the work output of the corresponding bimetallic effect. While the bimetallic effect is linear within a certain temperature range, shape memory thin film composites exhibit a non-linear behavior in combination with a hysteresis directly related to the hysteresis of the shape memory material.

18.3   Fabrication of shape memory composites

The shape memory component is deposited on to the substrate that forms the second component using thin film techniques, while crystallization can be obtained both during and after the deposition. Alternatively, the bimetallic compound of the composite can also be fabricated by thin film deposition. In this case sacrificial layers are used to (partially) release the shape memory thin film composite from the substrate. The evolution of stress in the shape memory composite during film post-annealing and further actuation can be seen in Fig. 18.1. During deposition compressive stress is being built up (1) which is released during the post-deposition heat treatment at 600 °C (2). Upon cooling a bimetallic tensile stress is evolved which is almost linear with temperature (3). At the martensitic transformation starting at M_S the tensile stress is partially released due to a suitable arrangement of the martensitic variants (4). Upon heating to the austenitic phase, the shape and the corresponding stress of the shape memory composite in the austenitic phase are restored in a temperature range dependent on the hysteresis of the shape memory component (5). The shape memory film of the composite based on TiNi can be fabricated using different techniques like DC or RF magnetron sputtering [2], ion beam sputtering [3] or laser ablation [4]. In binary TiNi materials, limitations arise due to their relatively low transition temperatures of less than 60 °C/140 °C (martensite/austenite finish temperature, respectively) in combination with a quite large hysteresis for TiNi films exhibiting no intermediate R-phase transition, and even lower temperatures (less than 50 °C)

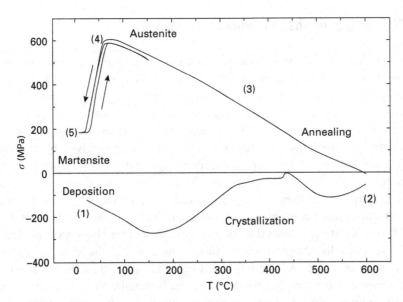

Figure 18.1   Development of stress in shape memory thin film composites during annealing ((1) to (3)) and during operation ((4) and (5)).

for Ni-rich films showing the R-phase transition with its narrow hysteresis. In order to reduce hysteresis in combination with large effect sizes, films of the ternary alloy Ti(NiCu) [5] can be implemented while (TiHf)Ni [6] alloys increase the hysteresis. To significantly increase the transition temperatures ternary Ti (NiPd) [7] films are the best choice. Typical transformation temperatures, their hysteresis and the transformation stress (effect size) for different TiNiX thin film composites are shown in Fig. 18.2.

Due to the transition between low stress in the low-temperature phase to high stress in the high-temperature phase the usual shape of these actuators is almost flat in the martensitic state and shows a significant curvature in the austenitic state. However, by designing composites of more than two layers more complex motions like a transition from convex to concave curvature or compensation of the overlaying bimetallic effect can be obtained (Fig. 18.3).

18.4   Bistable SMA composites

A general disadvantage of shape memory actuators is their high energy consumption, especially in applications which require long periods of the actuator being in the austenitic state. In the case of microsystems the heating of the total device is an additional problem which has to be addressed. For this purpose, there is an interest in actuators that only require energy for the change of the actuator state.

Bistable actuators can be obtained as a specific combination of a shape memory thin film composite and a suitable polymer. Figure 18.4 illustrates the functional

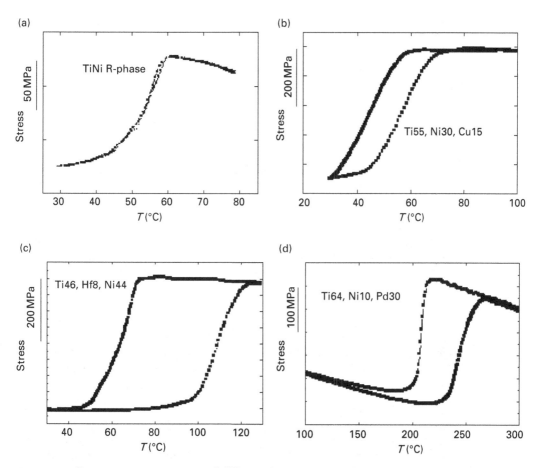

Figure 18.2 Stress–temperature curves of different shape memory thin composites (2 mm TiNiX/50 mm Mo composites).

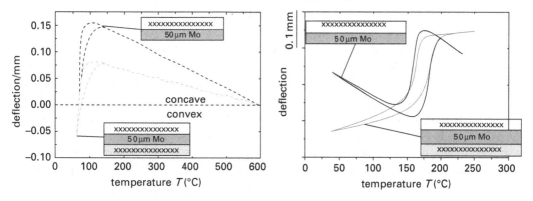

Figure 18.3 Stress–temperature curves of shape memory thin composites containing a third layer in order to achieve a deflection from the convex to concave bending (left) or to compensate for the bimetallic effect (right).

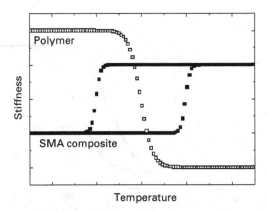

Figure 18.4   Schematic stiffness–temperature curves of a shape memory thin composite with large hysteresis and for a polymer showing a glass transition temperature in the temperature range defined by the hysteresis of the shape memory material. This adjustment is the essential requirement to achieve bistable behavior in shape memory thin film–metal–polymer composites.

principle of such a bistable shape memory thin film composite actuator [8]. The diagram illustrates the superposition of the temperature dependent stiffness of the polymer and the hysteresis of the shape memory composite, exhibiting shape memory two-way behavior. The essential condition for obtaining a bistable behavior is that the temperature of the glass transition (T_g) of the thermoelastic polymer lies within the hysteresis of the shape memory alloy. Upon heating, first the polymer becomes soft due to its glass transition at T_g before the shape memory thin film starts to undergo the transformation from martensite to austenite at the austenite start temperature A_s. Due to the low stiffness of the polymer above T_g, the actuator is able to bend and allows the total polymer-shape memory composite to transform into the shape defined by the austenitic state. Upon cooling, first the polymer becomes stiff below its glass transition temperature T_g before the shape memory material transforms into the martensitic state. If the thickness of the polymer is high enough, the stiffness of the polymer below T_g is sufficient to keep the total composite in the curved shape given by the austenitic phase even after its transformation to the martensitic state. Therefore, the actuator remains in its bended state. To switch the bistable actuator to the flat martensitic shape, the actuator has to be heated to a temperature between T_g and A_s. Above the glass transition T_g the polymer becomes soft and therefore releases the shape memory composite which is still martensitic and tends to go to the flat shape as long as the temperature is below A_s. Upon cooling, the flat shape is fixed again by the polymer. Due to this special bistable design, energy consumption is only needed during switching processes, while the temperature or duration of the heating pulse defines the final shape of the actuator.

The images in Fig. 18.5 indicate that combining the TiNiHf composites with a Lucryl polymer leads to an actuator with a bistable behavior having similar curvatures to normal shape memory thin film composites. The thickness of the

Figure 18.5   Comparison of normal (a),(b) and bistable (c),(d) shape memory thin film composites (10 mm TiHfNi/ 25 mm Mo): (a) shows the curvature of the martensitic phase taken at room temperature, (b) the austenitic phase taken at 120 °C, (c),(d) the corresponding bistable states both taken at room temperature. The lateral dimensions are 20 mm × 20 mm, the thickness of the polymer is 75 mm.

polymer layers surrounding the metal is 75 μm and the thickness of the SMA composite is 35 μm (25 μm Mo + 10 μm shape memory thin film). The bistable actuator can be heated either directly by Joule heating, contactless using an induction heater or by an environmental temperature change. For the design of the actuator the heating method has to be considered, especially for the selection of the corresponding materials since the direction of the heat flow during heating is opposed to the case of heating by the ambient temperature.

18.5   Phase-coupled SMA composites

Not taking into account any possible mixed states of martensitic and austenitic phases, a composite using one shape memory component is characterized by two (final) states which have different transformation temperatures upon heating and cooling due to the hysteresis of the shape memory component. If two shape

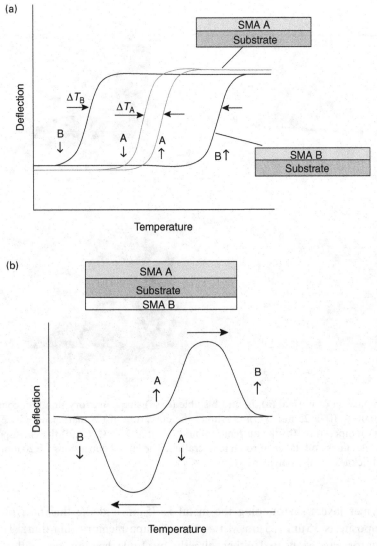

Figure 18.6   Deflection–temperature curves for two shape memory thin film composites with small (SMA A) and broad (SMA B) hysteresis, respectively, with a martensite–austenite transformation in the same temperature range (a) and the deflection–temperature–curve for a three layer composite showing the different transformation paths upon heating and cooling (b).

memory alloys with different transformation temperatures are integrated into one composite, in general four different states can be generated (martensitic/martensitic, martensitic/austenitic, austenitic/martensitic, austenitic/austenitic). Upon heating and cooling, the order in which the four stages occur can be different. This can be demonstrated for two shape memory materials, one having a narrow and the other a broad hysteresis with the narrow one lying within the broad one. Such shape memory thin film composites can be realized with shape memory alloys of the systems Ti(Ni,Cu) and (Ti,Hf)Ni (Fig. 18.6(a)). The results of the deflection versus temperature measurements of the individual composites TiNiCu/Mo and TiNiHf/Mo show that these alloys have a similar effect size and that the narrow hysteresis of the TiNiCu alloy lies within the broad hysteresis of the TiNiHf alloy. Fig. 18.6(b) shows that the shape of the actuator between the two final states (both martensitic or both austenitic) follows different paths upon heating or cooling. Thus this mechanism is able to provide wave-like or inch-worm behavior and can be used to generate travelling waves. The corresponding composite structure has been developed with alternating Ti52Ni34Cu14 and Ti47Hf11Ni42 films deposited on both sides of a Mo foil [9].

18.6 Applications of shape memory thin film composites

In general, both free standing films and thin film composites can be used for actuators in micro- or nanotechnology. If the application requires very large strains free standing films should be considered while high stresses or forces are developed by composites. The following part reviews a number of suggested shape memory thin film composite actuators and their applications.

The first applications of shape memory thin films were devices for microfluidics like microvalves [10] and micropumps [11] and for microgrippers [12] which can be used for microassembly tasks. Figure 18.7 shows microgrippers with TiNiCu/Mo thin film composite actuators on micropatterned silicon [9]. The large deflections of the actuators demonstrate the uniqueness of SMA thin film composites. In contrast to other materials, SMA composites are able to provide deflections which are of the same order of magnitude as the dimensions of the complete gripper system. For the assembling of microparts that have to be handled under high vacuum in the chamber of a scanning electron microscope, bistable microgrippers can be used. Due to the bistability, these grippers require an energy supply only to open and close the gripper. Additionally, an excessive heat flow from the SMA layer to the objects that have to be handled can be avoided. In contrast to microgrippers based on piezoelectricity or magnetic actuations, the influence of the actuator on the SEM resolution should be reduced due to the lower electromagnetic field strength required to operate the microgripper.

Based on shape memory thin film composites a novel mm-sized optical microscanner was developed that shows large deflection angles up to 170° in combination with maximum operational frequencies up to 100 Hz [13]. The performance of this microscanner is based on the strong change of bimorph stress in shape memory thin film composites upon relatively small temperature variations, while the effect size and frequency behavior depend on the cantilever dimensions. The built-in two-way behavior working as a reset spring enables a design of remarkable simplicity, which allows very simple fabrication processes.

Figure 18.7   Shape memory thin film composite (TiNiCu/Mo) microgripper fabricated on micromachined Si using a sacrificial layer technique.

Further potential applications are, e.g., related to active surfaces for tactile displays in medical applications [14] which require smart material systems with different material composites and complex thin film actuator design.

For applications in mobile communication high frequency components are required that allow the adjustment of the operating frequency. For this purpose, adjustable capacitors were suggested that are based on bistable shape memory thin film composites [8]. In this design the alternating deposition of shape memory layers on the substrate provides a parallel motion of the tip of the actuator which can thus result in a change of distance between the parallel capacitor plates. The complete actuator is embedded within a thermoplastic dielectric, whose melting temperature has to be adjusted in such a way that it lies within the hysteresis of the shape memory material. To change the capacitance, the system has to be heated above the melting temperature of the thermoplastic dielectric to allow any movement of the shape memory composite and thus a change of the distance between the capacitor plates. Depending on the accuracy of the temperature control, also intermediate positions and thus capacitances can be frozen in. A reset to the ground state (martensitic state) requires a heating of the device at a temperature which is above the melting temperature of the thermoplastic dielectric but below the austenitic start temperature. Further requirements are a shape memory transformation well above any ambient temperature and a complete encapsulation of the device.

Based on phase-coupled shape memory thin film composites, a design was developed which can work as a motion principle of a robot. For this purpose, the different shape memory alloys with different hysteresis width have to be patterned, as illustrated in Fig. 18.8, where the schematic cross-section of a possible robot leg is drawn as a superposition of the four different stages of a complete motion cycle. Every motion step corresponds to a different combination of austenite and martensite of the two shape memory alloys, which are marked by the four phases φ_1, φ_2, φ_3 and φ_4. Further details and more complex motions are described in [9].

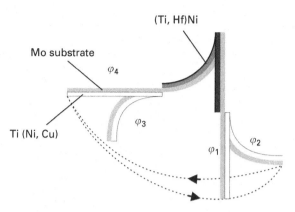

(Ti, Hf)Ni

Mo substrate

φ_4

φ_3

Ti (Ni, Cu)

φ_1 φ_2

Figure 18.8   Superposition of the cross-section of a robot leg at different phases of a motion cycle.

18.7   Summary

Shape memory thin film composites provide an actuator mechanism which is easy to fabricate, which can be integrated into MEMS batch fabrication, and which does not require any training of the shape memory material. By choosing different TiNi-based alloys the transformation temperature and hysteresis can be varied in a wide temperature range. For designing more complex composites, bistability as well as phase-coupled motions can be achieved which are potentially attractive for a number of applications.

References

[1] B. Winzek and E. Quandt, Thin Ti(Ni,Cu) film composites, *Proc. Actuator 2000*, Bremen, Germany, 2000, p. 172–176.

[2] J. D. Busch, A. D. Johnson, C. H. Lee and D. A. Stevenson, Shape-memory properties in Ni-Ti sputter-deposited film, *J. Appl. Phys.*, **68** (1990) 6224–6228.

[3] D. S. Grummon, S. Nam and L. Chang, Effect of superelastically reforming NiTi surface microalloys on fatigue crash nucleation in copper, *Mater. Res. Soc. Symp. Proc.*, **246** (1992) 259–264.

[4] K. Ikuta, M. Hayashi, T. Matsuura H. Fujishiro, Shape memory alloy thin film fabricated by laser ablation, *IEEE Proceedings on Micro Electro Mechanical Systems, MEMS'94*, Oiso, Japan, 1994, p. 355–360.

[5] S. Miyazaki, T. Hashinaga and A. Ishida, Martensitic transformations in sputter-deposited Ti-Ni-Cu shape memory alloy thin films, *Thin Solid Films*, **281–282** (1996) 364–367.

[6] A. D. Johnson, V. V. Martynov and R. S. Minners, Sputter deposition of high transition temperature Ti-Ni-Hf alloy thin films, *J. Physique IV*, **C8** 5 (1995) 783–787.

[7] E. Quandt, C. Helene, H. Holleck, *et al.*, Sputter deposition of TiNi, TiNiPd and TiPd films displaying the two-way shape-memory effect, *Sensors and Actuators A* **53** (1996) 434–439.

[8] T. Sterzl, B. Winzek, H. Rumpf and E. Quandt, Bistable shape memory composites for switches, grippers and adjustable capacitors, *Proc. Actuator 2002*, Bremen, Germany, 2002, p. 91–94.

[9] B. Winzek, S. Schmitz, H. Rumpf, *et al.*, Recent developments in shape memory thin film technology, *Mat. Sci. Eng. A* **378** (2004) 40–46.

[10] A. D. Johnson and E. J. Shahoian, Recent progress in thin film shape memory micro-actuators, *IEEE Proceedings on Micro Electro Mechanical Systems, MEMS'95*, Amsterdam, 1995, p. 216–220.

[11] W. L. Benard, H. Kahn, A. H. Heuer and M. A. Huff, A titanium–nickel shape memory alloy actuated micropump, *Proceedings of the 9th International Conference on Solid-State Sensors and Actuators, Transducers'97*, Chicago, 1997, p. 361–364.

[12] P. Krulevitch, A. P. Lee, P. B. Ramsey, J. C. Trevino, J. Hamilton and M. A. Northrup, Thin film shape memory alloy microactuators, *J. Microelectromech. Syst.*, **5** (1996) 270–282.

[13] D. Brugger, M. Kohl, B. Winzek and S. Schmitz, Optical microscanner based on a SMA thin film composite, *Proc. Actuator 2004*, Bremen, Germany, 2004, p. 90–93.

[14] R. Vitushinsky, F. Khelfaoui, S. Schmitz and B. Winzek, Metallic thin film composites with shape memory alloys for microswitches and tactile graphical displays, *Int. J. Appl. Electromag. Mech* **23**, 1–2 (2006) 113–118.

19 TiNi thin film shape memory alloys for optical sensing applications

Y. Q. Fu, W. M. Huang and C. Y. Chung

Abstract

This chapter focuses on the optical sensing applications based on TiNi films. When the TiNi film undergoes a phase transformation, both its surface roughness and reflection change, which can be used for a light valve or on–off optical switch. Different types of micromirror structures based on sputtered TiNi based films have been designed and fabricated for optical sensing applications. Based on the intrinsic two-way shape memory effect of free standing TiNi film, TiNi cantilever and membrane based mirror structures have been fabricated. Using bulk micromachining, TiNi/Si and TiNi/Si$_3$N$_4$ bimorph mirror structures were fabricated. As one application example, TiNi cantilevers have been used for infrared (IR) radiation detection. Upon absorption of IR radiation, TiNi cantilever arrays were heated up, leading to reverse R-phase transition and bending of the micromirrors.

19.1 Introduction

Optical MEMS, also called MOEMS (Micro-opto-electro-mechanical systems), generally refers to the optical and optoelectronic systems that include one or more micromechanical element [1, 2, 3, 4, 5]. The micromechanical elements in MOEMS are batch fabricated by micromachining techniques. They are smaller, lighter, faster and cheaper than their bulk counterparts, and can be monolithically integrated with the optical components. The digital micromirror device (DMD) from Texas Instruments, USA is a showpiece of optical MEMS [5]. It is an array of mirrors, which are independently addressable and can deflect light through a tristable range of motion (+10°, 0°, −10°), as illustrated in Fig. 19.1 [5]. A light source is directed onto the DMD while a signal is input to the device. Each individual mirror is placed in a binary mode. The mirrors that are "on" will reflect

Thin Film Shape Memory Alloys: Fundamentals and Device Applications, eds. Shuichi Miyazaki, Yong Qing Fu and Wei Min Huang. Published by Cambridge University Press. © Cambridge University Press 2009.

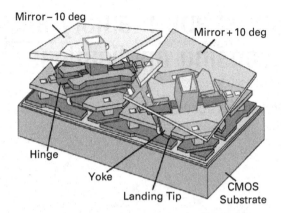

Figure 19.1 DMD device showing two mirrors in opposite tilted positions [5]. (Reproduced with permission from Texas Instruments, USA.)

Figure 19.2 Micrograph of a micromirror near the end of an optical fiber [6]. (Reproduced with permission from SPIE Press, USA.)

the light into the optical path and onto the projection surface. The mirrors in the "off" state will reflect the light into a baffle, and the associated pixel on the projection surface will appear dark [5]. These devices can be used in a variety of optical systems, including projection devices.

Optical switches are another optical MEMS application area. As one example shown in Fig. 19.2, a mechanical shutter blocks the light transmitted through a gap between fiber ends [6]. The mirror is on the end of a hinged and erect surface, which passes over a pivot to a plate parallel to the substrate. The translating action of the mirror can be done with a linear actuator that drives the mirror into the desired position.

Optical grating is also widely used in optical MEMS. One good example for the display market is the grating light valve (GLV, see Fig. 19.3) [7], which is made from parallel doubly supported beams of silicon, Si_3N_4 or SiO_2. By applying a

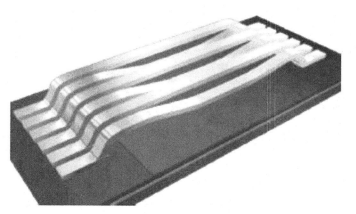

Figure 19.3   Silicon Light Machines' grating light valve [7]. (Reproduced with permission from Silicon Light Machines, USA.)

voltage, electrostatic force causes the beam to bend inward, and hence causes a diffraction of the incoming light, which can then be collected. In the "on" state, the ribbons form a flat mirror, while in the "off" state the suspended ribbons are pulled down by one quarter of a wavelength, so a grating is formed [7].

Of the many possible mechanisms of actuating a MEMS optical switch, several have emerged as the possible solutions for commercial products. Electrostatic and piezoelectric MEMS systems are the most common actuation designs due to the precisely controlled displacement, fast response and repeatability, but their displacement is limited. Magnetic actuation can generate large forces with high linearity, however, it is not commonly used in MEMS because of scalability and the complications of integrating the magnet into the MEMS structure. Another potential mechanism for the optical device design is electrothermal actuation based on compliant devices. The electrothermal compliant mechanisms are utilized because their flexibility and conductivity properties can create a constrained thermal expansion, flexible enough to satisfy the required displacement, and also stiff enough to support external loads [8].

MEMS applications of the TiNi based films have been focused on microactuators, such as microgrippers, micropumps and microvalves [9]. However, there are a few reports on optical application using a thin film SMA [10, 11]. Advantages of using shape memory actuation include: simple design mechanism, large transformation strain or displacement and low actuation temperature, which are quite attractive for optical mirror applications [9]. Figure 19.4 shows an illustration of a microlight valve design, in which a microfabricated TiNi diaphragm with a 0.26-mm diameter hole was used as a prototype light-valve [10]. The intensity of the transmitted light through the hole was increased by 10–17% after the diaphragm was heated and thus the center hole was opened.

In this chapter, the optical application of TiNi based thin films and different types of micromirror structures are introduced.

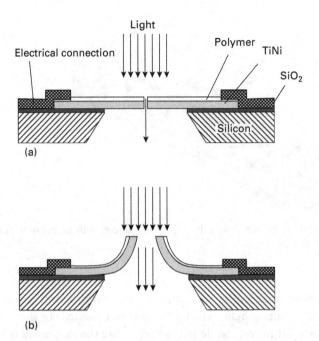

Figure 19.4   Microlight valve design: (a) off state at room temperature; (b) on state with TiNi film electrically heated so that light can pass through [10]. (Reproduced with permission from Optical Society of America, USA.)

19.2   Optical application based on the surface morphology change

When TiNi bulk materials or thin film SMAs undergo a phase transformation, both their surface roughness and reflection change, which can be used for optical applications. When a piece of bulk CuZnAl SMA was polished at 100 °C under the austenite phase and then cooled down to a lower temperature, the commonly observed surface relief appears due to the martensitic phase transformation, as shown in Fig. 19.5(a) [12]. When the material is heated, martensite transforms into austenite, and the surface of the sample becomes smooth (see Fig. 19.5(b)). The surface roughness of the sample at different temperatures was measured by a Wyko interferometer. At room temperature, the average surface roughness is about 119.7 nm because of the significant surface relief phenomenon. However, the roughness decreases to about 35.5 nm at a high temperature of 100 °C [12].

Ti50Ni50 film with a thickness of about 3 microns was prepared by the magnetron sputtering method. After deposition, the film was annealed at 600 °C for one hour for crystallization. A temperature controllable atomic force microscope (Shimadzu SFT 9800) was used to obtain the surface profile at different temperatures. Figure 19.6 shows a typical surface morphology of the TiNi film at room temperature (martensite) and 100 °C (austenite) obtained from AFM. At the

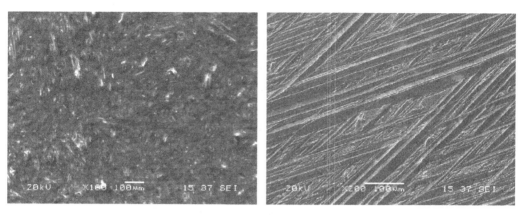

Figure 19.5 Surface morphology of CuZnAl single crystal at different temperatures [12]. (a) 20 °C; (b) 100 °C. (Reproduced with permission from SPIE, USA.)

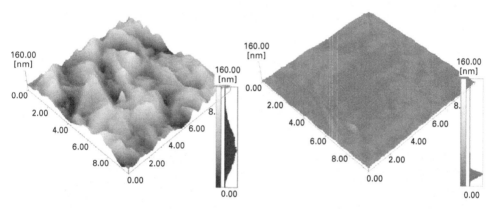

Figure 19.6 Surface of a TiNi shape memory thin film [13]. The bars on the right show the depth range. (a) 20 °C; (b) 100 °C. (Reproduced with permission from Institute of Physics, UK.)

high temperature of the austenite state, the TiNi film surface is smooth (with a roughness of 3 nm), whereas at low temperature of the martensite state, the surface roughness is very high (the measured roughness is about 22 nm) [13].

From the dramatic change of the surface roughness shown in Fig. 19.6, it is possible to fabricate microdisplay devices as illustrated in Fig. 19.7 and Fig. 19.8 using micromachining techniques, such as MEMS or laser processing [12]. Many micro-units or pixels can be made of thin film SMA. A microheater can be built up beneath these micro-units to control the individual transformation. Depending on the temperature of each unit, a uniform input beam can be converted into different patterns in the output direction, as shown in Fig. 19.7. One application example of the microdisplay device is shown in Fig. 19.8 [12].

Surface roughness shows a dramatic change in TiNi film, but it is not clear how significant is the change of reflection properties during the phase transformation. In order to investigate the reflection change upon thermal cycling, a spectrometer (SD2000 fiber optic spectrometer, Ocean Optics) was used. Figure 19.9 illustrates

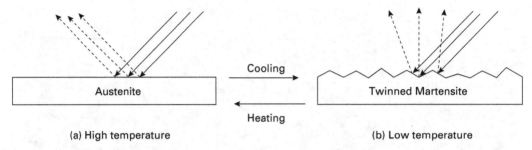

(a) High temperature   (b) Low temperature

Figure 19.7   Working principle of a microdisplay device [12]. (Reproduced with permission from SPIE, USA.)

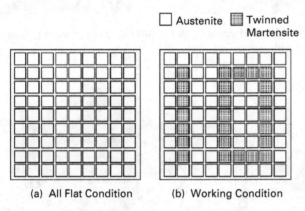

(a) All Flat Condition   (b) Working Condition

Figure 19.8   An illustration of proposed microdisplay device [12]. (Reproduced with permission from SPIE, USA.)

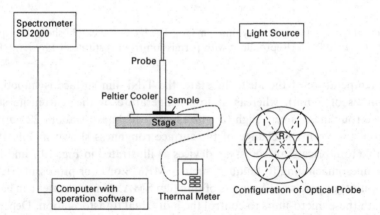

Figure 19.9   Experimental setup for the reflection measurement [14]. (Reproduced with permission from Institute of Physics, UK.)

the setup of the reflection measurement [14]. A tungsten–halogen light source was selected to provide a flat-top white light with a wavelength range from 360 nm to 2 µm. The reflection probe was located perpendicular to the sample. It is made up of seven individual optical fibers, six illumination fibers around one read fiber, with the

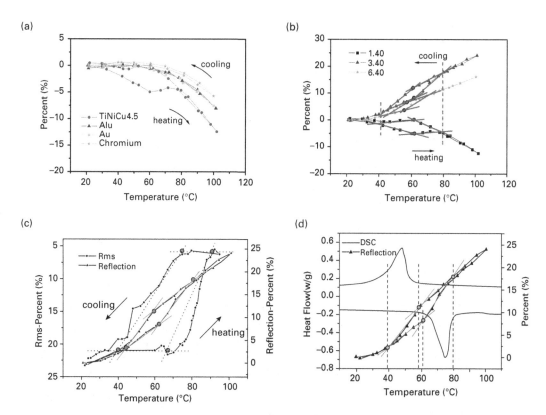

Figure 19.10 (a) Reflection change (taking the value at room temperature as reference) for a wavelength of 625.51 nm as a function of temperature in a thermal cycle at a distance of 4 mm; (b) comparison of reflection signals in TiNiCu film at distances of 1.4 mm, 3.4 mm and 6.4 mm; (c) comparison with roughness curve of TiNiCu film (at 3.4 mm); (d) comparison with DSC curve of TiNiCu film (at 3.4 mm) [14]. (Reproduced with permission from Institute of Physics, UK.)

same specifications (200 μm diameter, 0.22 numerical aperture and 250–800 nm optimization range). The white light passes through the six illumination fibres before reaching the sample surface. The reflection from the sample surfaces is collected by the spectrometer and analyzed. The testing sample used is a Ti50Ni47Cu3 film.

Figure 19.10(a) shows the reflection change (taking the value at room temperature as a reference) with a wavelength of 625.51 nm in a thermal cycle [14]. The distance between the probe and sample surface is 1.4 mm. The tests have also been performed on the reference materials (sputtered Al, Au and Cr films on the Si substrate) for comparison. As shown in Fig. 19.10(a), significant hysteresis is observed for the TiNiCu film. Obviously the hysteresis is caused by the change of surface conditions induced in the phase transformation. This hysteresis cannot be found in the intensity change curves of those reference films.

Figure 19.10(b) shows the reflection change of the TiNiCu film as a function of temperature with a distance of 1.4 mm, 3.4 mm and 6.4 mm between the probe

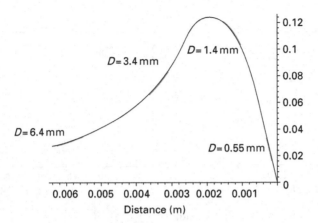

Figure 19.11 Reflectance changes upon distance change between the light source and sample surface from a geometric model [14]. (Reproduced with permission from Institute of Physics, UK.)

and sample surface. The hysteresis in the reflection signals can be found for all three curves. There are significant differences in background signals, indicating that the reflection is very sensitive to the displacement between the light source and the sample surface. This effect has been summarized in Fig. 19.11. The change of surface roughness has a strong influence on the reflection intensity, as shown in Fig. 19.10(c). It confirms that the change of surface roughness during the phase transformation is the dominant factor for the change of reflection in the TiNiCu thin film. Compared with the surface roughness measurement and DSC test (see Fig. 19.10(d)), the obtained phase transformation temperatures are very close [14]. This reveals a possibility for characterizing the shape memory behavior by optical measurement. From Fig. 19.10, it can also be concluded that there is an optimal distance for the reflection measurement. At a peak position, the influence of the specific configuration is minimal, and the reflectance change is more apparent. From the results shown in Fig. 19.10, it should be pointed out that during the phase transformation, the changes in the reflection seem less significant than those from the surface roughness changes.

Apart from surface relief morphology, there are other types of surface morphology changes, which show significant change in surface roughness, for example, surface wrinkling [15] as discussed in Chapter 7. In brief, the change in surface roughness and reflection during the phase transformation in TiNi thin film SMAs could be utilized not only as an alternative approach for characterization of SMAs, but also as novel mechanisms for optical applications.

19.3   Optical application based on free standing TiNi film

Freestanding TiNi based thin films possess an intrinsic two-way shape memory effect with large displacements: the film curls up at room temperature, but

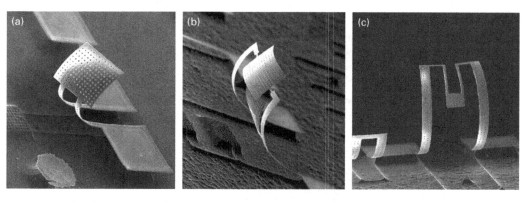

Figure 19.12 Cantilever based micromirror structure which can be actuated through the intrinsic two-way shape memory effect of the free standing film [16]. (Reproduced with permission from Institute of Physics, UK.)

Figure 19.13 TiNi micromirror structure actuated by four flexible beams: (a) SEM morphology; (b) and (c) micromirror actuation through electrically heating in the arm beams [16]. (Reproduced with permission from Institute of Physics, UK.)

becomes flat when heated and curls up again when cooling down to room temperature. The origin of the two-way shape memory effect observed in TiNi films can be attributed to the difference in sputtering yields of titanium and nickel, which produces a compositional gradient through the film thickness [9]. This two-way effect is suitable for application to optical mirror actuation. Figure 19.12 shows cantilever based micromirror structures of free standing film which can be actuated through the intrinsic two-way shape memory effect [16]. The performance of the released microactuators was demonstrated by simply heating up the structure on a hot plate to a maximum temperature of $100\,°C$. Vertical displacement up to 400 microns can be achieved with a beam length of 500 microns.

Figure 19.13 (a) shows another design of TiNi micromirror structure based on free standing TiNi film [16]. The micromirror is composed of a TiNi membrane cap as the mirror and four flexible beams with the corresponding TiNi electrical circuits and pads. The flexible beams are designed as the arms to support the cap, guide the out-of-plane motion and actuate the mirror. In operation, electrical current will be applied to the thermal element (TiNi electrodes), causing the

temperature increase in TiNi. The TiNi microbeams are bent up at room temperature due to the intrinsic gradient stress. When heated, the microbeams become flat after applying current through the TiNi beams, which results in a change of angle in the micromirror. This is clearly shown in Fig. 19.13(b) and (c).

Figure 19.14 shows a new design of micromirror structure [16]. The device consists of four arms with compliant spring structures to actuate the micromirror. Each arm is made up of two beams. One of them is wide and the other is narrow. When a current is passed through the arms, the higher current density in the narrow beam causes it to be heated up more than the wide beam, leading to both vertical (shape memory effect) and horizontal (thermal effect) movement through the difference in thermal expansion between the two beams. By actuating one arm, or a combination of different arms, the center micromirror can be actuated to different angles. Microactuators made from these freestanding TiNi films can provide large deflections based on simple designs. However, the force generated may not be so large, and the displacement cannot be precisely controlled.

19.4   Optical application based on bimorph structure

Bimorph TiNi/Si structures are more frequently used for optical applications. The advantage of using a TiNi/Si beam is that no additional bias structure is needed because the silicon beam can provide the bias force for pulling back to the original position after removing the electrical current [9]. The force generated is quite large, and the displacement can also be easily controlled.

19.4.1   TiNi/Si bimorph structure

A micromachined TiNi/Si cantilever can be used to actuate the micromirror, which is based on the significant beam bending upon thermal cycling (Fig. 19.15).

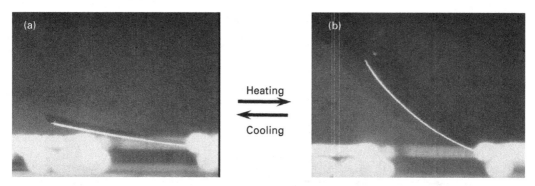

Figure 19.15   Tip deflection of cantilever structure with the change of temperature: (a) room temperature; (b) high temperature in austenite.

The first bimorph mirror example is a cantilever-based structure with a square Si cap (40 μm thick) as the top mirror (Fig. 19.16(a)) [17]. A V-shaped TiNi/Si beam structure (40 μm thick) acts as the actuating element. TiNi SMA has a suitable resistivity (80 μΩ·cm for martensite and 100 μΩ·cm for austenite), which enables it to be actuated electrically by Joule heating. In operation, electrical current is applied to the TiNi electrodes, resulting in an increase of temperature in the TiNi beam. Transformation from martensite to austenite causes the generation of large tensile stress, resulting in the bending of the V-shaped beam and thus the mirror cap (as shown in Fig. 19.16(b) and (c)).

The actuation performance of the microactuators was evaluated by passing a current through the patterns and resistively heating the film. The current was provided by a Keithley 224 A voltage/current power supply generating a square-wave voltage signal and this allowed actuation as a function of frequency. The deformation of the beam tip was measured using a CCD camera connected to a computer at different temperatures. Figure 19.17 shows the estimated deflection of the TiNi micromirror tip with the application of different powers by gradually increasing the current with a fixed voltage of 5 V [17]. When the current is less than 30 mA, the tip deflection does not show apparent change. With further increase in current, the tip deflection increases significantly (due to phase transformation and the shape memory effect) until above a current of 90 mA. Further increase in the current results in a slight decrease in tip deflection. Above a current of 140 mW, optical observation on the TiNi surface reveals that the film colour gradually changes, indicating the oxidation or deterioration of the film properties. The estimated maximum angle change is about 18°. The maximum frequency response detected by the naked eye is about 30 Hz [17]. Bulk TiNi devices typically have a frequency response of 0.1 to 2 Hz or less, mainly due to limited heat dissipation into the environment. The improved frequency response for the micromirror is due to the increased ratio of surface area to volume in TiNi films as compared to that of bulk material.

Figure 19.16 Cantilever-based micromirror structure (the first design) with an Si cap as the mirror (a) SEM morphology; (b) without electrical current; (c) after applying electrical current [17]. (Reproduced with permission from Institute of Physics, UK.)

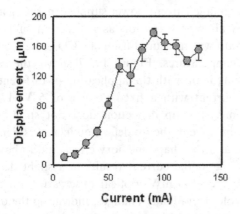

Figure 19.17 Tip displacement as a function of current for a V-shape cantilever TiNi/Si micromirror [17]. (Reproduced with permission from Institute of Physics, UK.)

Figure 19.18 shows the second example of TiNi/Si bimorph mirror design [17]. The micromirror is composed of three components: a membrane Si cap as the mirror body, three flexible Si beams and TiNi electrical circuits and pads. The Si membrane cap with a thickness of 40 microns is used as the top mirror. The

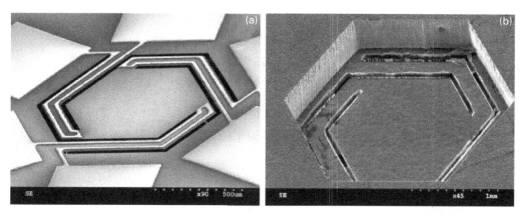

Figure 19.18 SEM images showing the second type of micromirror structure [17]. (a) Top view; (b) bottom view. (Reproduced with permission from Institute of Physics, UK.)

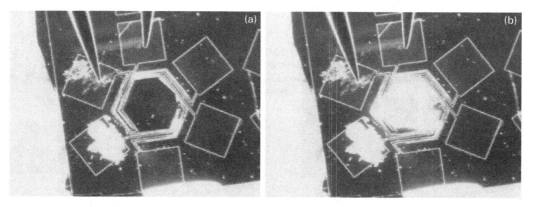

Figure 19.19 Tilting angle changes for the second type of micromirror [17]: (a) without electric current; (b) after applying electric current. (Reproduced with permission from Institute of Physics, UK.)

elbow-shaped flexible Si beams with TiNi electrodes are designed as the arms to support the Si cap, guide the out-of-plane motion and actuate the mirror. The TiNi/Si beam is flat at room temperature, but bends up after applying voltage to the TiNi electrodes, thus causing the angle change in the micromirror (see Fig. 19.19) [17]. Each elbowed beam is 100 μm wide and 40 μm thick. The distance between the Si cap edge and the elbowed beam is about 20 microns. The estimated maximum tilting angle for the second type of micromirror is about 10°. The problem with this type of micromirror is the rather limited tilting angle, because there are restrictions in the bending of tilting arms from the other two elbow beams. This design can also be used as a microvalve if the three arms are electrically actuated simultaneously. For both the mirror designs shown in Figs. 19.16 and 19.18, the Si membrane cap can be further designed to be coated with highly reflecting metallic layers such as gold, Cr, Al, etc., to improve its reflecting property.

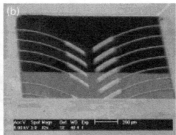

Figure 19.20 Bimorph TiNi/Si$_3$N$_4$ micromirror: (a) free standing Si$_3$N$_4$ cantilever; (b) bending down of TiNi/Si$_3$N$_4$ bimorph structure after amorphous TiNi film deposition; (c) cantilever bending up after crystallization [16]. (Reproduced with permission from Institute of Physics, UK.)

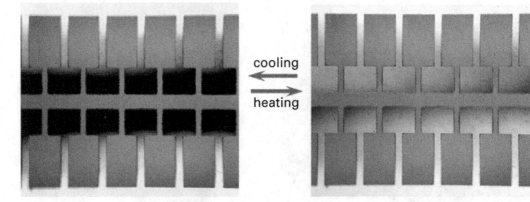

Figure 19.21 Observation of micromirror structure bending when heated using a hot plate up to 100°C [16]. (Reproduced with permission from Institute of Physics, UK.)

19.4.2   TiNi/Si$_3$N$_4$ microcantilever

Micromirror structures were also fabricated with a bimorph TiNi/Si$_3$N$_4$ system. A low stress Si$_3$N$_4$ layer of 2 microns was deposited on an Si wafer, then patterned into a cantilever mirror structure and etched using the RIE process with CF$_4$+O$_2$ (ratio of 9/1). The underneath Si was etched using KOH until the free standing Si$_3$N$_4$ structures were obtained. The fabricated Si$_3$N$_4$ mirror is shown in Fig. 19.20 (a) [16]. TiNi film was then deposited on the Si$_3$N$_4$ cantilever. After deposition, the cantilever beams bend down significantly due to large compressive stress in the amorphous TiNi film (see Fig. 19.20(b)). After annealing the films at 600 °C for one hour for crystallization, the bimorph structure bends up due to the bimorph thermal effect, as shown in Fig. 19.20(c). When heated using a hot plate, the cantilevers show the shape memory effect and become flat, as can be seen clearly in Fig. 19.21, forming a micromirror design [16].

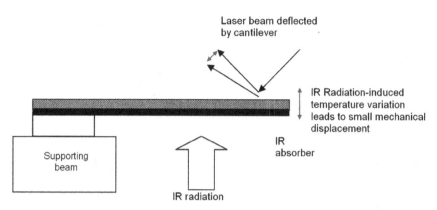

Figure 19.22   Design of IR sensor based on a TiNi thin film cantilever.

19.5   TiNi film for infrared image application

Infrared (IR) imaging has long been an extensively studied topic because of the applications in night-vision and security surveillance. In conventional IR imaging systems, the sensing element is usually a p-n junction photodetector fabricated from narrow bandgap semiconductors such as mercury cadmium telluride ($Hg_xCd_{1-x}Te$). However, thermal agitation, which leads to band-to-band transitions and creates a large amount of unwanted noise, has made such devices inoperative without cryogenic (77 K) cooling. The cost and inconvenience associated with these IR imaging systems have prompted a search for a better option that can work at room temperature. A new IR imaging approach can be designed which is based on the thermomechanical property of thin film SMA. A schematic diagram of the SMA IR sensor cantilever is shown in Fig. 19.22. When a TiNi cantilever undergoes shape change due to the martensitic transformation caused by the incident IR radiation, it will deflect. This can be detected by illuminating the cantilever using a laser beam or visible light.

In a preliminary feasibility study as shown in Fig. 19.23, 20 µm thick TiNi films with a width of 5 mm and length of 15 mm were used. A thin and light mirror was adhered to the TiNi film. A chopper was used in front of a halogen lamp, and the rotation speed of the chopper was varied so that the frequency of the IR light shining on the TiNi film could be adjusted. The laser beam shining on the mirror attached to the TiNi film was reflected to the optical detector and converted to a voltage signal. The frequency was increased from 1 Hz to 18 Hz and the signal received by the optical detector is shown in Fig. 19.24.

The detection reveals that the TiNi cantilever mirror responds and deflects to the varying intensity of IR radiation controlled by the chopper. The quick response of the TiNi film confirms itself a good candidate for IR imaging systems. The sensitivity (deflection per unit temperature rise) of the TiNi film determined was 0.07 °/K [18], while the corresponding sensitivity estimated using data from Perazzo et al. [19] in a bimaterial cantilever was only 0.006 °/K.

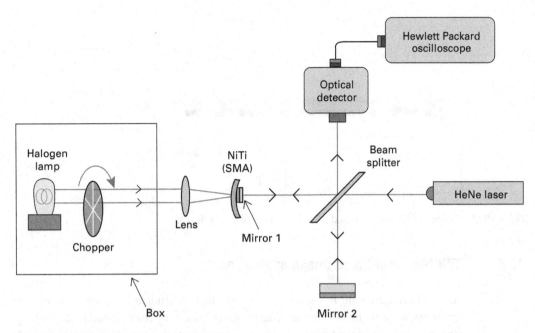

Figure 19.23   Schematic diagram of the IR imaging setup.

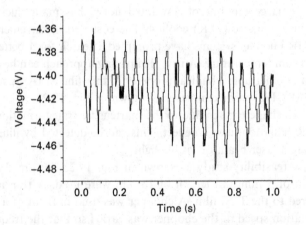

Figure 19.24   Interferometer signal output obtained at a chopper frequency of 18 Hz.

After the feasibility study, the SMA cantilever concept was further extended to digitized and miniaturized IR sensor microarrays [20]. Silicon plates coated with a thin layer of photoresist were used as the substrate. TiNi films 5 μm thick were deposited using a DC magnetron sputtering deposition technique. The films were then patterned by a standard photolithographic technique, and etched using HF/HNO$_3$/H$_2$O solution to obtain the free standing cantilever microarrays. The arrays were then adhered to the supporting frames made from 200 μm thick Si wafer. Three types of cantilever microarrays patterns with different width and

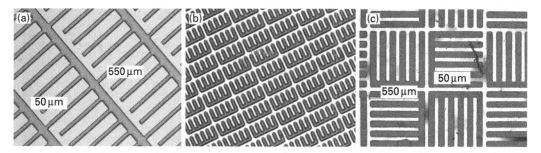

Figure 19.25   Three types of cantilever microarrays were made and characterized [20]: (a) type I, (b) type II, (c) type III. (Reproduced with permission from Elsevier, UK.)

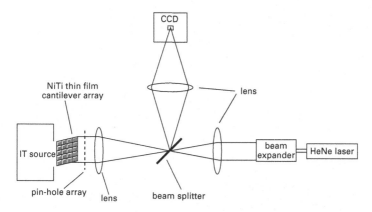

Figure 19.26   Schematic diagram showing the digital IR imaging system [20]. (Reproduced with permission from Elsevier, UK.)

lengths were made, as shown in Figs. 19.25(a) to (c) [20]. The microarrays were tested using an IR imaging system with a CCD camera, as shown schematically in Fig. 19.26 [20]. A beam of IR light was focused at the corner of the TiNi microarray. It was found that when type I and type III microarrays were illuminated with varying intensity of IR irradiation, the whole microarray vibrated seriously and failed to produce steady images. The type I and III cantilevers were found too long and not rigid enough.

On the other hand, steady optical images of type II microarrays were recorded successfully using a CCD camera. When the IR irradiation caused an approximate 8 °C increase in the temperature of the TiNi microarrays at one corner of the TiNi microarray (i.e., at the circled regions shown in Fig. 19.27), it led to a corresponding change in the image [20]. The intensity line profile of the image along the dotted line is plotted in Fig. 19.28. The intensity of the peaks decreased when the cantilevers were IR irradiated because the temperature increase led to the shape memory effect and deflection of the cantilevers.

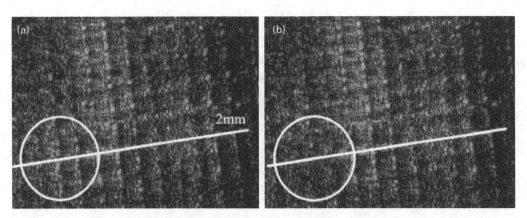

Figure 19.27 Optical images of the cantilever microarray obtained when (a) the IR source was off: room temperature; and (b) the IR light was focused at the circled region and the corresponding 8 °C temperature increase was due to IR illumination. (The total length of the line is about 2 mm) [20]. (Reproduced with permission from Elsevier, UK.)

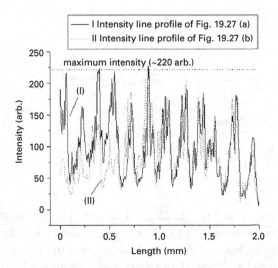

Figure 19.28 Intensity line profiles of TiNi at room temperature (I) and after IR irradiation (II) based on Fig. 19.27. [20]. (Reproduced with permission from Elsevier, UK.)

In summary, we have demonstrated the possibilities of the TiNi film based IR imaging system which utilized the thermomechanical properties to convert the temperature variation due to IR irradiation to a visible signal. However, the quality of the image formed is still not satisfactory. The blurred image shown in Fig. 19.27 is believed to be due to the "cross talk" between neighboring microcantilevers. Furthermore, the scattered light from neighboring microcantilevers also produced high background noise. It is believed that a light collimator can be added to reduce such background noise.

19.6   Summary

When TiNi film undergoes a phase transformation, both its surface roughness and reflection change, which can be used as a light valve or on–off optical switch for spatial light modulators. Surface roughness and reflection measurement could be alternative measurement methods for the phase transformation temperatures of thin film SMAs. Different types of TiNi based optical micromirror structures including free standing TiNi film, bimorph TiNi/Si and TiNi/Si$_3$N$_4$ structures were designed and fabricated in this study. The microbeams bend significantly with either heating the structure above 80 °C or applying voltage to TiNi electrodes (due to phase transformation and the shape memory effect), thus causing the changes in angles of the micromirrors. As one application example, TiNi cantilevers have also been used for infrared (IR) radiation detection. Upon the absorption of IR radiation, TiNi SMA thin film was heated up leading to a reverse R-phase transition. Illumination of the microarray can increase the temperature locally and cause bending of the micromirrors.

Acknowledgement

Help from Dr. Mingjie Wu, Dr. Min Hu and Prof. Hejun Du from Nanyang Technological University, Singapore, Mr. Mark K. C. Ng and Miss Amman P. M. Chan from the City University of Hong Kong is acknowledged.

References

[1]   R. A. Syms and D. F. Moore, Optical MEMS for Telecoms, *Materials Today*, **5**(2) (2002) 26–35.

[2]   T.-W. Yeow, K. L. E.   Law and A. Goldenberg, MEMS optical switches, *IEEE Commun. Mag.*, **39** (2001) 158–163.

[3]   M. C. Wu, O. Solgaard and J. E. Ford, Optical MEMS for lightwave communication, *J. Lightwave Technol.*, **24** (2006) 4433–4454.

[4]   W. Noell, P. Clerc, L. Dellmann, *et al.*, Applications of SOI-based optical MEMS, *IEEE J. Selected Topics Quant. Electron.* **8** (2002) 148–154.

[5]   L. J. Hornbeck, The DMD (TM) projection display chip: a MEMS-based technology, *MRS Bulletin*, **26** (2001) 325–327. www.ti.com/dlp/.

[6]   P. Rai-Choudhury, *MEMS and MOEMS Technology and Applications*, New York: SPIE Press (2000) p. 314.

[7]   G. Mandle, New projection technologies: silicon crystal reflective display and grating light valve, *SMPTE Motion Imaging Journal*, **114** (2005) 474–479. www.siliconlight.com.

[8]   A. Saxena and G. K. Ananthasuresh, On an optimal property of compliant topologies. *J. Structural and Multidisciplinary Optimization*, **19** (2000) 36–49.

[9]   Y. Q. Fu, H. J. Du, W. M. Huang, S. Zhang and M. Hu, TiNi-based thin films in MEMS applications: a review, *Sens. Actuat.*, **112** (2004) 395–408.

[10]   B. Sutapun, M. Tabib-Azar and M. A. Huff, Applications of shape memory alloys in optics, *Appl. Optics*, **37** (1998) 6811–6815.

[11]   M. Tabib-Azar, B. Sutapun and M. Huff, Applications of TiNi thin film shape memory alloys in micro-opto-electro-mechanical systems, *Sens. Actuat.*, **A 77** (1999) 34–38.

[12]   W. M. Huang, W. H. Zhang and X. Y. Gao, Micro mirror based on surface relief phenomenon in shape memory alloys. *Conf. Photonics Asia Symposium 2002*, Oct. 14–18, 2002, *Materials, Devices and Systems for Display and Lighting*, SPIE 4918, 2002, p. 155–161.

[13]   Q. He, W. M. Huang, M. H. Hong, *et al.*, Characterization of sputtering deposited TiNi shape memory thin films using a temperature controllable atomic force microscope, *Smart Mater. Struct.*, **13** (2004) 977–982.

[14]   M. J. Wu, W. M. Huang and F. Chollet, In situ characterization of TiNi based shape memory thin films by optical measurement, *Smart Mater. Struct.*, **15** (2006) N29–N35.

[15]   Y. Q. Fu, S. Sanjabi, Z. H. Barber, *et al.*, Evolution of surface morphology in TiNiCu shape memory thin films, *Appl. Phys. Lett.*, **89** (2006) 171922.

[16]   Y. Q. Fu, J. K. Luo, W. M. Huang, A. J. Flewitt and W. I. Milne. Thin film shape memory alloys for optical sensing applications, *J. Phy.: Conference Series*, **76** (2007) 012032.

[17]   Y. Q. Fu, J. K. Luo, M. Hu, H. J. Du, A. J. Flewitt and W. I. Milne, Micromirror structure actuated by TiNi shape memory thin films, *J. Micromech. Microengng.*, **15** (2005) 1872–1877.

[18]   H. P. Ho, C. Y. Chung, K. C. Ng, K. L. Cheng and S. Y. Wu, Novel far infrared imaging sensor based on the use of titanium-nickel shape memory alloys, *SPIE Conference on Smart Materials, Nano-, and Micro-Smart Systems*, Melbourne, Australia, 16–18 December 2002.

[19]   T. Perazzo, M. Mao, O. Kwon, A. Majumdar, J. B. Varesi, and P. Norton, Infrared vision using uncooled micro-optomechanical camera, *Appl. Phys. Lett.*, **74** (1999) 3567–3569.

[20]   P. M. Chan, C. Y. Chung and K. C. Ng, TiNi shape memory alloy thin film sensor micro-array for detection of infrared radiation, *J. Alloys and Compound*, **449** (2008) 148–151.

Index

Printed in the United States
by Baker & Taylor Publisher Services